KNOW YOUR OWN SHIP.

A SIMPLE EXPLANATION

OF

THE STABILITY, TRIM, CONSTRUCTION, TONNAGE, AND FREEBOARD OF SHIPS, TOGETHER WITH A FULLY WORKED OUT SET OF THE USUAL SHIP CALCULATIONS (FROM DRAWINGS).

SPECIALLY ARRANGED FOR THE USE OF SHIPS' OFFICERS, SUPERINTENDENTS, ENGINEERS, DRAUGHTSMEN, AND OTHERS.

BY

THOMAS WALTON,

LATE LECTURER ON NAVAL ARCHITECTURE TO SHIPS' OFFICERS AND STUDENTS IN THE GOVERNMENT NAVIGATION SCHOOL, LEITH, AND IN MIDDLESBROUGH SCIENCE SCHOOL;
AUTHOR OF
"CONSTRUCTION AND MAINTENANCE OF SHIPS BUILT OF STEEL."

With Numerous Illustrations.

FOURTH EDITION, GREATLY ENLARGED.

LONDON:
CHARLES GRIFFIN AND COMPANY, LIMITED;
EXETER STREET, STRAND.
1899.

[*All Rights Reserved.*]

This scarce antiquarian book is included in our special *Legacy Reprint Series*. In the interest of creating a more extensive selection of rare historical book reprints, we have chosen to reproduce this title even though it may possibly have occasional imperfections such as missing and blurred pages, missing text, poor pictures, markings, dark backgrounds and other reproduction issues beyond our control. Because this work is culturally important, we have made it available as a part of our commitment to protecting, preserving and promoting the world's literature.

PREFACE TO FOURTH EDITION.

DURING the three years which have elapsed since the First Edition of this work was issued, the Author has received numerous responses to the invitation then given for "suggestions and communications." These have come from ships' officers, engineers, and from students of naval architecture, working in shipyards. Though the original intention was to exclude the somewhat elaborate calculations involved in arriving at many of the conclusions dealt with in this work—displacement, moment of inertia, righting moment of stability, etc., etc.,—the communications received have proved beyond doubt that such calculations would be welcomed by a very large section of readers as a valuable addition to the book The author has been agreeably surprised to discover that many sea-going folk are not satisfied to merely understand and use the *results* of calculations given to them, but are determined to know and understand the *whole process of calculation*, by means of which these results are obtained, and, in consequence, very many hours have been spent in replying to queries of this nature. It will be evident that, as we come to deal with some of the more intricate calculations, a moderate amount of mathematical knowledge is required. However, it is hoped that by first dealing with the simple preliminary calculations in Chapter X., and then following on to the actual set of ship calculations, for which the necessary drawings are given, together with brief notes and explanations of the various steps taken, the reader will be able to trace the arithmetical application of the rules given.

Considerable additions have been made in Chapter III. on the subject of water pressures, and in Chapter VI. (Section III.) on the subject of * waterballast, to illustrate which numerous new diagrams have been inserted. Trim (Chapter VII.) is entirely new, and, it is hoped, will meet the requirements of those readers at whose suggestions it has been introduced. This chapter, in the form of a paper, was read before the Shipmasters' Society in London, in March 1898, and the request was there repeated that it should find a place in *Know Your Own Ship*.

The other sections of the book have been revised, and, as far as possible, brought up to date.

Suggestions involving details of ship construction will be fully dealt with in the companion volume, *Construction and Maintenance of Steel Ships*, undertaken by the Author at the special request of the publishers.

In its enlarged and improved form, it is hoped that this volume will meet with the same cordial acceptance which has been the chief characteristic of its history during its three years' existence.

* Extracted largely from a paper read by the Author before the Shipmasters' Society, London, November, 1897.

LONDON, *April*, 1899.

AUTHOR'S PREFACE TO THE FIRST EDITION.

EXPERIENCE is a wonderful teacher, though often a very slow one. In the course of time, it will instil into a seaman's mind a considerable knowledge of the capabilities and behaviour of his vessel under varying circumstances—her strength, her carrying capacity, her stability, or, in other words, her sea qualities.

This mode of obtaining knowledge is, however, far too costly for the intelligent seaman of to-day. He knows that many a good ship, and what is worse, many precious lives have been lost before it has been acquired, and all through pure—though it would be unjust to say, wilful—ignorance.

As regards the subject of stability, it has been said that it is useless to provide a captain with curves of stability, for he does not understand them, and if he did, they are of little use, all he requires being a statement of certain conditions of loading, beyond which he must not go, or his vessel will be unsafe. This is all very good in its way, but why does the captain not understand curves of stability, or, more broadly speaking, the subject of stability? Is not the answer this, that very little indeed has been done to provide the means of his obtaining such important information? Besides, if the curves are of little use, and a statement of conditions to ensure safety is sufficient, some credit must be given to the intelligence of the ship's officer of to-day. Let him not be put down as a machine to steer a ship, incapable of comprehending what may be reduced to a subject of comparative simplicity. On the other hand, let it not be supposed that lengthy experience in navigating ships necessarily means capability to so load any ship as to produce seaworthiness. The ignorance often displayed in loading and ballasting vessels, and the loss resulting therefrom, prove that such is not the case.

It is evidently often imagined that able, valuable, and instructive papers and lectures on vital subjects relating to ships, are only for, and interesting to, naval architects and experts, and that to ships' officers they are either unnecessary or else they have no interest in this means of instruction. A more foolish and unjust conclusion could scarcely be arrived at. If the stability of a vessel depends so much upon the loading, and officers superintend the loading of their own vessels, at any rate in foreign ports, if not always at home, the importance of an intelligent knowledge of the subject must be evident.

Moreover, is it not a fact that, with this knowledge, the ship's officer would be able to supply a great amount of valuable data to the shipbuilder and naval architect concerning the behaviour and capabilities of his vessel under every condition of weather and loading, which could not fail to be of immense value in the designing of future ships? By this means, the designer, who, as a rule, sees little of the ship he designs after a few hours' trial trip, generally under the most favourable circumstances, would then be the better able to produce a vessel more perfectly adapted to her requirements.

That such is the case is proved by the fact that some leading shipbuilding firms do everything in their power to encourage ship captains to give all the information they possibly can regarding the behaviour and performance of the vessels which they have built.

The best method of supplying this kind of instruction to seamen is believed to be by means of lectures thoroughly illustrated by diagrams and experiments.

During the past few years the writer has, in the Government Navigation School, Leith, N.B., given courses of lectures, specially designed for the instruction of seamen. The eagerness and enthusiasm displayed to obtain such information made it manifest that seamen are fully awake to its importance, and appreciate and understand its value. The able discussion which often followed these lectures, and the important points and features brought out in connection with personal experience, furthermore proved that seamen want no rule-of-thumb methods for guidance, but that a thorough grasp of the principles of the subject is required. This, together with the repeated requests which have been made to publish these lectures, has tempted the writer to do so in the form of this book. The old-fashioned method of "beginning at the beginning" of the subject has been adopted as the only trustworthy way of dealing with it; and while endeavouring to cover as much ground as possible, the aim has been to condense the matter as far as compatible with clearness, to present it in language easy enough to be understood by every seaman, and stripped of laborious mathematical formulæ, and at a cost which will place it within reach of seamen of limited means.

It is felt that no apology is necessary in presenting this book, for while the excellent work of Sir E. J. Reed and also that of Sir W. H. White are admirably adapted to the requirements of the naval architect and the shipbuilder, it is believed that no special attempt has been made to supply the need of seamen generally.

It is further hoped that the chapters on Construction, Tonnage, and Freeboard, while interesting and important subjects to seamen, will make the book acceptable to the shipowner, ship superintendent, ship draughtsman, and to those generally interested in shipping.

The Author's sincerest thanks are due to Messrs. W. Denny & Brothers, Dumbarton; to Messrs. J. L. Thompson & Sons, Ltd., Sunderland; and to Messrs. Ramage & Ferguson, Ltd., Leith, for permission to use curves of stability of vessels built by them; and also to J. Bolam, Esq., Head Master of the Government Navigation School, Leith, for the kindly sympathy and interest which he has manifested in the preparation of the book.

Suggestions and communications—especially from ships' officers and others desirous of promoting this branch of nautical instruction—will be gladly welcomed and acknowledged in future editions of this work, should these be called for.

LEITH, *March*, 1896.

EDITOR'S PREFATORY NOTE

THIS SERIES has been designed to meet the growing desire on the part of Officers in the Mercantile Marine for a MORE SCIENTIFIC INSIGHT into the principles of their profession, and the sciences upon which the Art of Navigation is founded. The treatises are, for the most part, WRITTEN BY SAILORS FOR SAILORS; and, where this is not the case, by authors who have special knowledge of the subjects dealt with and their application to the Sailor's life. The treatment will be thoroughly scientific, yet as free as possible from abstruse technicalities, and the style such as will render it easy for the young sailor to gain a knowledge of the elements of his profession by private reading and without difficulty.

E. B.

LONDON, *March*, 1896.

CONTENTS.

Chapter I.—Displacement and Deadweight.

Displacement—Displacement Curves — Deadweight — Deadweight Scale—"Tons per Inch Immersion"—Examples of Practical Application of "Tons per Inch Immersion" Curve—Coefficients of Displacement, ...   PAGES 1–11

Chapter II.—Moments.

Moments — Examples of Moments — Centre of Gravity — Effect upon Centre of Gravity of Moving Weights on a Ship, Vertically and Transversely, ...   12–18

Chapter III.—Buoyancy.

Buoyancy—Water Pressures — Reserve Buoyancy—Sheer—Value of Deck Erections—Centre of Buoyancy—Curves of Vertical and Longitudinal Centres of Buoyancy—Effect upon a Ship's Centre of Buoyancy of Immersing or Emerging Wedges of Buoyancy — Effect of Entry of Water upon Buoyancy—Camber, or Round of Beam—Testing Water Ballast Tanks, .   19–39

Chapter IV.—Strain.

Relation of Weight of Material in Structure to Strength—Strain when Floating Light in Dock—Relation between Weight and Buoyancy — Strain Increased or Decreased in Loading—Distribution and Arrangement of Material in Structure, so as to get Greatest Resistance to Bending—Types of Vessels subject to Greatest Strain—Strains among Waves—Panting Strains—Strains due to Propulsion by Steam and Wind—Strains from Deck Cargoes and Permanent Weights—Strains from Shipping Seas—Strains from Loading Cargoes Aground, .   40–50

Chapter V.—Structure.

Parts of Transverse Framing, and how Combined and United to produce Greatest Resistance to Alteration in Form—Sections of Material Used—Compensation for Dispensing with Hold Beams—Parts of Longitudinal Framing, how Combined and United to Transverse Framing to produce Greatest Resistance to all Kinds of Longitudinal Bending and Twisting—Forms of Keels and Centre Keelsons, and their Efficiency—Distribution of Material to Counteract Strain—Value of Efficiently-worked Shell and Deck Plating in Strengthening Ship Girder—Definitions of Important Terms—Illustration of Growth of Structural Strength, with Increase of Dimensions, by means of Progressive Midship Sections — Special Strengthening in Machinery Space—Methods of Supporting Aft End of Shafts in Twin-Screw Steamers—Arrangements to prevent Panting—Special Strengthening for Deck Cargoes and Permanent Deck Weights, and also to Counteract Strains due to Propulsion by Wind — Types of Vessels — Comparison of Scantlings of a Three-decked, a Spar-decked, and an Awning-decked Vessel—Bulkheads—Rivets and Riveting, . . . 51–111

Chapter VI. (Section I.).—Stability.

Definition — The Righting Lever — The Metacentre — Righting Moment of Stability—Conditions of Equilibrium — "Stiff" and "Tender"— Metacentric Stability—Moment of Inertia—Agents in Design influencing Metacentric Height—How to obtain Stiffness—Changes in Metacentric Height during the Operation of Loading—Stability of Objects of Cylindrical Form—A Curve of Stability—Metacentric Curves—How the Ship's Officer can Determine the Metacentric Height and then the Position of the Centre of Gravity in any Condition of Loading—Effect of Beam, Freeboard, Height of Centre of Gravity above Top of Keel, and Metacentric Height upon Stability — Wedges of Immersion and Emersion — Effect of Tumble Home upon Stability—Stability in Different Types of Vessels, ... 112–147

Chapter VI. (Section II.).—Rolling.

Rolling in Still Water—Relation of Stiffness and Tenderness to Rapidity of Movements in Rolling—Resistances to Rolling—Danger of Great Stiffness—Rolling among Waves—Lines of Action of Buoyancy and Gravity—A Raft, a Cylinder, and a Ship among Waves — Synchronism: how Produced and Destroyed — Effect of Loading upon Behaviour — Effect of Transverse Arrangement of Weights upon Rolling Motions—Alteration in Behaviour during a Voyage—The Metacentric Height—Fore and Aft Motions— Fore and Aft Arrangement of Weights, ... 148–165

Chapter VI. (Section III.).—Ballasting.

Similar Metacentric Heights at Different Draughts — Wind Pressure—Amount and Arrangement of Ballast—Means to Prevent Shifting of Ballast—Water Ballast—Trimming Tanks —Inadaptability of Double Bottom Tanks alone to provide an Efficient Means of Ballasting—Considerations upon the Height of the Transverse Metacentre between the Light and Load Draughts, and Effect upon Stability in Ballast—Unmanageableness in Ballast—Minimum Draught in Ballast—Arrangement of Ballast, ... 166–185

Chapter VI. (Section IV.)
Loading—Homogeneous Cargoes.

Alteration to Curve of Stability, owing to Change in Metacentric Height—Stability of Self-Trimming Vessels—Turret—Trunk, 186–192

Chapter VI. (Section V.).—Shifting Cargoes.

Variations in Stability on a Voyage, ... 193–197

Chapter VI. (Section VI.)
Effect of Admission of Water into the Interior of a Ship.

Admission through a Hole in the Skin into a Large Hold—Curves, showing Variation in Height of Metacentre with Increase of Draught — Buoyancy afforded by Cargo in Damaged Compartment — Longitudinal Bulkheads — Entry of Water into Damaged Compartment beneath a Watertight Flat—Entry of Water into Damaged Compartment beneath a Watertight Flat—Value of Water Ports—Water on Deck—Entrance of Water through a Deck Opening—Entry of Water into an End Compartment — Height of Bulkheads — Waterlogged Vessels, ... 198–212

Chapter VI. (Section VII.)

Sailing—Sail Area, etc., ... 213–218

Chapter VI. (Section VIII.)

Stability Information, ... 219–221

Chapter VI. (Section IX.)

Closing Remarks on Stability, ... 222–225

xvi   CONTENTS.

Chapter VII.—Trim.

Definition—Moment to Alter Trim—Change of Trim—Centre of Buoyancy of Successive Layers of Buoyancy at Successive Draughts—Longitudinal Metacentre—Longitudinal Metacentric Height—Moment to Alter Trim One Inch—Practical Examples showing how the Change of Trim is Ascertained, . . . 226-238

Chapter VIII.—Tonnage.

Importance to Shipowners from an Economical Point of View—Under-deck Tonnage—Gross Tonnage—Register Tonnage—Deductions for Register Tonnage—Importance of Propelling Deduction in Steamers—Deep Water Ballast Tanks—Deck Cargoes—Examples of Actual Ship Tonnages—Sailing Vessels—Suez Canal Tonnage—Yacht Tonnage, ... 239-250

Chapter IX.—Freeboard.

Definition—Method of Computing Freeboard—Type of Vessel—Nature of Deductions, and Additions to Freeboard—Examples of Estimating Freeboard for Different Types of Vessels, . . 251-267

Table of Natural Sines and Cotangents, etc., ... 268-270

Chapter X. (Section I.).—Calculations.

Useful Tables and Rules—Calculation of Weight of Steel Plate—Stanchion—Hollow Stanchion—Gallons in Fresh Water Tank—Tons in Coal Bunker—Rectangular Barge's Displacement and "Tons per Inch" Immersion—Simpson's Three Rules and Graphic Explanations—Calculation of Area of Deck or Waterplane—"Tons per Inch" Immersion of Ship's Waterplane—Ship's Displacement—Centre of Gravity of a Waterplane, Longitudinally or Transversely—Centre of Buoyancy, Vertically and Longitudinally—Moment of Inertia—Transverse Metacentre above Centre of Buoyancy—Centre of Gravity—Longitudinal Metacentre above Centre of Buoyancy—Alteration of Trim—Area of Section and Volume and Centre of Gravity of Wedge of Immersion or Emersion—Centre of Effort, ... 271-300

Chapter X. (Section II.).
A Set of Ship Calculations as Worked from Actual Drawings.

Displacement—Longitudinal Centre of Buoyancy—Vertical Centre of Buoyancy—Transverse Metacentre above Centre of Buoyancy, showing two Methods of Arrangement—Tons per Inch Immersion—Wetted Surface and Shell Displacement—Longitudinal Metacentre above Centre of Buoyancy—Results of Calculations upon Curves—Stability Calculation, . . . 301-324

Appendix A: Dynamic Stability and Oscillations among Waves, . 325-328

Appendix B: Test Questions, ... 328-332

INDEX, ... 333

KNOW YOUR OWN SHIP.

CHAPTER I.

DISPLACEMENT AND DEADWEIGHT.

CONTENTS. — Displacement — Displacement Curves — Deadweight — Deadweight Scale—"Tons per Inch Immersion"—Examples of Practical Application of "Tons per Inch Immersion" Curve—Coefficients of Displacement.

Terms.—" Displacement," "Deadweight," and "Tonnage" are terms often heard and used by those associated with ships and shipping in some form or other, but not always definitely understood. Their simplicity, however, renders them easy of explanation, and we shall, at the outset, devote our attention briefly to the two former—Displacement and Deadweight. The last —Tonnage—being a subject of larger dimensions, though simple in its character—will be reserved for a later chapter.

Displacement.—Any object floating in water displaces or dislodges a volume of water, and the weight of the displaced water is equal to the weight of the floating object. The prefix "dis" in the word "displacement" means "away from." Thus, displacement reveals its own meaning—viz., that which is placed out of its usual condition.

Displacement, in the technical sense in which it is applied to ships, or any other floating bodies, refers to the displacing of the water by the total or partial immersion of any object placed in it. The volume of water displaced may be measured in cubic feet or in tons, and the weight of water displaced is called the *Displacement*.

This fact may be simply illustrated by supposing a tank to be filled to the brim with water, and when in this condition let a box-shaped ship—3 feet long, 2 feet broad, and 2 feet deep—be placed in it, sinking to a depth of 1 foot. It is evident that if the tank were full before the ship was placed in it, some of the water must have overflowed (fig. 1).

Let a receiver be placed below the tank, with a reservoir 3 feet long, 2 feet broad, and 1 foot deep (exactly the same dimensions as the immersed part of the ship), attached to the bottom. It would be found that immediately the overflow had ceased, the reservoir would be just filled as shown by the dotted lines. This proves that the volume of water displaced is exactly

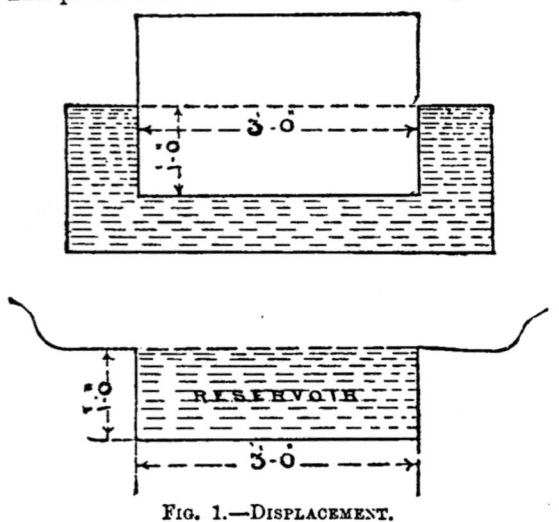

FIG. 1.—DISPLACEMENT.

equal to the volume of the immersed portion of the ship. Moreover, if the ship could be placed in one balance, and the displaced water in another, they would be found to exactly equal each other in weight. But supposing the ship were too large for any process of weighing to be adopted, the displacement could be ascertained by a very simple calculation.

A cubic foot of sea water weighs 64 lbs., and 35 cubic feet of sea water weigh 1 ton. Multiply together the length, breadth, and draught of the ship, and we get cubic feet; $3 \times 2 \times 1 = 6$ cubic feet. Therefore, since 1 cubic foot weighs 64 lbs., 6 cubic feet will equal $6 \times 64 = 384$ lbs. displacement, and this, while being the weight of water in the reservoir, is also the total weight of the ship in its present condition.

Again, supposing an unknown weight be placed in the ship increasing

DISPLACEMENT AND DEADWEIGHT. 3

the draught to 1 foot 6 inches, or 1·5 feet, then 3 × 2 × 1·5 = 9 cubic feet, 9 × 64 = 576 lbs. total displacement.

The displacement, when the ship was empty, was 384 lbs., therefore the weight added is 576 − 384 = 192 lbs.

In dealing with floating objects not box-shaped, such as real ships, although the same method of computing the displacement

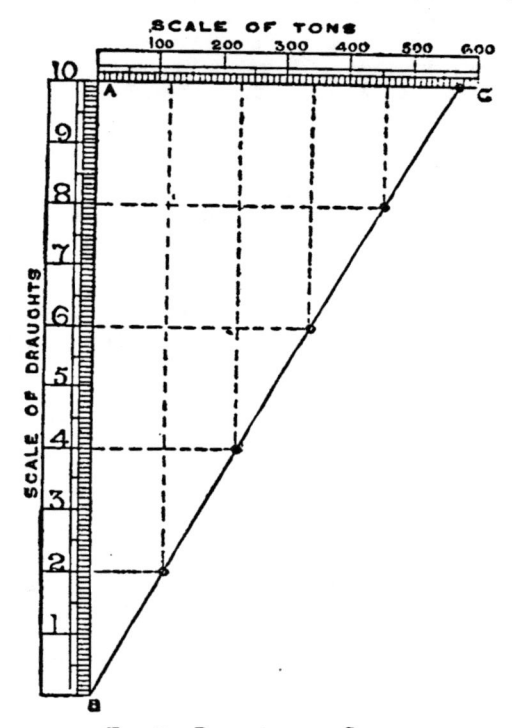

FIG. 2.—DISPLACEMENT CURVE.

cannot be adopted owing to the difference in form, yet the principle, that the weight of water displaced equals the total weight of the ship, remains the same. This important fact, as is evident, proves of immense value to shipbuilder and shipowner alike, for had the weight of a ship to be found by estimating the weight of every item in it—hull, engines, boilers, masts, etc. —separately, and adding them together, it will be seen how laborious the process, and how inaccurate the result, might possibly be.

The length, breadth, and draught of a ship cannot be multiplied together for displacement, but, by the application of a simple method known as Simpson's rules, the volume of the immersed portion of the ship can be ascertained, which, if considered as water, and divided by 35, will give the displacement in tons. (See Chapter X. for Simpson's Rules and Calculations.)

To make a separate calculation whenever the displacement is required at any particular draught would entail considerable labour and inconvenience. This is avoided by using what is termed a *Displacement Curve*, by means of which, in a moment, the displacement can be read off at any draught. It is constructed in the following manner (fig. 2).

Draw the vertical line A B, and upon it construct a scale at, say, $\frac{1}{2}$ inch = 1 foot, indicating the draughts up to the load line, exactly as read upon the stem or stern of a vessel, inserting in each foot space twelve equal divisions for inches. From the top of the line A B draw the horizontal line A C. Divide this line into $\frac{1}{2}$-inch spaces, each one representing 100 tons of displacement. Subdivide the spaces again into tenths, each of which will, therefore, represent 10 tons. Supposing our vessel be of box form, 100 feet long, 20 feet broad, and 10 feet draught, then by calculating the displacement at a series of draughts, say 2, 4, 6, 8, and 10 feet respectively, we are in a position to find a number of points by means of which the displacement curve is constructed.

The displacement at—

2 feet draught = 100 × 20 × 2 = 4,000 cubic feet
$\frac{4,000}{35}$ = 114·2 tons.

4 feet draught = 100 × 20 × 4 = 8,000 cubic feet
$\frac{8,000}{35}$ = 228·5 tons.

6 feet draught = 100 × 20 × 6 = 12,000 cubic feet
$\frac{12,000}{35}$ = 342·8 tons.

8 feet draught = 100 × 20 × 8 = 16,000 cubic feet
$\frac{16,000}{35}$ = 457·1 tons.

10 feet draught = 100 × 20 × 10 = 20,000 cubic feet
$\frac{20,000}{35}$ = 571·4 tons.

To construct the curve proceed as follows :—

DISPLACEMENT AND DEADWEIGHT. 5

Fig. 3.—Displacement Curve.

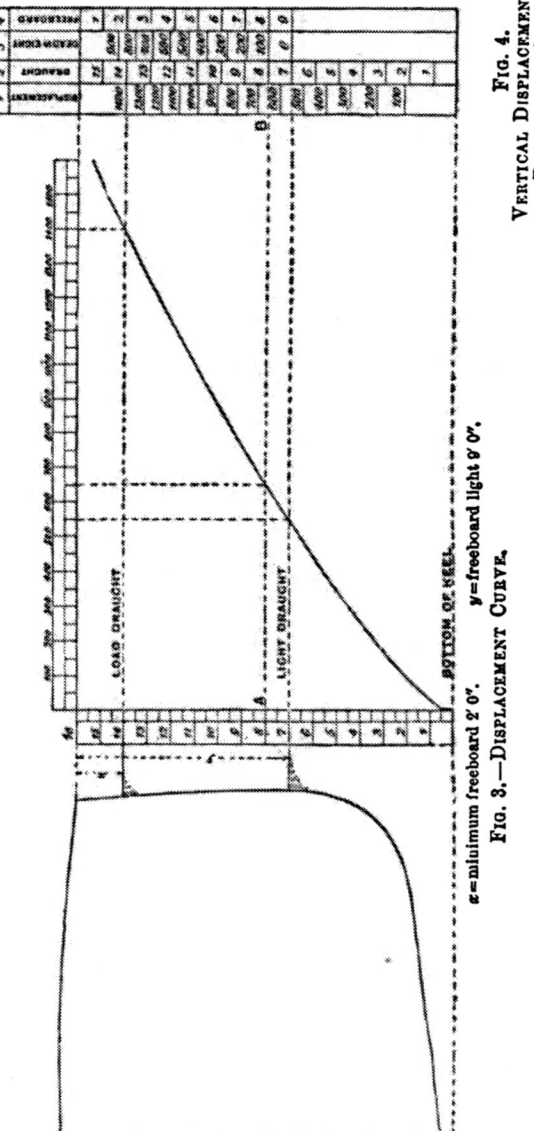

Fig. 4.
Vertical Displacement and Deadweight Scale.

Through the 2, 4, 6, 8, and 10 feet draughts on the scale draw lines parallel to the line AC. The displacement at the 2 feet waterline is found to be 114·2 tons. Through the point in the horizontal scale of tons representing this, drop a vertical line to the 2 feet waterline. The intersection of these lines gives the first point in the curve. Repeat the same operation for the other displacements at their respective waterlines. A line through the points of intersection gives the displacement curve. For a box-shaped vessel, it is a straight line as shown. Having this, the displacement can be read off at any intermediate waterline between the bottom of the keel and the load waterline. The displacement curve for an actual ship is constructed in the same way. Fig. 3 is an illustration of such an one.

The load waterline in this case is 14 feet above the bottom of the keel, which indicates on the curve 1400 tons of displacement. The mean draught of the vessel in her light condition is 7 feet, which reads from the curve 550 tons displacement, leaving a carrying capacity or deadweight of (1400 − 550 =) 850 tons. Let these terms be clearly understood. *The total weight of the ship in whatever condition and floating at any draught is equal to the displacement at that draught. Deadweight is carrying power only*, over and above the actual weight of the ship and her equipment. It, therefore, comprises cargo and bunker coal. *The deadweight of a ship floating at a particular draught is the difference between the displacements in the light condition, and at that draught.*

From the displacement curve an even simpler method of indicating displacement can be arranged, useful more especially to the officer of the cargo-carrying vessel. This is a *Vertical Displacement Scale*, and with it is usually combined a *Deadweight Scale* (see fig. 4).

Column 2 is a *scale of draughts* exactly similar to the scale of draughts on the displacement curve. Column 1 is a *scale indicating the displacements* corresponding to the draughts, the readings of which are identical with the readings from the curve, since the one is constructed from the other. For example, strike a horizontal line AB from the displacement curve to the vertical scale at, say, 8·feet draught. The reading from the curve gives 655 tons, which is the same on the vertical scale.

Column 3 is a *deadweight scale*. As already pointed out, deadweight is the difference between the displacement at any particular draught and the weight or the displacement of the vessel when light. In the above case the vessel floated light at a mean draught of 7 feet, which represents 550 tons on the displacement scale, while the deadweight stands at *nil*. The difference between the displacements at light draught and at 1 foot intervals above the light draught will equal the respective deadweights at these draughts.

Column 4 represents the *freeboard*. More about this will be found in Chapter IX. Suffice it for the present to state that by freeboard is meant the distance from the top of the weather decks at midships to the waterline at which the vessel floats. It is, therefore, measured from the deck downwards. In the above case the minimum freeboard was fixed at 2 feet.

Thus, by means of the vertical scale, a ship's officer can read at a glance the total displacement—the deadweight and the corresponding freeboard—at any particular draught.

Tons per Inch Immersion.—Another very useful curve, closely related to the subject of displacement, may be constructed. This is known as the *Tons per Inch Curve*. By "tons per inch" is meant the number of tons necessary to be placed on board or to be taken out of a vessel, to effect an increase or decrease of 1 inch in the mean draught. Thus the term "tons per inch" really means displacement per inch. It is found by calculating the displacement for a foot of the depth at the particular draught at which the vessel is floating, and dividing this by 12. The result is the increase or decrease of the displacement for 1 inch alteration in draught, or "tons per inch."

The area of the waterline in square feet is first found, and reckoning this to be a foot deep, the square feet are at once converted into cubic feet. These cubic feet divided by 35 give the number of tons per foot. Tons per foot divided by 12 give the required "tons per inch."

The formula may be written thus:—

$$\frac{\text{Area of waterline}}{35 \times 12} = \frac{\text{Area of waterline}}{420} = \text{"tons per inch."}$$

The "tons per inch" for the box vessel, for which the displacement curve (fig. 2) was constructed, would be:—

$100 \times 20 = 2000$ square feet area of waterline.
$2000 \times 1 = 2000$ cubic feet of displacement.
$\frac{2000}{420} = 4\cdot7$ "tons per inch."

It will be observed that since a box vessel is unchanging in horizontal section from the bottom to the top, the "tons per inch" will be the same at any draught, thus rendering the construction of a curve unnecessary. This is not so in the case of an ordinary ship. The waterplanes, from the top of the keel to the load waterline, all varying in form, necessitate the construction of a curve for readiness and convenience, so that the "tons per inch" may be ascertained immediately for any draught.

This curve is made in the same manner as the displacement curve. The "tons per inch" at the 4, 8, 12, and 16 feet heights above the top of the keel are 8·4, 9·7, 10·36, and 10·6 respectively. Vertical lines are dropped from these positions on the "tons" scale to their respective waterlines, the intersections of which give the points necessary for the construction of the curve (fig. 5). Its shape is somewhat different from the displacement curve for this reason. With every increase of draught the displacement must increase, and especially in the region of the load waterline, where the vessel is fullest in all ordinarily designed vessels, thus tending to make the curve continue to spread. On the other hand, the "tons per inch" increases rapidly until the vicinity of the load waterline is reached, and then the sides of the vessel, in the case of ordinary cargo steamers, being somewhat perpendicular, there is little variation in the area of the waterplanes, and here the "tons per inch" remains about the same, the curve contracting and bending to a vertical position, as shown in the illustration. As an example of how to read the "tons per inch" curve at, say, 6 feet 6 inches draught, strike the horizontal line A B to the curve. At the point of intersection draw a vertical line to the scale of tons, and there is indicated 9·1 "tons per inch."

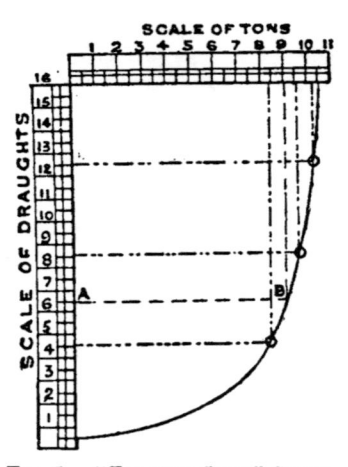

FIG. 5.—"TONS PER INCH" CURVE.

The use of a curve of "tons per inch" may be illustrated in a variety of ways.

For example, suppose a vessel to be floating at a certain draught, at which the "tons per inch" is 15. On calling at a port a moderate quantity of cargo has to be discharged, the weight of which is not exactly known. After discharging, the mean draught is found to have decreased 4½ inches, therefore the weight of the cargo discharged is 15 × 4½ = 67½ tons. Again, supposing a steamer floating at her load waterline, where the "tons per inch" is 15, to consume 100 tons of coal on a voyage, the decrease in draught would be $\frac{100}{15} = 6\cdot6$ inches approximately. Every seaman knows that on a vessel passing from salt to fresh water, an increase occurs in the mean draught, and a decrease when passing from fresh to salt water. The reason for this is, that salt and fresh water differ in density, and thus present different supporting qualities to objects floating in them. To support 1 ton of weight requires a displacement of 36 cubic feet of fresh water, while salt water, being denser and heavier, and more capable of affording support, will bear up a weight of 1 ton on a displacement of 35 cubic feet.

Sometimes where the depth of entrance to a dock is limited, it is very necessary to know what change of draught will occur in passing from salt to fresh water, or *vice versâ*. At the load draught this is marked on the ship's side by the Board of Trade, or one or other of the Registration Societies, if the vessel is classed. By the aid of the "tons per inch" curve, we may ascertain the change in draught for ourselves. Suppose a vessel to be floating at a certain draught where the displacement is 4,500 tons, and the "tons per inch" 20. Now, a cubic foot of sea water weighs 64 lbs., and a cubic foot of river water, which is chiefly fresh, about 63 lbs., the difference being $\frac{1}{64}$.* Since the total weight of the ship remains the same, the total weight of water displaced must remain the same also, though as it becomes fresh water it increases in volume, because it is $\frac{1}{64}$ lighter, measure for measure. As already stated, when floating in sea water, she displaces 4500 tons. Suppose her now to be floating in river water at the same waterline, her weight or displacement will be $\frac{1}{64}$ less.

$$\tfrac{1}{64} \text{ of } 4500 = 70\cdot3 \text{ tons.}$$

The "tons per inch" was 20 tons in salt water, but it also will be $\frac{1}{64}$ less.

$$20 \times \frac{63}{64} = 19\tfrac{11}{16} \text{ "tons per inch" in river water.}$$

The change in draught will therefore be—

$$70\cdot3 \div 19\tfrac{11}{16} = 3\cdot57 \text{ inches.}$$

* Fresh water may be taken at 62½ lbs. per cubic foot.

The formula may be shortened, and written thus—

$\frac{1}{64}$ of displacement ÷ $\frac{63}{64}$ of "tons per inch" = increase in draught;

therefore, $\frac{4500}{63 \times 20} = 3\cdot57$ inches.

Coefficients of Displacement.—In comparing the displacement or underwater form of one vessel with another, it is not sufficient to say that one is long and the other short, one broad and the other narrow, or one deep and the other shallow. Nor is a numerically correct idea conveyed by saying that one is fine and the other full or bluff. A more comprehensive means must be adopted, and this is attained by *coefficients*.

Suppose that out of a block of wood 6 feet long, 1½ feet broad, and 1 foot deep, the model of the underwater form of a vessel

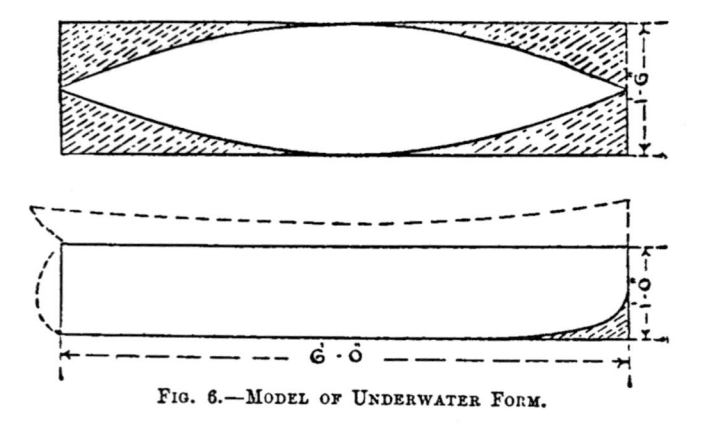

FIG. 6.—MODEL OF UNDERWATER FORM.

be cut out, as shown by fig. 6, the extreme dimensions of which are—length, 6 feet; breadth, 1½ feet; depth, 1 foot.

Before the block was cut, it contained $6 \times 1\frac{1}{2} \times 1 = 9$ cubic feet. The extreme dimensions of the remaining part in the form of the model are still the same—6 feet long, 1½ feet broad, 1 foot deep—but much of the volume of the block has been cut away, as shown by the hatched lines, leaving, say, 6 cubic feet, which is $\frac{6}{9} = \frac{2}{3}$, or, as it is generally written, ·66 of the whole block, and this is termed the *coefficient*, or, in other words, the

DISPLACEMENT AND DEADWEIGHT.

comparison of fineness. Thus the coefficient of fineness of any vessel is the fractional part (usually expressed in decimals) which the volume of the displacement bears to the circumscribed block.

·8 would be a very full vessel.
·7 to ·75, an average cargo steamer.
·65, a moderately fine cargo steamer.
·6, a fine passenger steamer.
·5, an exceedingly fine steamer, but an average for steam yachts.
·4, a very fine steam yacht.

By means of coefficients a comparison of the displacement or fineness between two or more vessels may be struck relatively to their circumscribing rectangular blocks.

Vessels of the same extreme dimensions, and the same coefficients of fineness, and, therefore, the same displacements, may vary considerably in form or design, which in turn may affect the speed.

Knowing the extreme dimensions of a vessel, and the coefficient of fineness, the exact displacement can easily be arrived at. For example, take a vessel 100 feet long, 20 feet broad, and floating at 8 feet draught, the coefficient of fineness being ·6.

The displacement would be—

$$\frac{100 \times 20 \times 8 \times ·6}{35} = 274·2 \text{ tons.}$$

CHAPTER II.

MOMENTS.

CONTENTS.—Moments—Examples of Moments—Centre of Gravity—Effect upon Centre of Gravity of Moving Weights on a Ship, Vertically and Transversely.

Moments.—Moment of a Force about a Point.—We may speak of a ship when inclined from the upright position as having a *moment* tending to bring her to the upright position again. This is usually termed a *righting moment*. On the other hand, it might be found that when the vessel was inclined to a certain angle, she possessed no inclination to return to her original upright position, but continued to heel until she capsized. In this case, we may say she possesses an *upsetting*, or *capsizing moment*.

This important term "moment" is easily understood. Everybody knows the meaning of simple distances like inches and feet, and of simple weights like pounds and tons. We must accustom ourselves to quantities which are got by multiplying distances by weights. Thus, we may have to multiply 5 feet by 8 tons. The product is called 40 *foot-tons*. If we multiply 5 feet by 8 pounds, the product is called 40 *foot-pounds*. If we multiply 5 inches by 8 tons, the product is called 40 *inch-tons*. Now weight is one kind of force, and other forces, such as pressures and resistances (say, wind or steam pressure and water resistance), are also conveniently measured in pounds or tons. And pounds or tons of any kind of force, such as pressure and resistance, may also be multiplied by inches or feet of distance, giving inch-tons, or foot-pounds. etc., as the case may be.

Now the quantity called the *Moment of a Force* (the only kind of moment we want just now) is got by taking some weight or other force and multiplying it by what may be conveniently called its *leverage*. This leverage is the perpendicular distance of the direction in which the force acts from some conveniently chosen point. The point often chosen for this purpose in a ship is the centre of gravity of the ship. For instance, 5 tons of wind pressure on sails multiplied by a leverage from the ship's centre of gravity of 30 feet would give a moment of force about that centre of gravity of 5 × 30 = 150 foot-tons. And a water pressure of 1000 tons multiplied by a leverage from the same centre of gravity of 3 inches would give a moment of force about that centre of gravity of $1000 \times 3 = 3000$ inch-tons.

Simple illustrations of moments of force may be got from levers.

Let A B (fig. 7) be a lever 5 feet long, supported at one end, A, as shown, and at the other having a weight of 4 tons suspended at right angles to the lever. In this condition there is a moment about A of 4×5 (weight multiplied by distance from centre of weight to point of support) = 20 foot-tons, tending to capsize, or break, or bend the lever at the point A, and the 20 foot-tons in this case may be called a capsizing moment,

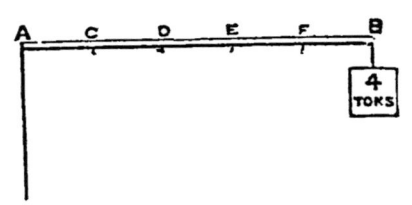

FIG. 7.—WEIGHT ACTING ON A LEVER.

or a bending moment. Let the lever be subdivided into foot intervals at the points C D E F. If the weight be now moved to the point F, which is 4 feet from A, the moment about A will be $4 \times 4 = 16$ foot-tons. In like manner, if moved

To the point E, the moment about A will be $4 \times 3 = 12$ foot-tons.
,, D, ,, ,, $4 \times 2 = 8$,,
,, C, ,, ,, $4 \times 1 = 4$,,
,, A, ,, ,, $4 \times 0 = 0$,,

The last, no foot-tons, as will be seen, is accounted for by the fact that the downward force of weight is acting in the same vertical line as the upward support, there being therefore no capsizing moment; or, as we are speaking of a lever or bar, no breaking or bending moment.

There is just another point bearing on moments which will also assist us in studying the structure and strains of ships. In most cases the leverage is the variable factor in influencing the amount of foot-tons, the weight remaining constant. It should now be noticed that the *moment is always greatest at the point of support, and when the weight is removed farthest from it.* Keeping the weight in the position shown at B, the tendency to fracture at the point C will be considerably less since the **leverage** is less. The moment at this point is $4 \times 4 = 16$ foot-tons, and it will continue to diminish until, at the point B, it has vanished completely.

But let us take another simple example illustrative of moments. Fig. 8 represents a lever supported at the point A, and with weights of 3 and 4 tons suspended at its extremities, at distances of 8 and 5 feet respectively from the point of support. First, what would be the tendency to fracture at the point A? On the side towards B there would be a moment of

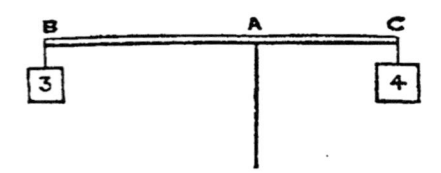

FIG. 8.—ACTION OF WEIGHTS ON A POINT OF SUPPORT.

$3 \times 8 = 24$ foot-tons, and on the side towards C, $4 \times 5 = 20$ foot-tons.

But suppose it is asked, "Is there a capsizing moment, and, if so, how much?" As we have already seen, there is a moment of 20 foot-tons on the side towards C and 24 foot-tons towards B, and since the moment towards B preponderates by $24 - 20 = 4$ foot-tons, the lever would therefore capsize towards that side.

Centre of Gravity.—But suppose we wished to find the point to which the support must be moved in order that the moments might balance one another, the lever remaining in a state of equilibrium or rest. This would be done by dividing the difference of the two moments, which was found to be 4, by the total weight $3 + 4 = 7$, 4 divided by $7 = \frac{4}{7}$ foot, and moving the support this distance towards B, the side possessing the greater moment. Let us prove this. The support, according to our calculation, will now be $8 - \frac{4}{7} = 7\frac{3}{7}$ feet from B, and $5 + \frac{4}{7} = 5\frac{4}{7}$ feet from C.

Towards B the moment is now $3 \times 7\frac{3}{7} = 22\frac{2}{7}$ foot-tons.
,, C ,, $4 \times 5\frac{4}{7} = 22\frac{2}{7}$,,

Being exactly equal to each other, the lever remains at rest. Now this *balancing point* or *centre of moments* is a very important point. For the point at which a body (acted on only by its own weight) will balance is called the *centre of gravity of that body*. And the point at which a system of bodies (acted on only by their own weight) will balance is called the *centre of gravity of that system of bodies.* An example of a system of bodies is a ship with or without cargo in her. And as such a body as a ship cannot be balanced experimentally so as to find the centre of gravity by trial, the calculation of the moments in question is employed upon the ship and the weights in her until their balancing point is found, and that balancing point is the centre of gravity of the ship and the load then in her, and the whole weight of ship and cargo may be supposed to act vertically downwards through that point.

For, proceeding further, in fig. 9, A B is a lever, with a weight of 2 tons at a distance of 2 feet from the support, and a weight of 4 tons at a distance of 4 feet from the support, both on the same side, and, similarly, weights of 2 and 4 tons at 2 and 4 feet respectively on the other side of the support. Then the moments must be the same, and it is evident that the lever is supported

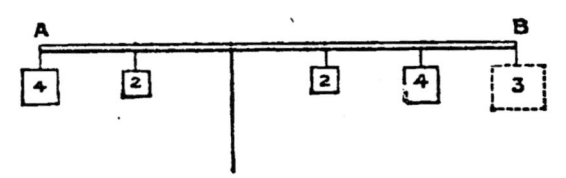

FIG. 9.—CENTRE OF GRAVITY.

at the centre of gravity of the total weight. Let a weight of 3 tons be now suspended from the extremity of the lever at B, and at a distance of 6 feet from the support, as shown by the dotted square. It is required to find the centre of gravity now. Since the moments preponderate towards B by $3 \times 6 = 18$ foot-tons, if this be divided by the sum of the weights, we shall get the distance the centre of gravity has moved, which is $\frac{18}{15} = 1\frac{1}{5}$ feet towards B.

But suppose that instead of a weight being added, the 2 ton weight on the side towards B be removed. Let us find the centre of gravity now. The moment towards A will preponderate by $2 \times 2 = 4$ foot-tons, and this, divided by the total weight, which is $2 + 4 + 4 = 10$, will equal $\frac{4}{10} = \frac{2}{5}$ foot = shift of centre of gravity towards A.

The reader is advised to make himself thoroughly familiar with the principle of moments, as illustrated in this chapter, as this will be found to be absolutely essential in order to deal successfully with the following chapters. Whenever several **weights** are connected by any means, as shown in the foregoing **examples,** the combined system of weights acts directly through one balancing point, and this is their Common Centre of Gravity. And, if this method of finding the common centre of gravity of a ship and its load be used, one first finds how far from stem or stern that centre of gravity lies, and how much it is shifted fore or aft by shifting weights fore and aft. And precisely the same sort of calculation may be used to find how far above the keel the same centre of gravity lies, and how much up and down it is shifted by shifting weights up and down in the ship. We have only in this case to measure our leverages vertically above and below the balancing point we are trying to find by assuming a point for its position, and multiply the weights by the vertical leverages to get the moments. We, therefore, now proceed to show how to apply these principles to actual ships.

Effect of Moving Weights on a Ship's Centre of Gravity. —Suppose a vessel be floating with her centre of gravity at the

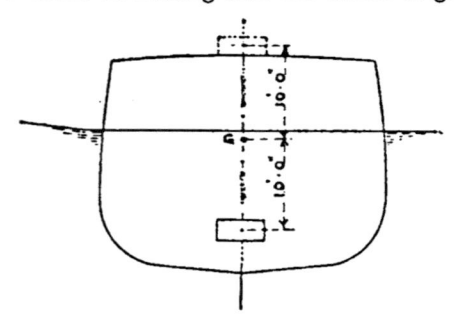

FIG. 10.—CENTRE OF GRAVITY OF A SHIP.

point G (fig. 10). (It may be said here that the position of the centre of gravity of the ship when light could be supplied by the builder, and this would limit the duty of the shipmaster to finding the position of the centre of gravity with various kinds of loading. But as it is not customary with all shipbuilders to do this, a simple experiment by which the shipmaster himself may find the centre of gravity will be given later on.) Her *total displacement*—that is, her *weight*—is 1000 tons. Let a weight of 30 tons, already on board, be raised from the hold and placed on the deck, at a distance of 20 feet from its former position, as shown. What effect will this have upon the centre of gravity? Since the centre of gravity is in the centre of the total weight it is evident that, *if a weight be raised, the centre of gravity must travel in the direction of the moved weight.* To find the exact distance the centre of gravity has moved, the

MOMENTS. 17

same rule is adopted as in the case of a vertical lever. *Multiply the weight moved by the distance it is moved, and divide the result by the total weight or displacement.*

$$\frac{30 \times 20}{1000} = \frac{600}{1000} = \cdot 6 \text{ foot,}$$

the distance the centre of gravity has been raised.

Again, suppose the weight, instead of being placed on deck, had been taken out of the vessel altogether. In this case, *multiply the weight by its given distance from the centre of gravity of the ship*, say, 10 feet, and *divide by the total displacement after the weight is removed* (1000 − 30 = 970 tons = displacement after weight is removed).

$$\frac{30 \times 10}{970} = \frac{300}{970} = \cdot 3 \text{ foot,}$$

the distance the centre of gravity has been raised.

Again, suppose, when the displacement is 1000 tons, a weight of 30 tons be placed on board 10 feet below the centre of gravity. *Multiply the new weight by its distance from the centre of gravity,* and *divide by the new displacement* (1000 + 30 = 1030 tons = new total displacement).

$$\frac{30 \times 10}{1030} = \frac{300}{1030} = \cdot 28 \text{ foot,}$$

the distance the centre of gravity has been lowered.

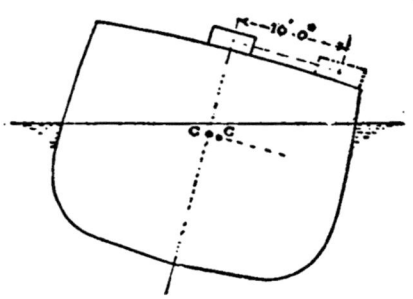

FIG. 11.—EFFECT OF WEIGHT MOVED ATHWARTSHIPS.

Let us now notice the effect of a weight moved athwartships (fig. 11).

The centre of gravity is in the position shown at G, and the total displacement is 1000 tons.

A weight of 20 tons already on board, on the centre of the upper deck, is moved 10 feet to starboard. The centre of gravity must move in the same direction, and in a line parallel to the line joining the centres of the weight in its original, and in its new position. To find the exact distance moved, *multiply the weight by the distance moved*, and *divide by the total displacement*.

$$\frac{20 \times 10}{1000} = \frac{200}{1000} = \cdot 2 \text{ foot.}$$

(G G^1), the distance the centre of gravity has travelled to starboard.

CHAPTER III.

BUOYANCY.

CONTENTS.—Buoyancy—Water Pressures—Reserve Buoyancy—Sheer—Value of Deck Erections—Centre of Buoyancy—Curves of Vertical and Longitudinal Centres of Buoyancy—Effect upon a Ship's Centre of Buoyancy of Immersing or Emerging Wedges of Buoyancy — Effect of Entry of Water upon Buoyancy — Camber, or Round of Beam — Testing Water Ballast Tanks.

Buoyancy.—Buoyancy means floating power. Under what conditions will a vessel float? Simply when its enclosed watertight volume is greater than its total weight in tons multiplied by 35, since it requires that 35 cubic feet of salt water be displaced before sufficient support is obtained to bear up 1 ton of weight. If the enclosed watertight volume of the vessel in cubic feet is less than its total weight in tons multiplied by 35, it will be evident that the vessel will sink. This we have learned from the chapter on "Displacement." But it may be asked, "Why does a vessel float?" or "What is the nature of the application of the pressure which is obviously produced by the water itself in order to sustain an object of greater or less weight floating upon it?" This we shall endeavour to explain.

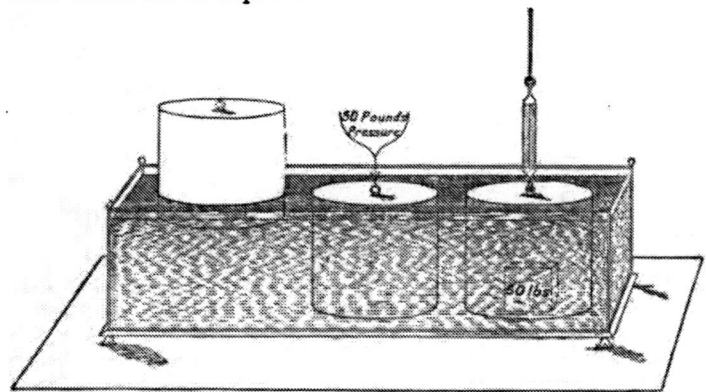

FIG. 12.—ILLUSTRATING WATER SUPPORT.

Water Pressure.—Fig. 12 represents a tank nearly full of fresh water. At the left-hand end of this tank is a hollow cylinder made of, say, sheet iron. It is completely watertight, being entirely closed at the ends, one of which is a watertight lid. By ordinary calculation, its volume is found to be 1 cubic foot, the area of one end being 1 square foot and the height 1 foot. When totally empty, with the ends closed, its weight is, say, $12\frac{1}{2}$ lbs. A flat piece of sheet iron, similar to that from which this cylinder is made, would sink if thrown into the tank, while the sheet iron cylinder shows no signs whatever of sinking, but floats as shown, with more than three-fourths of its volume out of water. In order to immerse this tank so that its uppermost surface is level with the water surface, there would require to be an application of considerable downward pressure, which pressure, if measured, would be found to be equal to a weight of 50 lbs. (See second position of cylinder.)

Or, supposing that, instead of a downward pressure applied to the outside, a weight of 50 lbs. had been placed inside the cylinder, and the whole suspended on a spring balance, the total weight registered will be $12\frac{1}{2} + 50 = 62\frac{1}{2}$ lbs. Suspended in mid air, it is clear that the entire support is afforded by the spring balance. While thus suspended, let the cylinder be gradually lowered into the water in the tank. (See third position of cylinder.) Immediately the bottom of the cylinder enters the water, and the immersion increases, the spring balance registers a reduction in weight, and this reduction continues in exact proportion to the rate of immersion, until, by the time that the cylinder is half immersed, it registers $31\frac{1}{4}$ tons, and finally, when the uppermost surface is flush with the water surface, the balance registers 0, and the cylinder is barely maintained at the water surface. The addition of the slightest weight would send the cylinder to the bottom. Such an experiment as this, proves that, while a floating object is subject to the same gravitation forces out of the water as in it, the downward pressure of its weight is balanced by an exactly equal upward pressure from the water itself. Now, these upward water pressures are of enormous importance to a floating ship, for not only, as we have seen, do they afford the support which keeps her on the bosom of the ocean, but are the means whereby, when she is forcibly inclined to a greater or less degree from the upright, she is enabled to regain her normal position, though, under other conditions, these same pressures may be the agents tending to capsize her when so inclined. (The stability aspect of the subject is dealt with in Chapter VI.). The foregoing experiment also enables us to estimate the nature of the application of the upward water pressures, for the spring balance clearly indicated that the amount of this pressure varies directly as the depth.

BUOYANCY. 21

Thus, taking a rectangular object floating in salt water, as shown in fig. 13, it can be similarly demonstrated that, for every square foot of area on the bottom of the box, there is an

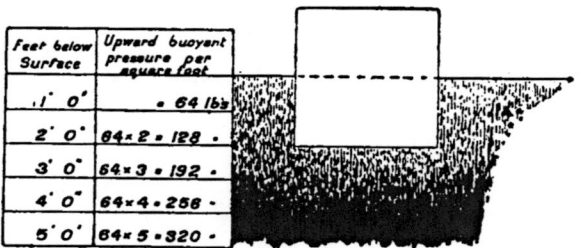

Feet below Surface	Upward buoyant pressure per square foot
1' 0"	= 64 lbs
2' 0"	64 × 2 = 128
3' 0"	64 × 3 = 192
4' 0"	64 × 4 = 256
5' 0"	64 × 5 = 320

FIG. 13.—ILLUSTRATING WATER PRESSURE INCREASING IN PROPORTION TO DRAUGHT.

upward pressure of 64 lbs. at 1 foot depth; at 2 feet depth the pressure is twice 64 lbs., and so on, in proportion to the depth. An attempt has been made to illustrate this by drawing the lines indicating the water pressures denser as the depth increases.

Our next step is to ascertain the direction of the application of these water pressures. Take a vessel similar to that shown in fig. 14, whose sides are perforated with a considerable number of small holes. The vessel is filled with water, and it is found that from every hole the water squirts out in a direction square to the surface of the vessel, as shown.

Or take the same vessel empty, and plunge it into the water. Here, again, the water is seen to squirt into the inside of the vessel in a direction perpendicular to the surface of the vessel. From these two illustrations it is evident that water contained in a vessel, large or small, exerts its pressure upon every unit of area on the inside, in lines of action perpendicular to the surface. On the other hand, when it is immersed, the pressure on its immersed surface is also perpendicular to that surface.

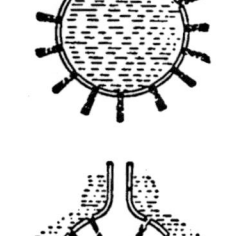

FIG. 14.—WATER PRESSURE.

So far, both by observation and the foregoing experiments, we have arrived at the conclusion that a floating object, whether it be the cylinder or a log of wood, or a 20,000 ton ship, is supported, or enabled to float by upward pressure given by the water

22   KNOW YOUR OWN SHIP.

itself, and, moreover, that the water pressure is exerted in lines of force perpendicular to the surface immersed—longitudinally, transversely, obliquely, and vertically, as shown by the lines of water pressures in fig. 15, tending to crush in the sides, ends, and bottom, or to push the vessel out of the water. This is understood all the more clearly when it is remembered that, immediately any floating object is taken out of the water, the water rushes in on all sides,

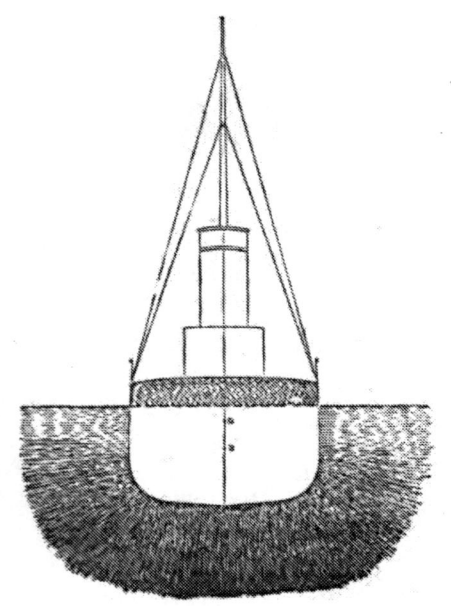

FIG. 15.—SHOWING LINES OF WATER PRESSURE.
Volume immersed = buoyancy.   Volume above water level = reserve buoyancy.
G = centre of gravity.   B = centre of buoyancy.

and, more quickly than can be seen, the cavity formerly occupied by the object is filled, and the water is again at its uniform level.

Thus, while a ship is floating, the water has still the inclination, if the phrase may be so used, to occupy the space filled by the immersed volume of the vessel. Why the water does not succeed in crushing in the exterior of the vessel will be dealt with in another chapter on "Structure." Why it does not succeed in thrusting the ship out of the water has already been explained by the fact that there is the downward pressure of the weight of the ship itself. Thus a vessel will sink to a draught such that the

BUOYANCY. 23

sum of the *upward* pressures of buoyancy exactly equal the weight of the ship. To balance a ship weighing, say, 5000 tons, in mid air, by means of supports, such as pillars of iron, would be no easy matter. A ship is a large, bulky object, and with such an enormous weight, a very careful estimate of the strength of the supports and their positions would have to be made in order to ensure the necessary conditions being fulfilled. But in water, the support and balance are perfect. There are exactly 5000 tons of supporting pressures from the water to exactly support the 5000

FIG. 16. – SHOWING LINES OF WATER PRESSURE.
(For explanation see fig. 15.)

tons weight of the ship, and the balance is perfectly effected by the centre of the support (or the resultant of all the upward pressures), and the centre of the weight being exactly in the same vertical line, which condition is absolutely essential in order to preserve an exact balance : or, in other words, the centre through which the resultant of the buoyancy pressures is exerted (B, fig. 15), which is found to occupy a position in the centre of the displacement or immersed volume, and the centre of gravity (G) must be in the same vertical line in order to ensure a condition of equilibrium.

The net result is that, in a ship floating at rest, whether upright or inclined, we have two equal opposing forces, exactly neutralising each other, and therefore producing no motion whatever.

Figure 16 shows the same vessel, forcibly inclined, and, as a result, the lines of force of gravity and buoyancy are no longer exerted in direct opposition to each other, G being in exactly the same position in the ship, while B has moved into the centre of the immersed volume. A ship, or any floating object, on being launched, or by any means placed in water, immediately, by these natural laws, places herself in such a position as we have described; indeed, she must do so before a condition of rest can be maintained. Such a position is not necessarily the upright, considered either transversely or longitudinally, but, under certain circumstances, due to form and the distribution of weight constituting the ship and cargo, may necessitate the vessel inclining to port or starboard, or trimming by the stem or the stern. While the total buoyancy pressures and weight of a floating ship are equal to each other, it must not be imagined that these buoyancy pressures measure the total pressure upon the immersed hull of a ship.

FIG. 17.—SHOWING THE TOTAL PRESSURE ON THE IMMERSED HULL OF A SHIP.

Figure 17 shows a box vessel floating, partially immersed. The lines of water pressures are shown vertical on the bottom and horizontal on the sides and ends (perpendicular to immersed surface).

The horizontal pressures afford no support whatever, simply tending to push the vessel in the direction of their thrust. This, however, is balanced by the opposing pressures on the opposite side. The vertical surfaces of floating objects are exposed only to horizontal, crushing pressures; such surfaces only as are horizontal or oblique are exposed to buoyancy pressures. So that a ship's immersed surface endures all the upward pressures providing buoyancy, or floating power, equal in amount, therefore, to her own weight, and all the horizontal pressures in addition, which afford no buoyancy, but simply tend to crush in the sides and ends of the vessel. It is clear, then, that to estimate the amount of buoyancy or upward pressure by using the total area of immersed surface would be entirely wrong. It must be distinctly understood that only upward pressures, either vertical or oblique, afford support, all others producing only crushing strains upon the vessel's immersed exterior.

Remembering that supporting pressures increase in proportion to the depth, a difficulty sometimes arises in understanding the nature of the pressures in wholly immersed objects. For it is reasoned that, if it be true that upward pressures increase with depth, how comes it that, as soon as a vessel becomes too heavy to remain supported at the surface of the water, it never finds a place of rest until it reaches the bottom? The answer is simple enough. It is quite true that, as the object descends in sinking, the pressure on the bottom surface increases in direct proportion to the depth, and remembering that water pressures are always perpendicular to the immersed surface, the pressures on the upper surface of the sinking object also increase in direct proportion to the depth. Thus, while it is perfectly true that, as the object sinks, it is enduring an increased pressure upon the whole of its external surface in direct proportion to the depth, the increased upward pressure is balanced by a proportionately increased downward pressure.

Since the law is so strictly enforced that no vessel shall proceed to sea loaded beyond her load watermark, except under penalty, it is of the greatest importance that the responsible authorities should exercise the greatest care in taking the correct load draught in the case of a vessel which has loaded aground. For it not unfrequently happens that a vessel registers a less mean draught immediately after becoming afloat than she did when she was aground. After the attention already given to buoyant pressures, the reason for this may be readily surmised. Suppose we could take one of the cylinders shown in fig. 12, and place the bottom surface in close contact with the bottom of the tank, so that neither air nor water could possibly get beneath it, we should then find that, though water surrounds it on all sides, even to its top surface, no buoyancy pressures are experienced by the cylinder, and therefore it has not the least floating power, even though it is in its light condition, weighing only $12\frac{1}{2}$ lbs.

But, more than this, we know that the atmospheric pressure at the earth's surface is 15 lbs. per square inch. Hence, if neither water nor air is capable of getting under the bottom surface of the cylinder, not only is there no buoyant pressure, but there is an atmospheric pressure of 15 lbs. on every square inch of the upper outside surface of the cylinder, as well as 15 lbs. per square inch on the internal surface of the cylinder, assuming, naturally, that it is filled with air, and not a vacuum. There is, therefore, a pressure of 15 lbs. per square inch on the inside bottom surface of the cylinder. And as this bottom surface contains an area of 1 square foot, or 144 square inches, the total pressure is $144 \times 15 = 2160$ lbs., or nearly a ton, and against this there is not the slightest upward pressure from the water.

Suppose, however, that the bottom of the cylinder is only partially in close contact with the bottom of the tank, so that neither water nor air could intervene over an area of, say, 10 per cent. of the bottom. The buoyancy pressure would now be $\frac{90}{100}$ of 64 lbs. (supposing the water to be salt) = 57·6 lbs., while the atmospheric pressure is $144 \times \frac{10}{100} \times 15 = 216$ lbs. The buoyancy pressure is 57·6 lbs., and the atmospheric pressure 216 lbs.; the difference, 158·4 lbs., represents the preponderating downward pressure, indicating that the cylinder will not float.

It is true, these are theoretical illustrations, though, in some degree, somewhat similar conditions are found to exist in the case of actual ships. It will depend upon the nature of the bottom upon which the ship is lying, and the amount of the bottom surface in close contact with the bottom upon which she rests. This is the only explanation which can be given for such an occurrence as described at the beginning of this feature of buoyancy, where a ship registers a less draught after floating than immediately before when she was aground.

Reserve Buoyancy.—Let it be imagined that a vessel floating at the waterline where the "tons per inch" is, say, 12 tons, has an additional weight of 60 tons placed on board. The result will be that the weight of the vessel having increased 60 tons in excess of the buoyancy, she must become further immersed. But how far? Well, since we have observed that 1 ton of buoyancy equals 1 ton of displacement, she will therefore sink until she has displaced 60 tons more water, which is an increase of $\frac{60}{12}$, or 5 inches in the draught.

If, on the other hand, 60 tons be taken out of the vessel, the buoyancy will now be 60 tons in excess of the displacement, the result being that the vessel will rise out of the water until the buoyancy is reduced by 60 tons, and the draught is decreased by $\frac{60}{12}$, or 5 inches.

Thus, we have here one of the conditions of a vessel floating at rest in still water—namely, that *the total weight of the vessel equals the total weight of the water displaced*, or the *buoyancy*. The buoyancy of the immersed portion of the vessel represents that which is requisite to keep her afloat. The buoyancy of all enclosed watertight space above the waterline is therefore surplus buoyancy, or safety buoyancy, or, as it is more commonly termed, *Reserve Buoyancy* (see figs. 15, 16, and 18).

Sheer.—It is not sufficient that a vessel have just enough buoyancy to keep her afloat, for if she had only this, every wave would submerge her. It is the surplus buoyancy that gives her rising power, and, as we shall see in Chapter VI., provides righting power when inclined. The great advantage of sheer (fig. 18),

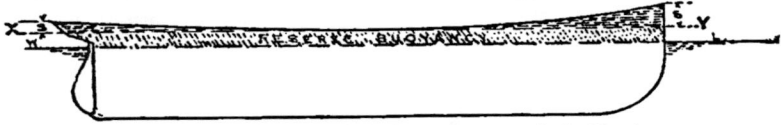

S=Sheer. *Volume above line X, Y = Reserve buoyancy obtained by sheer.*
FIG. 18.—SHEER.

which gives surplus buoyancy at the ends of a vessel, will now be evident, for every time the vessel pitches into the trough of a sea, she immediately displaces more water than her weight, and is, therefore, thrown up again.

Value of Deck Erections.—The Board of Trade recognises this to such an extent as to recommend a certain amount of sheer, according to the type of the vessel. When this amount is exceeded, a reduction in the freeboard is allowed, and when the sheer is less an addition is made to the freeboard (see chapter on "Freeboard"). The value of poops and forecastles, especially the latter, and particularly if efficiently closed at the ends, will now be clearly understood, since they all add to the buoyancy, and at those parts of the vessel where it is greatly needed. Though not to the same extent in efficiency, yet all bridges with watertight ends, and deck houses and hatches—in short, all enclosed watertight erections afford reserve buoyancy.

Centre of Buoyancy.—Now, just as the Centre of Gravity of the weights ranged on a lever can be ascertained, as explained in the previous chapter, so the centre of action of the numerous forces of buoyancy may be found; and since the forces acting upon any body may always be supposed to act directly through the centre of action, the value of this point will be readily granted, especially when we come to deal more closely with the subject of Stability.

However, instead of speaking of the centre of action of the forces of buoyancy, this point is, for brevity, termed the *Centre of Buoyancy,*[*] and, moreover, it is found that the centre of buoyancy is the centre of displacement. Thus, to find the centre

[*] B in figs. 15 and 16.

of buoyancy it is simply necessary to calculate the centre of displacement, as the two names indicate exactly the same point. The centre of buoyancy being the centre of the displacement, it must vary in position with every variation of draught, so that it becomes necessary to arrange a convenient method of readily ascertaining the centre of buoyancy at any draught. This is done by calculating the centres of buoyancy at several draughts parallel to the load waterline, and constructing curves. By means of a calculation of moments, using horizontal areas of waterplanes, instead of weights as in the case of the lever examples in the previous chapter, the positions of the vertical centres of buoyancy are found at the required draughts. In a similar manner the longitudinal centres of buoyancy at several draughts are calculated, and by using vertical areas of transverse sections of displacement at regular intervals fore and aft instead of weights, the several longitudinal centres of buoyancy may be found. See Chapter X. for examples of calculations.

C.B. = Centre of buoyancy.

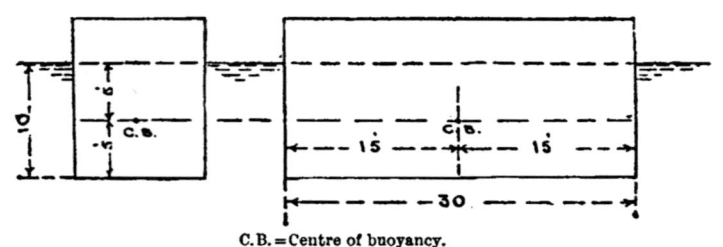

FIG. 19.—CENTRE OF BUOYANCY OF BOX-SHAPED VESSEL.

Curves of Vertical and Longitudinal Centres of Buoyancy.—Thus, to find the actual centre of buoyancy, it is necessary to construct two curves, one for the *vertical centres of buoyancy*, to give vertical position, and the other for the *longitudinal centres of buoyancy*, giving longitudinal position. The intersection of the two lines is the point required.

These curves, at the expense of little time and trouble, can be supplied to the ship's officer by the shipbuilder or naval architect.

To construct curves for a box-shaped vessel would be unnecessary, as it is evident that the vertical centres of buoyancy must always be at half the draught (see fig. 19), and the longitudinal centres of buoyancy at the middle of the length if floating with the bottom parallel to the waterline.

Thus, at 10 feet draught, the vertical centre of buoyancy is 5 feet down from the waterline, at 6 feet draught it is 3 feet down from the waterline, and so on.

BUOYANCY. 29

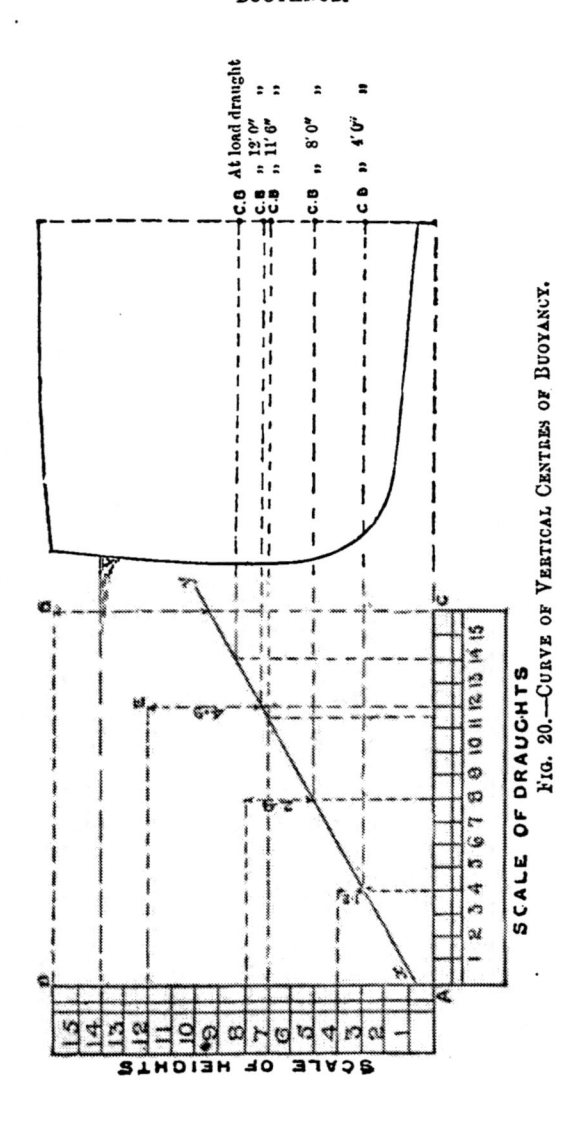

FIG. 20.—CURVE OF VERTICAL CENTRES OF BUOYANCY.

Curve of Vertical Centres of Buoyancy.—The heights of the vertical centres of buoyancy used in the construction of the curve in fig. 20 are for a vessel about 200 feet long, and with a draught of 14 feet when fully loaded. Suppose the vertical centres of buoyancy at the 4, 8, 12, and 16 feet draughts are found to be 1·2, 2·9, 4·9, and 6·6 feet respectively below their respective waterlines. To construct the curve proceed as follows:—Draw the lines A B and A C at right angles to each other. Let A B represent a scale of heights and A C a scale of draughts. Through the 16 feet height in the scale A B draw a horizontal line as shown, and through the 16 feet draught in A C draw a vertical line intersecting the other at D. From the point D set down the distance of the centre of buoyancy below the 16 feet waterline = 6·6 feet. Proceed in the same manner for the 12 feet waterline. Through the 12 feet height draw a horizontal line, and through the 12 feet draught draw a vertical line intersecting the other at E. From E set down the distance of the centre of buoyancy below the 12 feet waterline = 4·9 feet. In a similar manner the centres of buoyancy at the 8 and 4 feet waterlines may be set off. Through these points draw the line $x\,y$, which is *the curve of the vertical centres of buoyancy* required. By means of it the height of the vertical centres of buoyancy above the bottom of the keel may be read off at any draught.

It will be understood that the curve constructed for any particular vessel will be of little use for any other vessel unless of exactly the same form and proportion in the immersed portion of the hull. But for the sake of example, let the curve in fig. 20 be for a yacht with a rising keel and drawing 10 feet forward and 13 feet aft. This represents a mean draught of $\frac{10 + 13}{2}$ = 11 feet 6 inches. At 11 feet 6 inches on the horizontal scale of draughts, set up a vertical line until it intersects the curve of the centres of buoyancy. From this point draw a horizontal line until it meets the vertical scale of heights, and there we read 7 feet, which is the height of the vertical centre of buoyancy above the bottom of the keel. Had the vessel been an ordinary cargo one, floating on even keel, and drawing 11 feet 6 inches fore and aft, the height of the centre of buoyancy would have been the same. As has already been stated, this point, taken by itself alone, is of little use to anyone. It is only when used in relation to other points, with which we shall deal, that it possesses importance.

Curve of Longitudinal Centres of Buoyancy.—The longitudinal centres of buoyancy for the same vessel of which fig. 20 is the curve of the vertical centres of buoyancy are found to be

BUOYANCY. 31

98·7, 99·2, 99·8, and 100 feet from the afterside of the stern post at the 16, 12, 8, and 4 feet waterlines respectively.

The curve would be constructed in the following manner (fig. 21):—

Draw the vertical line A B, and upon it construct a scale of draughts. From B draw the horizontal line B C, and upon it construct a scale of feet, long enough to include the greatest distance of the centre of buoyancy from the stern post, which in this case is 100 feet. 98·7 feet on the line B C gives the first point in the curve. At 12 feet draught draw a horizontal line, and through the point in the scale indicating 99·2 feet drop a vertical line. The point of intersection gives the second point in

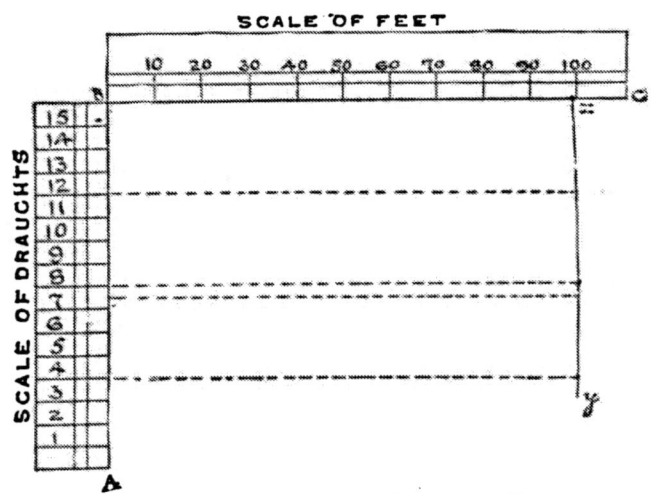

FIG. 21.—CURVE OF LONGITUDINAL CENTRES OF BUOYANCY.

the curve. Proceed in the same manner with the 8 and 4 feet waterlines, obtaining the points indicated by the large dots. Through these points draw the line $x\,y$, which is *the curve of the longitudinal centres of buoyancy*. Supposing we are asked to read off the longitudinal centre of buoyancy at, say, 7 feet 6 inches draught, we draw the horizontal line from the 7 feet 6 inches height in the scale of draughts until it intersects the curve, and from this point we strike a vertical line to the scale of distances, and there is indicated 99·6 feet from the afterside of the stern post. From this curve we can see that since the vessel is 200 feet long, the displacements of the fore and after bodies are exactly equal at the 4 feet draught, as the vessel's centre of support (centre of buoyancy) is at the middle of the length. Then, as the draughts increase, we notice that the centre of buoyancy travels a little towards the stern, showing that the after body is slightly fuller than the fore body, increasing in this respect up to the load waterline. Now, we have simply reduced our ship to a huge lever, balanced practically at the centre when floating at 7 feet draught. Suppose in this condition the vessel weighs 550 tons, this being her displacement. She is then to be additionally loaded in the following manner:—

50 tons are placed 20 feet abaft of the longitudinal centre of buoyancy.
30 ,, ,, 60 ,, ,, ,, ,,
20 ,, ,, 30 feet forward ,, ,, ,,
40 ,, ,, 70 ,, ,, ,, ,,

What effect will this have? First of all, we know that the draught will be increased. This could be found before the weights are placed on board, by adding the weights together, and referring to the displacement scale in figs. 3 and 4.

$550 + 50 + 30 + 20 + 40 = 690$ tons, which reads 8 feet $4\frac{1}{2}$ inches mean draught.

There may be another effect. If the moments of the weights preponderate ahead or astern of the centre of buoyancy, then the vessel will trim* by the head or the stern, as the case may be. Let us see—

Moments on after side of Centre of Buoyancy.	Moments on fore side of Centre of Buoyancy.
50 × 20 = 1000 foot-tons.	20 × 30 = 600 foot-tons.
30 × 60 = 1800 ,,	40 × 70 = 2800 ,,
Total, 2800 ,,	Total, 3400 ,,

Then, since the moments preponderate on the fore side by $3400 - 2800 = 600$ foot-tons, the vessel will trim by the stem.

But suppose we now wish to know what weight must be placed at, say, 25 feet aft of the centre of buoyancy to bring the vessel again on even keel—that is, with the new load line parallel to the keel. This would be discovered by dividing the foot-tons in excess by the distance the new weight has to be placed from the original centre of buoyancy. The result will be the required weight—

$$\frac{600}{25} = 24 \text{ tons.}$$

* By *trim* is meant the difference between the draught at the stem, and the draught at the stern.

It will now be seen that the moments on each side of the centre of buoyancy are equal.

Vessels passing from salt or sea water to fresh or river water increase in draught. The reason of this has already been explained.

Effect of Wedges of Buoyancy on Ship's Centre of Buoyancy.—Another thing which may be observed is, that not only do some vessels change draught in passing from sea to river water, but that they also change trim. The reason of this will be evident when it is known that in some vessels the fore body is fuller, and has more displacement than the after body. Therefore, in increasing in draught, the longitudinal centre of buoyancy will travel forward (since it must remain in the centre of displacement), and the fore end of the vessel, having more support, will sink less than the after end. Hence, the change of trim.

In lecturing before ships' officers, the question has more than once been asked: Is it possible, with a vessel trimming by the stern and the centre of the disc on the load waterline, to place any more cargo on board and yet not submerge the disc, and, consequently, not increase the draught?

To such a question it is certainly possible to answer "yes," but such a vessel would be so exceptional in her design that the answer is practically "no."

For on examination of the lines of ordinary vessels, it is generally found that the fore body is slightly fuller than the after body, but this fulness takes place usually on the lower lines. At the region of the load line, however, the greatest fulness is usually aft, so that, imagining the vessel to be floating first at the waterline W L (fig. 22), and then by shifting weights forward

FIG. 22.—WEDGES OF BUOYANCY.

to float at A B, the wedge y would generally be of greater volume than x.

We have already seen that before a vessel will float at rest at any waterline, the weight of displacement and the buoyancy must equal each other. If by any means the weight of displacement be increased, the vessel will increase in draught; if by any means the buoyancy be increased, the vessel will rise out of the water, and the draught will be decreased.

c

Thus it follows that if the fore wedge x be less in volume than the aft wedge y, the draught would actually be slightly increased since the buoyancy is less. If x and y be equal, no change will take place in the draught, and only when x is greater than y would any reduction in the mean draught be observed. Such is not likely to occur except in ill-designed vessels, for the effect upon the speed by the production of great resistance would certainly outweigh the consideration of carrying a trifle more deadweight at such a cost.

Again, suppose the vessel to be floating at the waterline A B, and weights to be then shifted aft, so that the waterline is now at W L. If the volume of the immersed wedge y is more than the emerged wedge x, the draught will be somewhat decreased, and if y be less than x, an increase in draught would occur. We have observed that the vertical centre of buoyancy is the vertical centre of the displacement, and that with every transverse movement of the vessel (whether by means of external force in river or dock, or under the influence of wind or waves) there is a corresponding movement of the centre of buoyancy into the new centre of displacement. We shall now see how this new position may be found, and in this again the study of moments (Chapter II.) comes to our assistance.

Fig. 23 is a cylinder 10 feet in diameter, 20 feet long, and floating at 5 feet draught. The displacement in this condition would be $\frac{1}{2} \times 10^2 \times \cdot 7854 \times 20 = 785 \cdot 4$ cubic feet. B is the centre of buoyancy when floating upright. Let the cylinder be inclined to an angle of 20°. Observe clearly what takes place. W L was the original waterline; after the inclination the waterline is W′ L′, so that the wedge A, which was previously actual buoyancy, has come out of the water and become reserve buoyancy, and the wedge B, which was formerly reserve buoyancy, has become actual buoyancy. (Let g and k be the centres of buoyancy of each of the wedges.) This simply amounts to shifting the wedge of buoyancy, with g as its centre, to the position of the wedge, with k as its centre, a distance of about $6\frac{1}{2}$ feet. Let the wedge, with g as its centre, equal 87 cubic feet, and, as just stated, the distance from the centre of buoyancy of the emerged wedge to the centre of buoyancy of the immersed equal $6\frac{1}{2}$ feet. As the volume of displacement must be the same, at whatever angle of inclination, it follows that the wedge, with k as its centre, will equal 87 cubic feet. Now the new centre of buoyancy of the whole figure must have travelled in the direction in which the actual buoyancy was shifted; that is, to starboard of its original position in the figure —viz., to B[1], and in a line parallel to the line joining g and k, the centres of the wedges. The exact distance may be found by

BUOYANCY. 35

multiplying the volume of the wedge of buoyancy moved, 87 *cubic feet, by the distance moved*, 6½ *feet*, and *dividing by the total volume of displacement;* or, in other words, *the volume of buoyancy.*

$$\frac{87 \times 6\frac{1}{2}}{785\cdot 4} = \cdot 72 \text{ feet, centre of buoyancy moved to starboard.}$$

In coming to vessels of ship form, the principle of finding the shift of the centre of buoyancy is exactly the same, but there is

Cylinder floating upright.

Cylinder inclined to 20°.

Wedges of immersion and emersion.

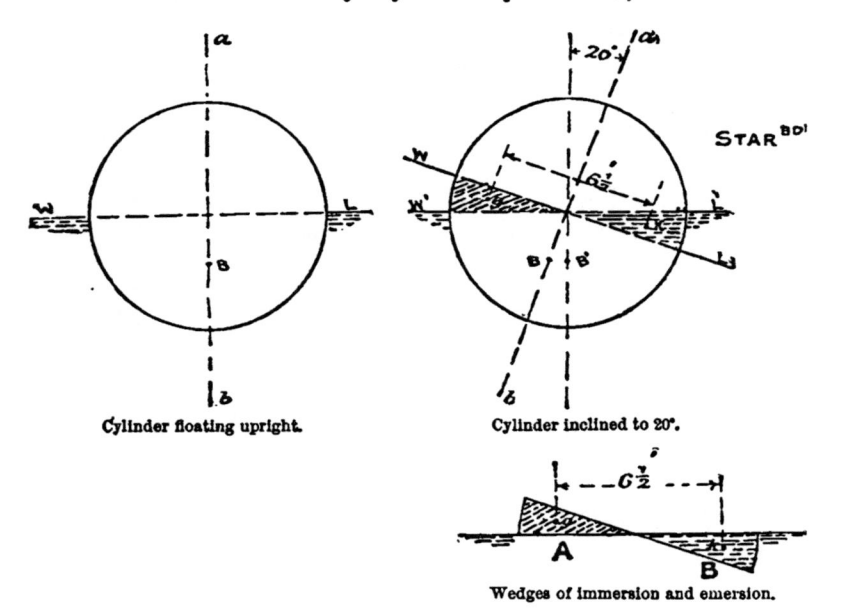

FIG. 23.—WEDGES OF BUOYANCY IN A CYLINDER.

a difference in the wedges. In one respect, however, the two wedges in any one ship do not differ, and that is, that the volume of the wedge of immersion is always equal to the volume of the wedge of emersion. In cylindrically-shaped vessels, whose water-line passes through the centre of the cylinder, revolving as they do on the centre of their diameter, not only are the wedges equal in volume, but identical in shape, any sections of the wedges all fore and aft being exactly similar. This is not so with vessels of ship form, for in these the wedges of immersion and emersion vary very considerably in form, especially towards the ends, and even more so still if the angle to which the vessel is inclined be great.

A little thought or personal observation of the actual form of a ship's hull will make this the more clear.

Thus, while it is a simple matter to find the centre of gravity of the wedges of immersion and emersion of a floating cylinder—whose waterline passes through the centre of the cylinder—it entails more work to find the corresponding points in an actual ship; and, while involving a considerable amount of labour, is not a matter of serious difficulty. (See Chapter X. for samples of Calculations.)

Effect of Entry of Water upon Buoyancy.—There is still another aspect of the subject of Buoyancy which calls for some attention. Suppose, first of all, that by some means or other a quantity of water enter the hold of a vessel. What will be the effect? This will all depend upon how the water entered. Let us imagine that the sea breaking over the bulwarks entered by means of some deck opening—hatchway, for instance—the outside shell plating of the vessel being intact and perfectly watertight. The result of water finding ingress in this way would exactly resemble the result of loading cargo, the water being deadweight at the rate of 35 cubic feet to 1 ton. Should water continue to enter the vessel, the draught would gradually increase until, if it happened that the total weight of the vessel, cargo, and water in her were more than the maximum possible displacement, she would naturally sink. If, on the other hand, when the hold was filled, the total weight of the vessel and the water in her were less than the maximum possible displacement, she would remain afloat—that is, disregarding the effect such might have upon the stability of the vessel, and also taking it for granted that little change of trim took place; for otherwise, should the water find ingress into a large hold towards the end of a vessel, this might result in her going down by the head or stern, as the case might be. But let us take an entirely different case. Supposing the outside plating in the way of some hold be damaged below the load waterline, the sea would consequently rush in. If the hold were empty, it would fill, unless prevented, up to the level of the outside sea level, with water. But this is different from the former case, where water was poured into the hold from above. There the water acted as deadweight, but not so in this example, as there is now free communication between the water in the hold and the sea outside. What has happened is this: The vessel has been robbed of the total buoyancy, both actual and reserve, of this compartment, even though it be found that, after increasing in draught, she still floats with considerable freeboard. The empty space from the water in the hold to the top of the hatchway, as stated, is no longer reserve buoyancy, and the vessel has lost the entire buoyancy of this compartment; and if it happens that the total buoyancy of the other intact compartments of the vessel is greater than the total displacement, as it was before the structure was damaged, and before the water entered, the vessel will float. If, on the contrary, the vessel possesses less remaining buoyancy than this, she will be entirely immersed, and will sink. This will perhaps be somewhat clearer if illustrated by a box vessel, as shown in fig. 24. The vessel is divided into three watertight

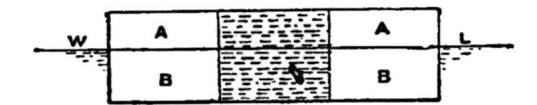

FIG. 24.—EFFECT ON BUOYANCY OF ENTRY OF WATER INTO A DAMAGED COMPARTMENT.

compartments by two watertight bulkheads, and floats at the waterline, W L. The centre compartment is damaged, and its actual and reserve buoyancy entirely lost, the sea having free entrance. If the sum of the volumes of A A (the reserve buoyancy) and B B (parts of the original actual buoyancy) be at least equal to the original actual buoyancy—that is, to the whole of the volume below the original waterline—it is possible for the vessel to float; if it be less, she will inevitably sink.

The reader must bear in mind that the inflow of water into such a hold as this has added nothing whatever to the weight of the ship, although the draught has increased; the weight, or the total displacement, remains exactly the same, but the entire empty space in the damaged hold is to be left out altogether, and ignored in reckoning upon the actual and reserve buoyancy. In increasing her draught, the vessel has simply taken from the reserve buoyancy in the other watertight and intact compartments a volume equal to the volume of that part of the damaged compartment which was previously below the waterline, and which was, therefore, previous to the accident, in use as actual buoyancy.

Again, suppose this accident happened to the vessel when her hold was filled with cargo, say timber, for example. How does she stand now? The water will flow into the hold, and occupy what space it can. This will naturally be very small indeed, since the hold is already practically full. As the water cannot possibly occupy the space taken up by the timber, it has to be content with what remains. Therefore, the volume below the original waterline of all the corners and crevices not occupied by the timber is lost actual buoyancy, and the vessel will sink until she has taken from the reserve buoyancy a volume equal to the volume of the space unoccupied by the timber, and below the original waterline, which, in most cases, would be comparatively little.

Camber or Round of Beam.—Vessels classed at Lloyd's require that all weather or uppermost decks have a round upon them, or camber, of at least $\frac{1}{4}$ inch to 1 foot of beam. Thus, a vessel of 40 feet beam will have a camber of $\frac{1}{4}$ inch × 40 = 10 inches, and will, therefore, be 10 inches deeper at the middle of the breadth at midships than at the sides. One important advantage to be gained by this is an addition to the reserve buoyancy, and, little as it may seem, its importance is so recognised, that, if the stipulated amount of camber be reduced, an increase of freeboard is demanded; if the amount be exceeded, a reduction in the freeboard is allowed. (See chapter on "Freeboard.")

Testing Water Ballast Tanks.—All water ballast tanks should be tested, in order to ensure that all joints and connections of plates and angle bars and the caulking be thoroughly watertight. This is done by means of water pressure. An iron pipe of the required length is fitted vertically into the top of the tank which is about to be tested. Water is then pumped into the tank until it is forced out at the top of the pipe. To accomplish this, considerable pressure has to be applied, which is in direct proportion to the height of the pipe.

Supposing the pipe to be 1 square inch in sectional area and 20 feet long from the crown of the tank, at the moment the tank is just full there is no pressure whatever upon the crown, but as soon as the water overflows at the top of the pipe it is evident that there must be a pressure at the bottom of the pipe of $\frac{20 \times 62\frac{1}{2}}{144}$ = 8·6 lbs. (fresh water). The pressure upon the inside of the crown of the tank must also be 8·6 lbs. per square inch or 20 × 62$\frac{1}{2}$ = 1250 lbs. per square foot.

Lloyd's requirements for the testing of tanks in vessels classed by them is as follows:—

Double Bottoms.—To have a head of water at least equal to the extreme draught—that is, the pipe previously mentioned must extend to the height of the maximum load line.

Deep Tanks and Peak Tanks.—To have a head of water at least 8 feet above the crown of the tank.

Fore and After Peak Bulkheads without Peak Ballast Tanks.—These are required to be tested by filling the peaks with water to the height of the load line. Other Bulkheads and Decks may have their watertightness tested by playing a hose upon them with a good head of water.

CHAPTER IV.

STRAIN.

CONTENTS.—Relation of Weight of Material in Structure to Strength—Strain when Floating Light in Dock—Relation between Weight and Buoyancy—Strain Increased or Decreased in Loading—Distribution and Arrangement of Material in Structure so as to get Greatest Resistance to Bending—Types of Vessels Subject to Greatest Strain—Strains among Waves—Panting Strains—Strains due to Propulsion by Steam and Wind—Strains from Deck Cargoes and Permanent Weights—Strains from Shipping Seas—Strains from Loading Cargoes Aground.

Relation of Weight of Material in Structure to Strength.— It would be rather absurd to commence the study of the structure of ships—whether they be steamers, sailing ships, or yachts—before first having some knowledge of the strains which, under varying circumstances, they would most probably have to bear. To build a ship capable of enduring, without damage to its structure, every possible strain which might be brought to bear upon it, however excessive, would necessitate the introduction of such an amount of heavy material into its structure as to render it greatly deficient in its carrying capacity. It does not even follow that the vessel with the heaviest material is necessarily the best, or even the strongest, ship, but rather the one with the lightest material so combined as to give the maximum strength and efficiency, and sufficient to cover the strains which in all likelihood, under reasonable circumstances, would have to be endured. This is the aim of all such classification societies as Lloyd's, Bureau Veritas, The British Corporation, etc.

Ships are built on a combination of two systems of framing —viz., longitudinal and transverse.

Longitudinal framing includes all those parts in the framework of a vessel which run in a fore and aft direction, whose function is to afford longitudinal strength.

Transverse framing includes all those parts in the framework of a vessel whose function is to give transverse or athwartship strength.

As has been already indicated, the strongest ship is only obtained when these two systems have been intelligently woven together, the strength of the one co-operating with the strength of the other—that is, in relation to the work which they have

STRAIN. 41

to do. When this is accomplished the whole is then covered by a skin in the form of a shell-plating and decks, and by this means the skeleton or framework of the ship is still further united and strengthened.

Strain when Floating Light.—Looking at an ordinary cargo steamer floating in the dock in her light condition, and lying at rest at her moorings, one would almost imagine at first sight that she is perfectly free from strain. But on investigation this is found not to be so. As has been previously shown, any object placed in water, whether it be a ship or a log of wood, will sink until it has displaced a volume of water equal in weight to itself. Or, in other words, before the object will remain stationary, and at rest at any waterline, the downward

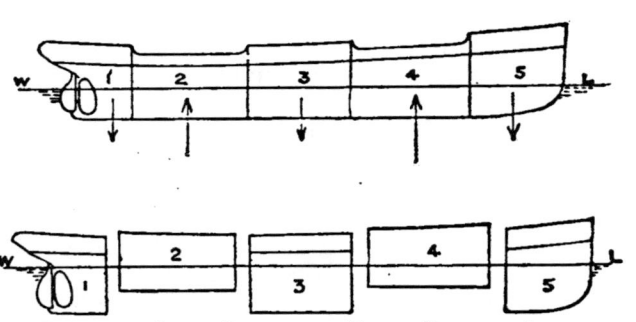

FIGS. 25 and 26.—STRAINS ON VESSEL FLOATING LIGHT.

pressure of the weight of the object floating must be exactly balanced by an equal upward pressure of the water.

Relation between Weight and Buoyancy.—Supposing the vessel shown in fig. 25 be 200 feet long, and classed at Lloyd's, she would require four watertight bulkheads, one at each end of the engine and boiler space, a collision bulkhead at the fore end, and another bulkhead at the aft end. The weight of the vessel light being, say, 500 tons, if placed in the water would displace a volume of water 500 tons in weight, and would thus remain stationary, say, at the waterline, W L.

But suppose the vessel could be divided off into five separate parts at the four watertight bulkheads, and each part floated separately, as shown in the sketch (fig. 26). It will now be seen that the draughts vary for each part, and none of them float at the original waterline. A little observation will soon explain this. The total weight of all the parts is exactly the same, and therefore the total water displaced is the same; but throughout the length of the vessel there is an unequal distribution of weight and buoyancy. Thus, for example, referring to the sketch, parts 1 and 5—the ends of the vessel—being very fine, and yet of considerable weight, which is increased by the poop and forecastle, in order that they may be balanced by the buoyancy they will have to sink to the draught as shown. In parts 2 and 4 we have the vessel rapidly increasing in internal capacity or fulness, and small in weight in comparison with the volume of the enclosed space, the result being that these parts float at a less draught than previously. In No. 3 we have the fullest part of the vessel with greatest floating power. But in this compartment is concentrated the weight of engines and boilers, which tends to increase the draught upon what it is in the combined ship. Thus we see that throughout the length there is a series of upward and downward vertical strains, as shown by the arrows, tending to alter the form of the vessel longitudinally. Note that these strains *tend* to alter the form of the vessel. This affords one consideration for the naval architect or shipbuilder, in constructing an efficient ship—viz., that there is sufficient strength to prevent any such alteration in form taking place, or even any sign of such strain being endured.

Strains in Loading.—Now it can easily be seen how these strains may be considerably increased when loading; for instance, if in the case of a miscellaneous cargo the heavy weights be placed towards the ends of the vessel where the buoyancy is least, the tendency of the ends of the vessel to droop would be greatly aggravated. The endeavour should be to distribute the heavy weights of the cargo so as to produce a balancing effect between the forces of weight and buoyancy, and thus avoid great local excess. By this means it is possible to reduce the strain even from what it is in the light condition. In considering strains at sea, the evil of bad loading will be seen still more clearly.

In addition to these vertical strains to which the vessel is subject while lying at rest, there are collapsing strains acting upon every portion of the immersed skin of the ship (see fig. 15, Chapter III.). For, be it remembered that the forces of buoyancy act in perpendicular lines to the immersed surface. Thus, while there is an upward pressure tending to thrust the vessel out of the water, there are also horizontal and oblique forces tending to crush in the sides of the ship.

It will be obvious that the greater the immersed girth of the vessel, the greater the strain. Thus, the strain is greatest at midships, and on each side of midships; and towards the ends, as the vessel becomes finer, it gradually diminishes. Considerable as strains in still water may be under certain circumstances, on investigating the strains experienced at sea we shall see how enormously they are increased.

First, observe the strains endured by a ship in the condition shown in fig. 27. Here the vessel is supported at midships on the summit of a wave, the extremities being practically unsupported. The ships may now be compared to a hollow girder with weights ranged miscellaneously throughout its length,

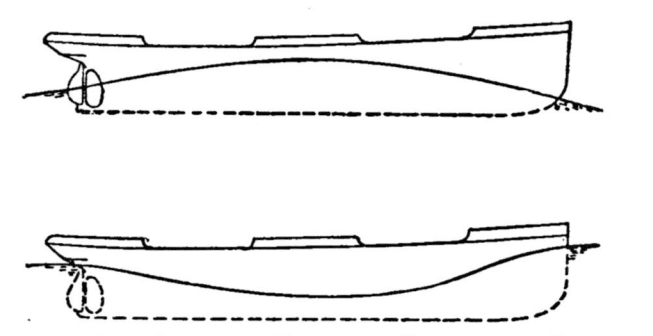

FIGS. 27 and 28.—STRAINS OF VESSELS ON WAVES AND IN TROUGHS.

and supported only at the centre, the result being a severe hogging strain tending to make the ends droop.

Distribution of Material to Resist Bending.—The question now arises, How should the material employed in the construction of the vessel be distributed so as to withstand this tendency to bend?

In fig. 29, let A B be a bar of iron or steel 100 feet long, supported at the middle of its length, and with a weight of 10 tons attached to each end. This would not be an exact illustration of a loaded vessel supported at midships upon a wave as in fig. 27, but it will form a fair approximation to the strains experienced by a vessel when in a light condition with large peak ballast tanks full, or it will show the evil of loading a vessel with the heaviest cargo at the ends, and will serve to illustrate the principle it is wished to make clear. Let the sectional area of the bar be the same throughout its length. The bending moment of each weight would be 10 × 50 = 500 foot-tons, and the tendency of the bar would be to bend or break at the point of support since the strain is greatest at this point. The tendency to break at 10 feet on each side of the support would be 10 × 40 = 400 foot-tons, and at 20 feet from

44 KNOW YOUR OWN SHIP.

the support 10 × 30 = 300 foot-tons, and so on, the tendency to break diminishing towards the ends as the leverage decreases.

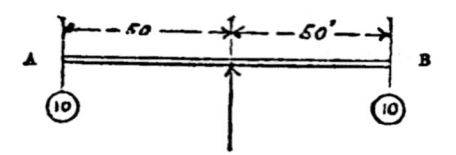

FIG. 29.—STRAIN ON A BAR LOADED AT EACH END.

The bending moment at any section from the centre to the end of the bar might be graphically illustrated in the following manner:—

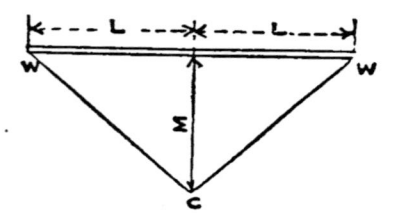

FIG. 30.—DISTRIBUTION OF BENDING MOMENT.

Let W = weight hung at each end of bar.
 ,, L = half-length of bar (that is, length from centre to end of bar).
 ,, M = maximum bending moment (which occurs at centre of bar).

Make M equal weight multiplied by leverage—W × L = 10 × 50 = 500 foot-tons at the middle of the bar.
Join C W.

Let M be set off to any arbitrary scale; then by using the same scale the bending moment can be measured at any intermediate position between the support and the weight. In a similar manner the strains experienced by a vessel supported at the middle of her length upon a wave are greatest in the region of the half length amidships. The structural arrangements introduced, and the great value of such erections as long bridges over the middle of the length in affording strength to resist those bending strains, will be shown at a later stage.

Now, supposing the bar in figs. 29 and 30 to be 4 square inches in sectional area, the question may be asked, Is it possible to arrange the material, preserving the same sectional area so as to get greater efficiency in resisting longitudinal bending? Let the bar be rolled out so as to make it 8 inches by ½ inch in section (same sectional area as previously), the length remaining the same. It will easily be seen that if the bar (or plate, as it now is) be supported as before with the wide 8-inch side horizontal, and the weights attached to its extremities, that it possesses less resistance to bending than when the bar was square in section.

But should the bar be supported with its wide 8-inch surface vertical, it will be found that its resistance to bending has been increased beyond what it was in the square section of the bar. Let us make a brief examination of this difference of resistance to bending with the same sectional area.

Let fig. 31 represent the 8-inch by ½-inch plate, placed with

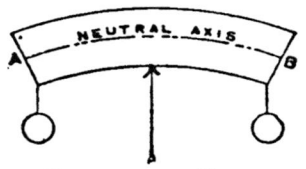

FIG. 31.—BENDING RESISTANCE OF BAR PLACED VERTICALLY.

the 8-inch side vertically, and let A B be the neutral axis. (The *neutral axis* is an imaginary line of no strain passing longitudinally through, say, the centre of depth of the plate. Since the plate is 8 inches deep, it will be 4 inches from each edge.) Supposing a weight be attached to each end of the plate, observe what must take place before it can bend. On the upper edge there must be considerable expansion or elongation, and on the lower, contraction or compression. The wider the plate is made, the greater will be the resistance to tensile strain on the upper edge, and compressive strain on the lower. It will also be found that the nearer the neutral axis is approached, the less elongation and compression are required to result in the same amount of bending, and at the neutral axis there is neither elongation nor compression.

Now, theory and experiment agree in showing that the stretching and compressive stresses at the top and bottom edges are now (8 inches by ½ inch, wide side vertical) only one-fourth of what they were when the bar was square in shape (2 inches by 2 inches), and only one-sixteenth of what they are when the bar is 8 inches by ½ inch, but laid with its thin edge vertical (the formally stated law is that *the resistance of the bar to this kind of stress varies directly with the square of the depth of the bar, and directly as its breadth*).

The resistance to bending, moreover, will be greatly increased if the vertical plate be turned into a girder by attaching a bulb to its upper and lower edge, as shown by A in fig. 32, or by adding strengthening flanges to the upper and lower edges, where the tensile and compressive strains are greatest; for example, see

FIG. 32.—BENDING RESISTANCE OF GIRDERS.

fig. 32, B, C, D, E, and F. (Stress, Strain, and Strength are dealt with much more exhaustively in the author's companion volume to *Know your own Ship.*)

Types of Vessels subject to Greatest Strain.—The principles which apply to a plate or girder apply equally well to a ship, for, after all, a ship is simply a huge, hollow girder, and from the foregoing reasoning the following deductions may be made:—

1st. That vessels of great length, and therefore subject to excessive bending moment among waves, require more longitudinal strength than short ones (not necessarily more transverse strength).

2nd. That long, shallow vessels possessing less resistance to tensile and compressive strains require additional longitudinal strength.

3rd. That in all vessels more strength is required in the region of midships, while a reduction may be gradually effected towards the ends.

Strains among Waves.—In studying the case of a ship in the condition shown by fig. 28, it is observed that since the ends are supported on waves, and the midship part, containing engines, boilers, bunkers, etc., is to a large extent unsupported, the vessel, especially if a long one, will endure a severe sagging strain tending to make her droop amidships. There are cases on record where long, shallow vessels have actually fractured through the middle and sunk. The vessel may now be compared in some measure to a bar, supported at the ends with a weight in the centre, thus—

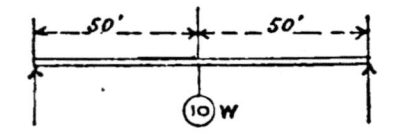

FIG. 33.—BAR WEIGHTED IN THE CENTRE.

and if the weight be, say, 10 tons, and its distance from each support 50 feet, then by the following diagram the bending moment may be illustrated:—

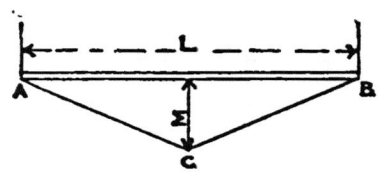

FIG. 34.—DISTRIBUTION OF BENDING MOMENT IN BAR.

Here, 10 tons in the middle gives 5 tons pressing on each support, and the bending moment in the middle (= pressure × leverage) = 5 tons × 50 feet = 250 foot-tons = M.

Set off M (250) to any scale, and join C B and C A. By using the same scale, the bending moments can be measured at any intermediate position between the support and the weight. Here again in this illustration it is shown that the longitudinal bending moment is greatest at the middle of the bar corresponding to the midship portion of the ship, decreasing towards the ends, where it vanishes altogether. This provides another reason why vessels should have greater longitudinal strength amidships than elsewhere.

Moreover, owing to the rapid transit of the waves, and, therefore, the unequal distribution of weight and buoyancy, the

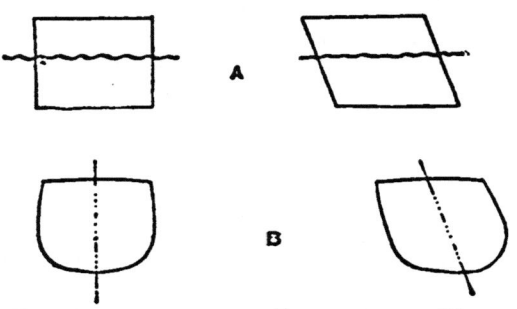

FIG. 35.—STRAINS DUE TO ROLLING AMONG WAVES.

vessel is subject to a succession of severe and sudden strains. It will also be seen that in rolling among waves, there is a great tendency for a vessel to alter in transverse form. Under such circumstances, she may be compared to a box, as shown in fig. 35.

If a series of irregular, collapsing strains be put upon the exterior of the box, the tendency is not so much to fracture in the positions shown by the wavy lines, but to work at the corners. Exactly the same thing takes place with a ship at sea as she rolls among the waves. The strain tends to have the effect shown in B, fig. 35.

Thus in determining the size and arrangement of all material in the construction of ships, there must be sufficient longitudinal strength to resist all longitudinal, bending forces, and sufficient transverse strength to resist all collapsing forces or alteration to transverse form of the nature already described, with a reasonable margin for safety.

As has been already shown, it is not necessary to have as much sectional area of material towards the ends of the vessel as near amidships, since the strain is less; but, nevertheless, in comparison with the strain which has to be borne, both sections should be equally able to withstand such strains as come upon them, for it should be remembered that the vessel is no stronger than her weakest part.

Thus far, only strains affecting the ship as a whole have been considered, but there are several other strains it is necessary to take into account which only affect the vessel locally.

1. *Panting Strains.*—The fore end of the vessel, especially when of a bluff form of stem and driven at a high speed, being the first part to pass through the water, naturally suffers great head resistance, tending to make this part pant or work in and out.

2. *Strains due to Propulsion by Steam.*—These strains may be divided into (1) strains owing to weight of engines and boilers, and (2) strains due to vibration of shaft, etc.

3. *Strains due to Propulsion by Sail.*—In vessels with lofty masts and much sail area, great strain is transmitted to the hull through the masts by the force of the wind and the action of rolling.

4. *Strains owing to heavy permanent weights carried,* such as winches, windlass, cranes, anchors, guns, etc.

5. *Strains from Deck Cargoes.*—It is customary with many vessels, such as those engaged in the "Baltic" trades, to carry coal or some other British export on the outward voyage, and to return with cargoes of timber, and in order to get the vessel down to her load waterline large deck cargoes are carried. Many of the shipmasters engaged in such trades are not unacquainted with the fact that, owing to the heavy deck weight, these vessels are severely strained, and sometimes take a set or sort of twist, and this is only discovered after the cargo has been removed, and perhaps not until even a few days later still, when the vessel,

sometimes with a considerable report, frees herself from her strained condition with a severe trembling from stem to stern. On examination it is found that very many of the rivets in the heads of the hold stanchions or pillars have been sheared, and considerable damage done to the beam knees. This is an abundant proof that vessels intended to carry heavy deck cargoes require special strengthening.

Such damage as that just explained might often be obviated by wedging or shoring the space between the top of the hold cargo and the beams, thus assisting the beams in enduring the strain of the deck cargo.

* 6. *Strains from the shipping of seas* against poop or bridge fronts. Bridge and poop fronts with closed ends are subject to sudden and severe strains owing to the shipping of heavy seas, which will evidently spend their force against these bulkheads. Such parts, therefore, require special attention, and it is only possible to secure the maximum allowance on the freeboard when these parts are most efficiently constructed.

7. *Strains from Loading Cargoes Aground.*—Vessels engaged in trades where it is known that they will lie aground during the operation of loading and unloading, require special stiffening on the bottom. And so on, cases might be enumerated where, special strain having to be endured, special strengthening must be provided.

Excessive strains, such as those borne by a vessel when run ashore, as frequently happens, so that one end only is water-borne, or where the vessel, possibly laden with a heavy cargo, is laid across a sandbank, with ends unsupported when the tide has left

---

* Strains from the shipping of heavy seas on deck are often more severe than is generally supposed, as shown by the following illustration. A fine new steel steamer, of nearly 4000 tons gross tonnage, was crossing the Atlantic in bad weather, in the beginning of the year 1899, when she shipped a heavy sea over the port side just in front of the bridge. The water fell with terrific force upon the fore main hatch and the deck. The hatch coamings on the port side were burst away, and the sea poured down into the fore main hold, damaging a considerable amount of cargo. Beams and stanchions were sprung and bent, the rivets in the beam knees sheared, and the deck considerably damaged. The sea, in sweeping over the starboard side, carried nearly the whole of the bulwarks with it. All this damage was caused by the shipping of a single sea, for the vessel shipped very little water afterwards, which gave the crew an opportunity of temporarily closing up the damaged hatchway. The vessel was comparatively new, being built to Lloyd's highest class in 1898, yet it seems quite certain that if she had shipped another sea of a like nature, she would have foundered, and contributed to swell the list of vessels "unheard of" or "missing." Too much attention cannot be paid to the thorough protection of all deck openings, both hatchways and engine casings, for it is very probable that the loss of most of those vessels which are never again heard of is caused by accidents of a similar nature to that just described.

her, the shipbuilder cannot attempt to cover, but all strains, such as those already mentioned, may be thoroughly provided against.

A vessel in dry dock, unless carefully shored, may be considerably damaged. Cases of dry docking, where the bilges have drooped, have occurred through carelessness or ignorance. Having briefly enumerated the chief strains borne by a vessel under various circumstances, the reader will now be better prepared to understand why exceptional strength is introduced into the structure, either considered as a whole, or in particular parts only.

CHAPTER V.

STRUCTURE.

CONTENTS.—Parts of Transverse Framing, and How Combined and United to Produce Greatest Resistance to Alteration in Form—Sections of Material Used—Compensation for Dispensing with Hold Beams—Parts of Longitudinal Framing, How Combined and United to Transverse Framing to Produce Greatest Resistance to all Kinds of Longitudinal Bending and Twisting—Forms of Keels and Centre Keelsons, and their Efficiency—Distribution of Material to Counteract Strain—Value of Efficiently-Worked Shell and Deck Plating in Strengthening Ship Girder—Definitions of Important Terms—Illustration of Growth of Structural Strength, with Increase of Dimensions by means of Progressive Midship Sections—Special Strengthening in Machinery Space—Methods of Supporting Aft End of Shafts in Twin-Screw Steamers—Arrangements to Prevent Panting—Special Strengthening for Deck Cargoes and Permanent Deck Weights, and also to Counteract Strains due to Propulsion by wind—Types of Vessels—Comparison of Scantlings of a Three-Decked, a Spar-Decked, and an Awning-Decked Vessel—Bulkheads—Rivets and Riveting.

Transverse Framing.—The parts of the structure of a ship affording resistance to transverse strains, according to the usual mode of construction, are included in the combination known as *transverse framing*. A complete transverse frame comprises a *frame bar*, a *reverse bar*, a *floor plate*, a *beam*, and a *pillar*,

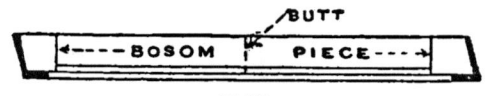

FIG. 37.—ANGLE BAR.

efficiently united. According to the size of the vessel, the spacing of the transverse frames varies from about 20 to 26 * inches from stem to stern. Let us take for our example an ordinary merchant steamer, a midship section of which is shown in fig. 36.

* *A leading Liverpool Shipping Company is adopting a System of spacing the Transverse frames 36 inches apart.*

52 KNOW YOUR OWN SHIP.

The *frame bar* extends continuously from keel to gunwale in this type of vessel. Should there, from any cause, be a break in the length of the frame, the strength should be preserved by lapping the parts to make the connection, or by fitting angle butt straps, or other efficient means of compensation should be adopted (see fig. 37, and section A B in fig. 36). This, indeed, is a rule which should be rigorously observed throughout the construction of a ship, that *wherever a structural part is weakened, the strength be fully recovered by compensation in some form or other.* If the frames meet or butt on the keel, as they usually do, they are connected by pieces of angle bar about 3 feet long, fitted back to back, which, in addition, provide a substantial means of connection to the shell plating. The usual form of frame is the plain angle bar (see A, fig. 38).

It will be noticed that one flange of the bar is longer than the other; the long flange always points into the interior of the ship, and the short one is attached to the shell plating. Since the greater the girth of the vessel, and thus the greater the collapsing strain, it is evident that the vessel needs more transverse strength in the region of midships, where it is fullest, than towards the ends. It is, therefore, usual to make the frames one-twentieth of an inch thicker for three-fifths of the length at amidships than at the ends. The frame bar is made stronger and more rigid, and, therefore, the better able to keep out the ship's side by means of a *reverse bar*, which is similar in section, but smaller in size. It is riveted to the back of the frame (see B, fig. 38), and being

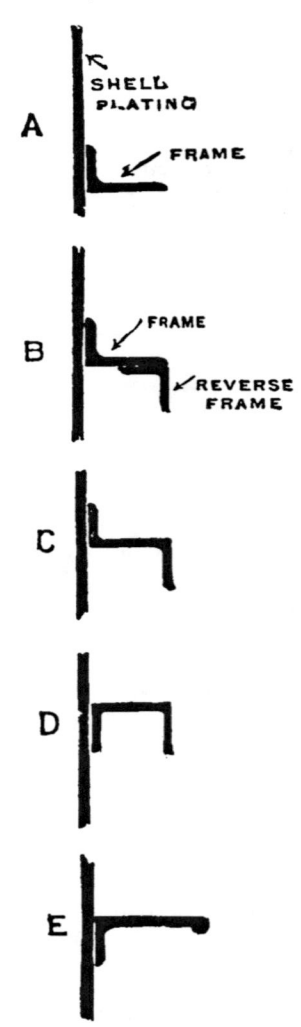

FIG. 38.—FRAME BARS.

on the side of the frame hidden from view in fig. 36, it is there shown in dotted lines. The reverse bar does not always extend to the same height as the frame. As will be seen further on, its height is governed by the transverse dimensions of the vessel, for the greater the girth and beam, the greater the need for transverse strength. Across the bottom of the ship, and extending well up the bilge, is a deep plate called a *floor plate*. On the lower edge of this plate is attached the large flange of the frame bar, and across the other side on the upper edge is bent and riveted the reverse angle after it leaves the frame (see section A B, fig. 36). The floor plate, being now converted into a girder, affords great stiffness and strength to the bottom of the ship.

Instead of the usual frame and reverse frame there are other sections of bar iron or steel which may be used. For example, there is the Z-bar (see C, fig. 38), which is a combination of the frame and reverse frame rolled in one section, thus saving the necessity of riveting these two together. It is very strong, and is extensively used in the building of ships for the Royal Navy.

A very similar bar to this, sometimes used for framing large vessels, is the channel iron section shown in D (fig. 38).

This also saves the riveting of a reverse bar. When this section is used, it is generally dispensed with towards the ends of the vessel, and the ordinary frame and reverse angle bar substituted, as it is difficult to bend and bevel the channel bars as required at the ends of the vessel. There is also the bulb angle section (see E, fig. 38).

This is sometimes adopted in vessels where no sparring is required in the holds, thus permitting some kinds of cargo to be trimmed right against the shell, an advantage being gained in cubic capacity. By this means also the reverse bar may be dispensed with when the bulb angle is made strong enough to equal the frame and reverse bars together.

The beams form an important part of transverse framing, uniting, as they do, the upper extremities of the frame bars and holding them in position, thus forming the foundation for decks. In addition, they complete the transverse section of the hollow girder into which we have resolved our ship. It would be useless to secure sufficient structural strength in the various bars forming the transverse skeleton of a vessel, unless at the same time every attention were given to the efficient connection of these parts to one another. We have already noticed how the frame butts are strengthened and connected, but there is still the connection of the frame and the beams. This is done by welding or riveting to each extremity of the beam a *knee plate* (fig. 39), which is fitted into the bosom of the frame.

The British Corporation recognises the necessity of an efficient connection at this part, by compelling all ships built under its survey to have knees three times the depth of the beam, and one-and-three-quarters the depth at the throat, to all beams in the way of the main deck. Lloyd's require the depth of the knee to be two-and-a-half times the depth of the beam, and one-and-a-half times the depth of the beam at the throat for steam vessels, while for sailing ships over 36 feet broad the knees must be three times the depth of the beam.

Like frame bars, the beams may vary in sectional form. Under iron or steel decks it is usual and better to fit angle or angle bulb beams to every frame, and under wood decks, on alternate frames, beams of the following sections may be adopted:—

(1) Butterly Bulb (see A, fig. 40).

(2) Bulb Plate, and double angles riveted to its upper edge (see B, fig. 40). Or,

(3) Channel Bar (see D, fig. 38).

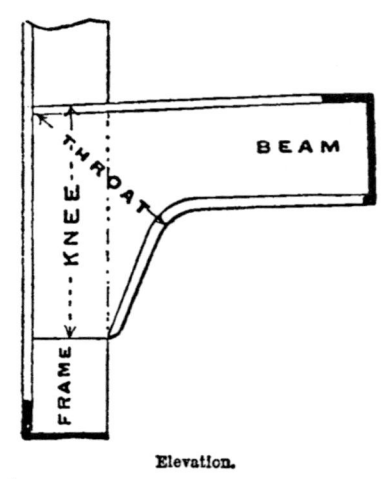

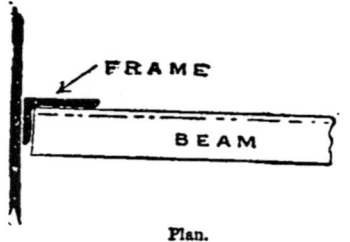

FIG. 39.—BEAM KNEE.

FIG. 40.—BEAMS.

However, these forms may be considerably modified according

STRUCTURE. 55

to the length of the beams, since the length of the beams determines the size of bar to be used. Vessels may have two, three, four, or more tiers of beams, according to their depth.

Compensation for Dispensing with Hold Beams.—Now, let us suppose a shipowner is about to have a vessel built with a depth of 16 feet. He finds, if he intends to class the vessel at

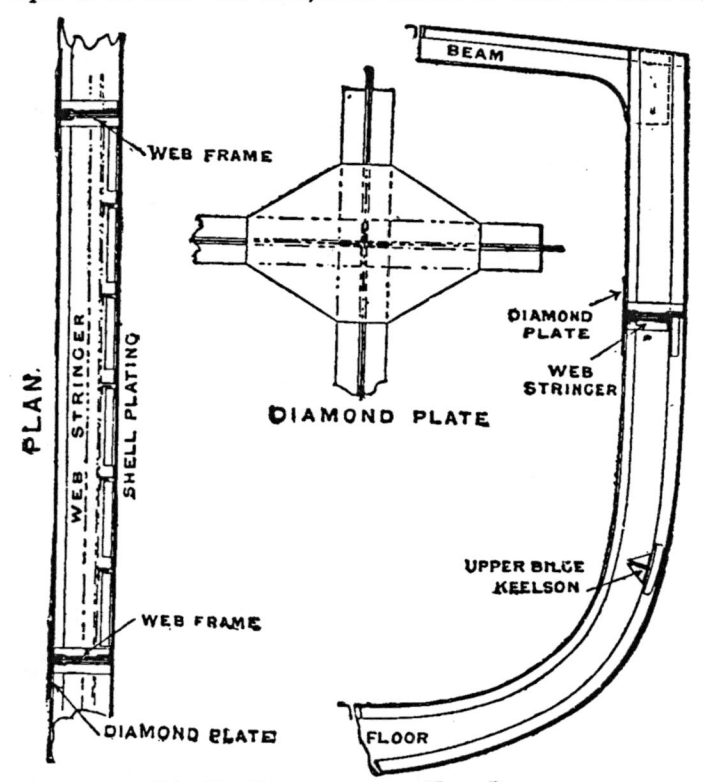

FIG. 41.—SUBSTITUTES FOR HOLD BEAMS.

Lloyd's, that she requires hold beams fastened to every tenth frame, and at the extremities of the beams a stringer plate (see fig. 36) securely attached to beams and shell. But as these would interfere with the stowage of cargo he intends to carry, let us see what alternative he may adopt in order to dispense with the hold beams. If he wishes, he may fit transverse web frames (see fig. 41) at distances of eight frame spaces apart all fore and aft; this compensates for loss of beams.

In addition, there must be fitted a longitudinal web frame, or web stringer as it is called, all fore and aft, between the transverse web frames, and securely attached to them by angles on the upper and the lower sides, and also by means of an efficient diamond plate on its front edge. This compensates for the loss of the stringer plate on hold beams. The next thing to be done is to have the upper extremities of the web frames firmly tied together and held in position. This is done by fitting extra strong beams across the vessel, attached to the web frames by extra strong deep knees. Since the function of the beam and stringer plate is to tie and stiffen the sides of the ship, the web frame being a transverse plate girder, and the web stringer a longitudinal plate girder, the one firmly united to the other enables them to do their work together, and thus serve the same purpose. It will be observed that the web frame shown in fig. 41 is a continuation of the floor plate, and also that both it and the web stringer are stiffened by double angles on their inner edges.

Another alternative is to stiffen the ship at every frame space all fore and aft by fitting together two large angles (as shown in fig. 55). In conjunction with special hold stringers, this method, known as "Deep Framing," makes a substitute for both hold beams and reverse frames.

There is still another part in the transverse framing to be noticed—viz., the *pillars*. They are riveted to the beams, and usually to the girder on the top of the floors, or some other part of the bottom of the vessel. They bind the upper and lower parts of the structure together, and perform the function of a strut and a tie by holding the beams and the bottom of the ship in their right positions relatively to each other; and thus by uniting the two great horizontal flanges of the ship girder, the deck and the bottom, they enable them to act in unison in resisting longitudinal strains.

When considering the subject of strains, we noticed that in rolling among waves a vessel has the tendency to alter her transverse form, and to work at the corners. After observing the combination constituting transverse framing, we shall now be able to see how these parts unite in offering resistance to the alteration of transverse form.

1. *At the Bilge Corners.*—Here we have the floor plate curved up the bilge to a height of twice its depth at the middle line of the vessel, thereby supporting the bilge in the form of a web at every frame.

STRUCTURE. 57

2. *At the Deck Corners.*—In this case we see the efficiency developed by the deep beam knees in giving support in the form of webs at every beam. Fig. 55 shows a similar web in a double-bottomed vessel.

3. Furthermore, there is the great assistance provided in the form of the beams themselves holding the sides of the vessel rigid, the beams being in their turn supported by the pillars.

Longitudinal Framing.—The longitudinal framework of the vessel is made up of the *keel, keelsons,* and *stringers.* These may partake of a variety of forms, with a view of which we shall briefly deal, together with the means adopted for binding them all together in order to secure an efficient and strong framework. The number of keelsons and stringers depends upon the size and proportions of the vessel.

Keel.—The keel shown in fig. 36 is known as the ordinary bar keel. It is made up of long lengths of bar iron connected by means of scarphs, the length of which should be sufficient to secure a good connection (see fig. 42). If the vessel be classed at Lloyd's,

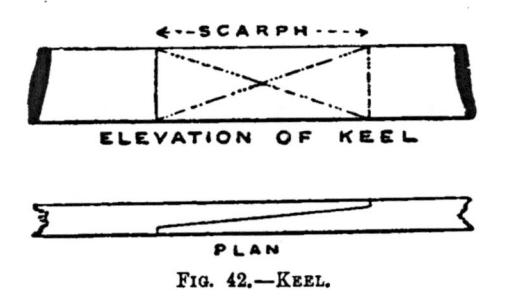

FIG. 42.—KEEL.

the scarphs will be nine times the thickness of the keel, and if with the British Corporation, three times the depth of keel.

The same method of connection unites the keel to the stem and the stern posts. These connections may also be made by welding, but this is seldom done.

A superior arrangement of this kind of keel is the one known as the side bar keel (see E, fig. 43).

It consists of a deep plate extending down from above the top of the floors to the bottom of the keel, the thickness of the keel part being made up by attaching two side bars or slabs of iron, one on each side of the lower extremity of the centre plate. The whole is then riveted to the two strakes of shell plating which cover the keel, and are called the *garboard strakes* (see fig. 36). A thoroughly strong result is thus secured. Holes

58   KNOW YOUR OWN SHIP.

are cut in the centre plate at the top of the keel to allow the *heel piece*, or frame back bar, as it is sometimes called, to be fixed in position. The butts or connections of these plates comprising the keel must be kept well clear of each other, and separated by at least two frame spaces, wherever practicable.

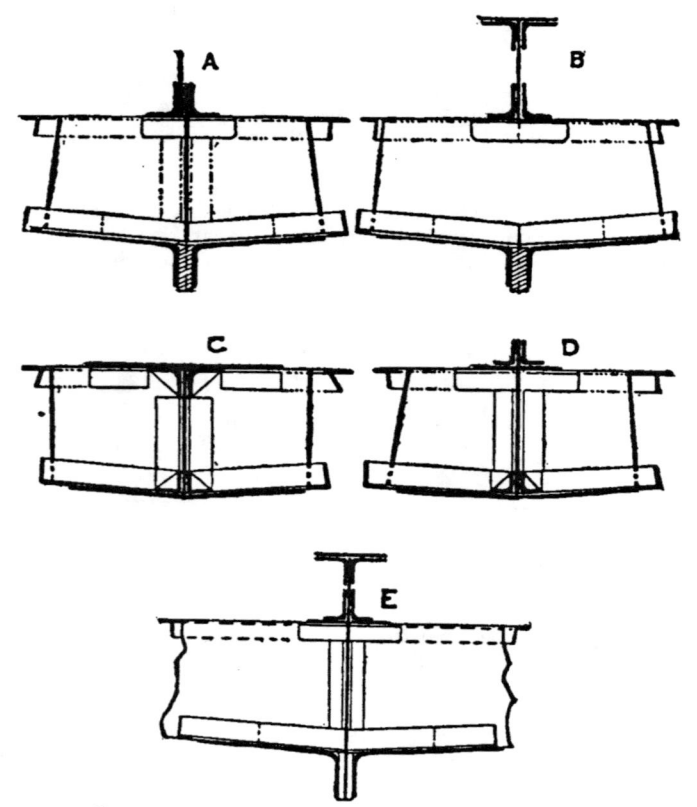

FIG. 43.—KEELS AND KEELSONS OF VARIOUS FORMS.

The centre plate being carried up above the top of the floors forms part of the centre keelson. Two horizontal plates are then attached to the floor plates, one on each side of the centre plate, the connection being made by means of the reverse angles. To the upper edge of the centre keelson plate two angles are riveted, and also two others on the top of the horizontal plates. The combination now forms a splendid backbone to the whole ship. It will be noticed in fig. 36, as also in A, B, C, D, and E (fig. 43), that a short piece of angle bar called a *lug piece* is attached to the top of the floors on the opposite side to the reverse angle, thereby ensuring a doubly strong connection between the keelson and the transverse framing, for unless the longitudinal and transverse framings are thoroughly united, their separate strength is of little value to the ship as a whole, and they would thus fail in their chief function.

Keelsons.—The commonest form of centre keelson consists of a single plate standing upon the top of the floors with double angles riveted to its upper and lower edges, as shown in B (fig. 43). In addition, a plate called a *rider plate* is riveted on the top of the two uppermost angles. The great disadvantage of this keelson, especially in large, heavy vessels, is, that it affords no resistance whatever to buckling of the floors, and thus it has often happened that when vessels of this form of construction have grounded upon an uneven bottom, the keel has been sprung up, and consequently the floors having nothing between them to stiffen them vertically at their deepest part, have buckled. This lack of stiffening between floors is the great defect of all keelsons standing simply upon the top of the floors.

A very good kind of centre keelson is that shown in A (fig. 43). Here we have the deficiency in the previous keelson remedied. Its parts are as follows:—First, there is the deep bulb plate, with angles on its lower edge, attached to the top of the floors. Between one of the angles and the bulb plate an intercostal plate is let down between all the floors on to the top of the keel, and secured to the floors by vertical angle bars, as shown.

A now unfamiliar, though very efficient, form of centre keelson is sometimes to be seen in old vessels. A sketch of the same is shown in C (fig. 43). It consists of a continuous centre plate extending from the top of the floors to the top of the keel, the latter being a broad, thick plate known as a flat-plate keel. On the top of the floors a thick, broad plate is laid, and attached to the vertical keelson plate by large, double, continuous angles, as shown. Since the entire centre keelson is continuous, it follows that the floor plates must butt on either side of it, the connections between the two being made by double angles. The vertical plates comprising the centre keelson are connected by double butt straps (see fig. 73), treble riveted. The horizontal plate is also riveted to the reverse angles on the top of the floors, and in addition to a short lug piece fitted on the top edge of the floor opposite to the reverse bar. This form of centre keelson is usually adopted in double bottoms (see fig. 55).

In D (fig. 43) we have another modification of a centre keelson with a flat plate keel. The centre plate is continuous, and extends above the top of the floors sufficiently high to take two large angle bars which are riveted to two horizontal plates shown on each side of the centre plate, and also to the top of the floors.

Keelsons and stringers are fitted for the purpose of giving longitudinal stiffness to the vessel, and also in order to tie or unite the transverse framing, so that, when strain is brought to bear upon any particular part, it is transmitted to the structure as a whole.

Keelsons and stringers are all forms of girders (see fig. 36), varying both in number and size, according to the dimensions and structural requirements of the vessel.

Those longitudinal stiffeners located along the bottom of the vessel between bilge and bilge are called *keelsons*; above the bilge they are termed *stringers*.

Stringers.—It will be noticed that wherever a tier of beams is fitted in a vessel, a broad, thick plate, called a stringer plate, is attached to its extremities, and connected with the shell by a strong angle bar. This bar, called a *shell bar*, is fitted intercostally between the frames if below the weather deck, and to the reverse frames extending above the beams a continuous angle bar is riveted, and also to the stringer plate. If the deck is an iron or steel one, the plate at the end of the beams is still called a stringer plate. It is always thicker than the adjoining plating, and, being firmly connected with the beams and shell, forms a splendid longitudinal stiffener to the vessel, acting in conjunction with the beams and transverse framing in keeping out the sides of the vessel to their proper position and shape, and in resisting longitudinal twisting strains.

If the beams are widely spaced and no deck is laid, the stringer plate is supported by means of knees or bracket plates underneath.

We observed at the beginning of this chapter that not only is the transverse strain greatest, but the longitudinal also, in the region amidships, and is gradually reduced towards the ends of the vessel, thus showing that a reduction may be made in the thickness of the material used in the construction towards the ends. This applies generally throughout the vessel, for, be it remembered, excessive strength is useless.

Distribution of Material to Counteract Strain.—With such an able means of conveying instruction to our minds as the eye, it seems very probable that, with short explanatory notes, a few sketches, showing both the arrangement and growth of the framing and plating, ranging from the smallest to large types of merchant vessels, will prove of more value, and will perhaps be plainer than pages of printed matter. Before doing this we must not omit to notice that although the first aim is to secure the greatest possible efficiency by a judicious combination of longitudinal and transverse framing, yet immense strength is added by an efficiently worked skin, or *shell plating*, as it is more commonly termed. Some parts of this outside plating are capable of rendering more service to the structure than others.

For example, the ordinary bar, or *hanging keel*, as it is often termed, has its only connection to the vessel by means of the strakes, or rows of plating, called the garboard strakes, on either side of the keel. The absurdity of connecting a thin plate to a thick bar with large rivets, widely spaced, will be easily understood, and thus the garboard strake is made thicker than its adjacent plating. Moreover, where no heel pieces are fitted, connecting the lower extremities or heels of the frames on one side of the vessel to those on the other side, the garboard strakes accomplish this by securing the heels of the frames firmly to the top of the keel. It also adds stiffness to the bottom flange of the ship girder.

We have already seen the advantage of strengthening the upper and lower edges or flanges of a girder in increasing its efficiency to resist longitudinal bending; and since a ship, as previously stated, is simply a huge, hollow girder, any method of deepening it vertically (ship's side plating, etc.), or increasing the strength of its upper and lower flanges, must add to its longitudinal strength. Hence it is compulsory, if a vessel be classed, to have its uppermost strake of outside plating, called the *sheer strake*, and sometimes the strake next below, increased considerably in thickness (see fig. 36). Also on the bottom of the girder, in the region of the bilge, one or two of the strakes are thickened in long vessels. Midway between the bilge and the sheer strake—approximately in the region of the neutral axis—where the strains vanish, the thickness of the plating is least. The value of long bridges extending over the midship length of a vessel, increasing the depth of the girder at the very place where the bending strain is greatest, must be evident. Indeed, in two- and three-decked vessels of over thirteen depths to length, Lloyd's require that they have a substantial erection extending over the midship half length. A complete or partial steel deck over the middle of the length, together with the beams, also affords great strength to the ship girder in increasing the efficiency of its upper flange.

Definition of Important Terms.—It is necessary at this stage that a few terms be clearly understood.

1. *Length between Perpendiculars.*—For vessels with straight stem this is taken from the fore part of the stem to the after part of the stern post. Should the vessel have a clipper or curved stem, the length is measured from the place where the line of the upper deck beams would intersect the fore edge of the stem, if it were produced in the same direction as the part below the cutwater (fig. 44).

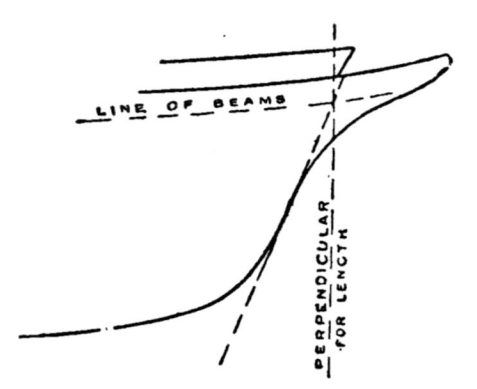

FIG. 44.—LENGTH BETWEEN PERPENDICULARS.

2. "*Lloyd's Length.*"—Lloyd's length is the same as the foregoing, except that the length is taken from the after side of the stem to the fore side of the stern post.

3. *Extreme Breadth.*—This is measured over the outside plating at the greatest breadth of the vessel.

4. *Breadth Moulded.*—This is taken over the frames at the greatest breadth of the vessel.

5. *Depth Moulded.*—This is measured in one-, two-, and three-deck vessels at the middle of the length from the top of the keel to the top of the upper deck beams at the side of the vessel.

In spar- and awning-decked vessels, the depth moulded is measured from the top of the keel to the top of the main deck beams at the side of the vessel.

6. *Lloyd's Depth.*—This is somewhat different. We have seen that they require a round up upon the weather decks, of a

quarter of an inch to one foot of beam. This round up is added to the moulded depth, and gives Lloyd's depth. With this modification it is otherwise the same as No. 5.

In designing our series of midship sections illustrating the arrangement, amount, and development in structural strength in progressive sizes of vessels, we will consider, for the sake of example, that the vessels are to be classed at Lloyd's.

The size and spacing of all transverse framing—frames, reverse frames, floor plates, pillars—are regulated by numbers obtained entirely from transverse dimensions, as follows:—

Add together (measurements being taken in feet) *half the moulded breadth, the depth (Lloyd's), and the girth of the half midship frame section of the vessel, measured from the centre line at the top of the keel to the upper deck stringer plate.* By referring to the tables in Lloyd's rules, the sizes of these parts of the structure, corresponding to the sum of these dimensions, may be found. The number for three-deck steam vessels is produced by the *deduction of 7 feet from the sum of the measurements taken to the top of the upper deck beams.*

The sizes of all longitudinal framing—keel, keelsons, stringers, as well as thickness of outside and deck plating, stem bar, and stern frame—are governed by the number obtained by *multiplying Lloyd's first number for frames, etc., by the length of the vessel.*

Vessels of extreme dimensions require special stiffening above that ordinarily needed by the numbers obtained as above, and special provision is made for this in the rules.

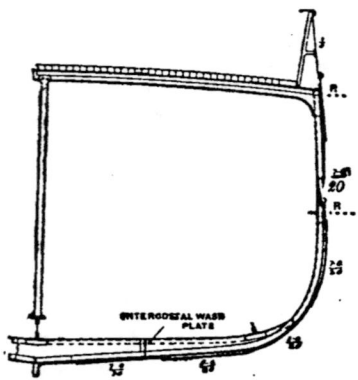

Under 13 feet depth.   12½ depths in length.

FIG. 45.—DIMENSIONS OF FRAMEWORK AND PLATING FOR VESSELS LESS THAN 13 FEET IN DEPTH.

Lloyd's Numerals.

½ Girth	21·8
½ Breadth, ...	12·0
Depth, ...	12·88
1st No., ...	46·63
Length, ...	161
2nd No., ...	7507

Frames, $3 \times 3 \times \frac{6-5}{20}$, spaced 21 ins.
Reverse frames, $2\frac{1}{2} \times 2\frac{1}{2} \times \frac{5}{20}$.
Floors, $13 \times \frac{6-5}{20}$.
Centre keelson, $11 \times \frac{9-7}{20}$.
Keelson continuous angles, $3\frac{1}{2} \times 3 \times \frac{6}{20}$.

Beams, $5\frac{1}{2} \times \frac{8}{20}$ — bulb plate with Double angles, $2\frac{1}{2} \times 2\frac{1}{4} \times \frac{5}{20}$.
Sheer strake, $32 \times \frac{11-8}{20}$.
Garboard strake, $31 \times \frac{9-8}{20}$.
Stringer plate, $36 \times \frac{8}{20} - 19 \times \frac{6}{20}$.
Gunwale angle bar, $3 \times 3 \times \frac{7}{20}$.
Keel, $7 \times 1\frac{3}{8}$.

Additions for Extreme Length.

To thickness of sheer strake, $\frac{1}{20}$ is added for ¾ L amidships.
To bilge keelson, a bulb plate is added for ⅔ L amidships.
To thickness of 2 strakes at bilge, $\frac{1}{20}$ is added for ½ L amidships.

Abbreviations.

L means length, thus

¾ L amidships	,,	three-fourths length amidships.
R	,,	reserve frame height.
$\frac{8-7}{20}$	,,	thickness reduced from $\frac{8}{20}$ to $\frac{7}{20}$.

Note.—All sizes of plates and angle bars are given in inches.

STRUCTURE. 65

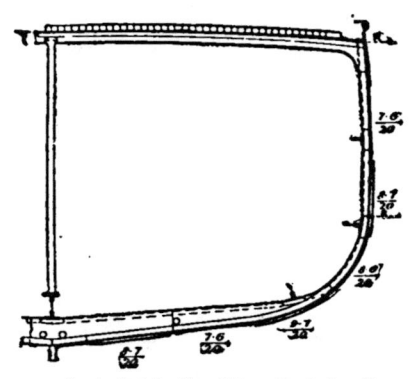

Under 14 feet depth.   12¼ depths in length.

FIG. 46.—DIMENSIONS OF FRAMEWORK AND PLATING FOR
VESSELS LESS THAN 14 FEET IN DEPTH.

Lloyd's Numerals.

½ Girth,	24
½ Breadth,	13·5
Depth,	13·83
1st No.,	51·33
Length,	173
2nd No.,	8880

Frames, $3 \times 3 \times \dfrac{6-5}{20}$, spaced 21 inches.

Reverse frames, $2\frac{1}{2} \times 2\frac{1}{2} \times \frac{6}{20}$.

Floors, $14\frac{1}{2} \times \dfrac{6-5}{20}$.

Centre keelson, $12 \times \dfrac{9-7}{20}$.

Keelson continuous angles, $3\frac{1}{2} \times 3 \times \frac{6}{20}$.

Beams, $6\frac{1}{2} \times \frac{9}{20}$ — bulb plate with $\Big\}$ on alternate frames.
 Double angles, $2\frac{1}{2} \times 2\frac{1}{2} \times \frac{6}{20}$.

Hold pillars, $2\frac{1}{2}$.

Sheer strake, $33 \times \dfrac{12-8}{20}$.

Garboard strake, $32 \times \dfrac{9-8}{20}$.

Deck stringer plate, $38 \times \frac{9}{20}$ — $20 \times \frac{8}{20}$.

Gunwale angle bar, $3 \times 3 \times \frac{7}{20}$.

Keel, $7\frac{1}{4} \times 1\frac{7}{8}$.

Additions for Extreme Length.

To thickness of sheer strake, $\frac{1}{20}$ is added for $\frac{2}{5}$ L amidships.
To bilge keelson, a bulb plate is added for $\frac{3}{5}$ L amidships.
To thickness of 2 strakes at bilge, $\frac{1}{20}$ is added for $\frac{1}{2}$ L amidships.

E

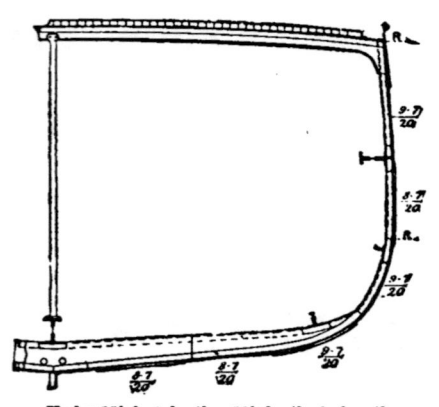

66 KNOW YOUR OWN SHIP.

Under 15½ feet depth.   12½ depths in length.

FIG. 47.—DIMENSIONS OF FRAMEWORK AND PLATING FOR VESSELS LESS THAN 15½ FEET IN DEPTH.

Lloyd's Numerals.

½ Girth,	26·5
½ Breadth,	14·5
Depth,	15·33
1st No.,	56·33
Length,	192
2nd No.	10815

Frames, $3\frac{1}{2} \times 3 \times \frac{6-5}{20}$, spaced 22 inches.

Reverse frames, $2 \times 2\frac{1}{2} \times \frac{6}{16}$.

Floors, $16 \times \frac{7-6}{20}$.

Centre keelson, $12 \times \frac{10-8}{20}$.

Keelson continuous angles, $4\frac{1}{2} \times 3 \times \frac{7}{10}$.

Hold pillars, $2\frac{5}{8}$.

Beams, $7 \times \frac{7}{16}$ – bulb plate with } on alternate frames.
Double angles, $3 \times 3 \times \frac{6}{20}$.

Sheer strake, $34 \times \frac{12-8}{20}$.

Garboard strake, $32 \times \frac{9-8}{20}$.

Deck stringer plate, $40 \times \frac{9}{20} - 22 \times \frac{7}{20}$.

Gunwale angle bar, $3\frac{1}{2} \times 3\frac{1}{2} \times \frac{7}{16}$.

Keel, $7\frac{1}{2} \times 2$.

Additions for Extreme Length.

To thickness of sheer strake, $\frac{2}{20}$ is added for $\frac{2}{3}$ L amidships.
To strake below sheer strake, $\frac{1}{20}$ is added for $\frac{1}{2}$ L amidships.
To bilge keelson, a bulb plate is added for $\frac{4}{5}$ L amidships.
To thickness of 2 strakes at bilge, $\frac{1}{20}$ is added for $\frac{1}{2}$ L amidships.

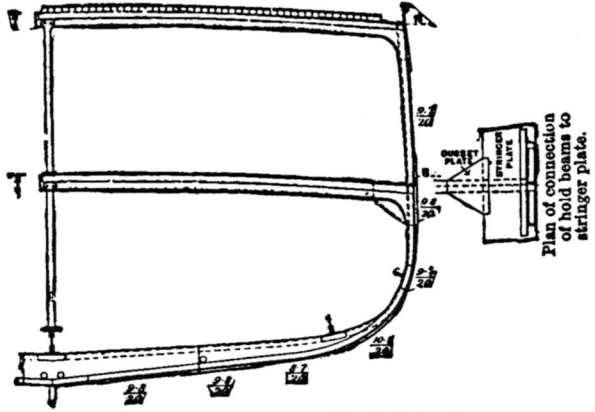

Under 16½ feet depth. 12¼ depths in length.

FIG. 48.—DIMENSIONS OF FRAMEWORK AND PLATING FOR VESSELS LESS THAN 16½ FEET IN DEPTH.

Lloyd's Numerals.

½ Girth,	28·3
½ Breadth,	15·5
Depth,	16·33
1st No.,	60·13
Length,	205
2nd No.,	12326

Frames, $3\frac{1}{2} \times 3 \times \frac{7-6}{20}$, spaced 22 inches.

Reverse frames, $3 \times 2\frac{1}{2} \times \frac{6}{20}$.

Floors, $17\frac{1}{2} \times \frac{8-7}{20}$.

Centre keelson, $13 \times \frac{10-8}{20}$.

Keelson continuous angles, $4\frac{1}{2} \times 3\frac{1}{2} \times \frac{7}{20}$.

Main deck beams, $7\frac{1}{2} \times \frac{7}{20}$, bulb plate with } on alternate frames.
  Double angles, $3 \times 3 \times \frac{7}{20}$.

Hold beams, $8\frac{1}{2} \times \frac{8}{20}$, bulb plate with } on every 10th frame.
  Double angles, $4 \times 3 \times \frac{7}{20}$, with covering plate

Main deck stringer plate, $44 \times \frac{9}{20} - 24 \times \frac{7}{20}$.

Hold stringer plate, $27 \times \frac{7}{20} - 21 \times \frac{6}{20}$.

Gunwale angle bar, $4 \times 4 \times \frac{7}{20}$.

Sheer strake, $35 \times \frac{12-8}{20}$.

Garboard strake, $33 \times \frac{10-9}{20}$.

Keel, $7\frac{1}{2} \times 2\frac{1}{4}$.

Hold pillars, $2\frac{3}{4}$.

Additions for Extreme Length.

To thickness of sheer strake, $\frac{1}{20}$ is added for $\frac{3}{4}$ L amidships.
To strake below sheer strake, $\frac{1}{20}$ is added for $\frac{1}{2}$ L amidships.
To bilge keelson, a bulb plate is added for $\frac{3}{4}$ L amidships.

Under 23 feet depth. 12½ depths in length.

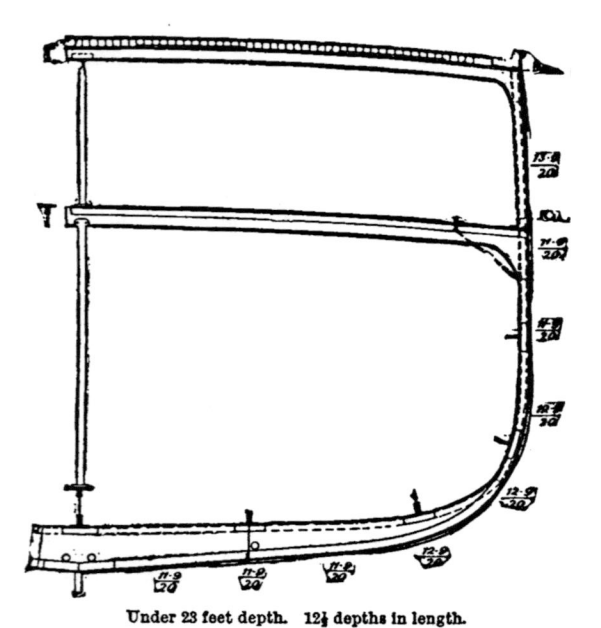

Fig. 49.—Dimensions of Framework and Plating for Vessels less than 23 feet in depth.

STRUCTURE. 69

Lloyd's Numerals.

½ Girth,	38
½ Breadth,	19
Depth,	22·83
1st No.,	79·83
Length,	286
2nd No.,	22831

Frames, $5 \times 3 \times \dfrac{8-7}{20}$, spaced 24 inches.

Reverse frames, $3\frac{1}{2} \times 3 \times \frac{8}{20}$.

Floors, $24 \times \dfrac{10-8}{20}$.

Centre keelson, $20 \times \dfrac{13-11}{20}$.

Keelson continuous angles, $6 \times 4 \times \frac{9}{20}$.
Intercostal keelson plate, $\frac{8}{20}$.
Complete steel deck, $\frac{7}{20}$ thick on main deck.

Main deck beams, $6\frac{1}{2} \times 3 \times \dfrac{9-8}{20}$, bulb angle on every frame.

Hold beams, $10\frac{1}{2} \times 1\frac{8}{8}$ bulb plate, with } on every 10th frame.
Double angles, $4\frac{1}{2} \times 4 \times \frac{9}{20}$

Main deck stringer plate, $41 \times 1\frac{8}{20} - 35 \times \frac{8}{20}$.
Hold stringer plate, $38 \times \frac{9}{20} - 29 \times \frac{8}{20}$.
Gunwale angle bar, $4\frac{1}{2} \times 4\frac{1}{2} \times 1\frac{8}{8}$.

Sheer strake, $42 \times \dfrac{15-10}{20}$.

Garboard strake, $36 \times \dfrac{12-11}{20}$.

Keel, $10 \times 2\frac{1}{2}$.
Hold pillars, $3\frac{1}{2}$.

Additions for Extreme Length.

To thickness of sheer strake, $\frac{1}{20}$ is added for ¾ L amidships.
To strake below sheer strake, $\frac{1}{20}$ is added for ½ L amidships.
To bilge keelson, a bulb plate is added for ⅜ L amidships.
To thickness of 3 strakes at bilge, $\frac{1}{20}$ is added for ¼ L amidships.

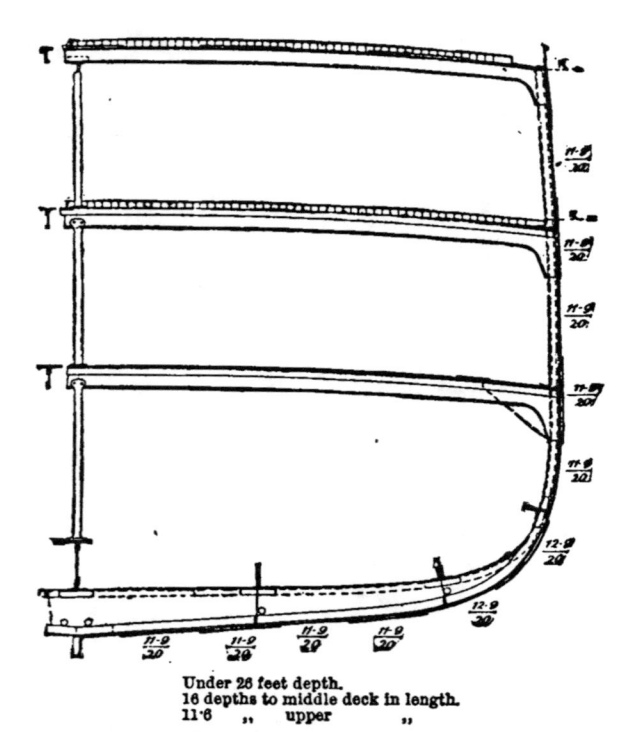

Under 26 feet depth.
16 depths to middle deck in length.
11·6 ,, upper ,,

FIG. 50.—DIMENSIONS OF FRAMEWORK AND PLATING FOR VESSELS LESS THAN 26 FEET IN DEPTH.

STRUCTURE. 71

Lloyd's Numerals.

½ Girth,	. . . . . .	40·75
¼ Breadth,	. . . . . .	20·5
Depth,	. . . . . .	25·83
		87·08
		7
1st No.,	. . . . . .	80·08
Length,	. . . . . .	301
2nd No.,	. . . . . .	24104

Frames, $5 \times 3\frac{1}{2} \times \frac{8-7}{20}$ spaced 24 inches.

Reverse frames, $3\frac{1}{2} \times 3\frac{1}{2} \times \frac{8}{20}$.

Floors, $24\frac{1}{2} \times \frac{10-8}{20}$.

Centre keelson, $25 \times \frac{14-12}{20}$.

Keelson continuous angles, $6\frac{1}{2} \times 4 \times \frac{9}{20}$.
Intercostal keelson plate, $\frac{7}{20}$.
Complete steel deck, $\frac{7}{20}$ thick on main deck.

Upper deck beams, $7\frac{1}{2} \times 3 \times \frac{10-9}{20}$ bulb angles on every frame.

Middle deck beams, $10 \times 1\frac{5}{8}$, bulb plate with ⎫ on every alternate
  Double angles, $3\frac{1}{2} \times 3\frac{1}{2} \times \frac{7}{20}$ ⎭ frame.

Hold beams, $11 \times 1\frac{1}{16}$, bulb plate with ⎫ on every tenth frame.
  Double angles, $5 \times 4 \times \frac{9}{20}$ ⎭

Upper deck stringer plate, $43 \times 1\frac{5}{8} - 36 \times \frac{8}{20}$.
Middle deck stringer plate, $62 \times 1\frac{5}{8} - 36 \times \frac{8}{20}$.
Hold stringer plate, $40 \times \frac{9}{20} - 31 \times \frac{8}{20}$.
Gunwale angle bar, $4\frac{1}{2} \times 4\frac{1}{2} \times 1\frac{5}{8}$.

Sheer strake, $42 \times \frac{15-10}{20}$.

Garboard strake, $36 \times \frac{12-11}{20}$.

Keel, $10 \times 2\frac{5}{8}$.
Hold pillars, $3\frac{5}{8}$.

Additions for Extreme Length.

To thickness of sheer strake, $\frac{2}{20}$ is added for ¾ L amidships.
To bilge keelson, a bulb plate is added for ⅔ L amidships, and
  an intercostal plate for ½ L amidships.
To thickness of 2 strakes at bilge, $\frac{1}{20}$ is added all fore and aft.
Centre keelson increased in depth.
To side keelson, a bulb is added for ½ L amidships.
To bilge stringer, an intercostal plate is added for ¾ L amidships.

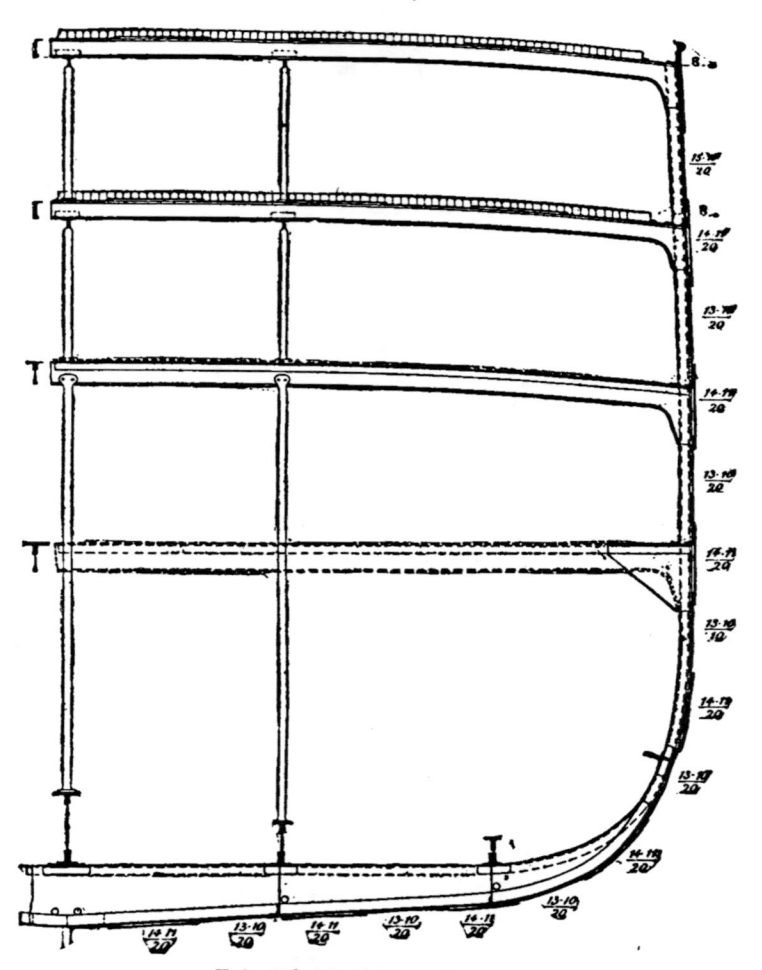

Under 36 feet depth (also under 39 feet).
15·5 depths to middle deck in length.
12·5 ,, upper ,,

FIG. 51.—DIMENSIONS OF FRAMEWORK FOR VESSELS LESS THAN 36 FEET IN DEPTH.

Note.—The scantlings are for the vessel under 36 feet depth, all of which are shown in clear, black lines. The fig. is drawn to the under 39 feet depth in order to show the introduction of the new tier of beams indicated by dotted lines. Such a vessel would require both additional topside and bottom strengthening.

STRUCTURE.

Lloyd's Numerals.

½ Girth,	52·5
½ Breadth,	25
Depth,	35·83
	113·33
	7
1st No.,	106·33
Length,	448
2nd No.,	47635

Frames, $6 \times 3\frac{1}{2} \times \dfrac{10-9}{20}$, spaced 25 inches.

Reverse frames, $4\frac{1}{2} \times 3\frac{1}{2} \times \frac{13}{20}$.

Floors, $32 \times \dfrac{10-8}{20}$.

Centre keelson, $36 \times \dfrac{14-13}{20}$.

Keelson continuous angles, $6\frac{1}{2} \times 4\frac{1}{2} \times \frac{13}{20}$.
Intercostal keelson plates, $\frac{9}{20}$.
Foundation plate, $18 \times \frac{13}{20}$.
Complete steel deck to upper deck, $\frac{7}{20}$.
Complete steel deck to middle deck, $\frac{7}{20}$.

Upper deck beams, $9 \times 3 \times \dfrac{12-11}{20}$, bulb angle on every frame.

Middle deck beams, $9 \times 3 \times \dfrac{12-11}{20}$, bulb angle on every frame.

Lower deck beams, $12 \times \frac{11}{20}$, with }
  Double angles, $3\frac{1}{2} \times 3\frac{1}{2} \times \frac{9}{20}$ } on every 2nd frame.

Upper deck stringer plate, $64 \times \frac{13}{20} - 51 \times \frac{9}{20}$.
Middle deck stringer plate, $64 \times \frac{11}{20} - 51 \times \frac{9}{20}$.
Lower deck stringer plate, $56 \times \frac{13}{20} - 44 \times \frac{8}{20}$.
Gunwale angle bar, $5 \times 5 \times \frac{11}{20}$.

Sheer strake, $46 \times \dfrac{15-12}{20}$.

Garboard strake, $36 \times \dfrac{15-14}{20}$.

Keel, $12 \times 3\frac{1}{4}$.
Hold pillars, 4.

Additions for Extreme Length.

Sheer strake doubled for whole width, for ¾ L amidships.
To strake below sheer strake, $\frac{1}{20}$ is added for L amidships.
To upper deck stringer plate, $\frac{1}{20}$ is added for ¾ L amidships.
To side keelson, continuous plate is added for ¼ L amidships.
To bilge keelson, continuous plate is added for ½ L amidships.
To bilge stringer, intercostal plate is added for ¾ L amidships.
  Centre keelson increased in depth.

Relation of Strength to Dimensions—Notes on "Midship Sections."—In considering the subject of strains, it was found that both longitudinal and transverse strains decreased towards the ends of the vessel, being greatest on each side of midships. Naturally, therefore, in turning to the transverse sections (figs. 45 to 51), we expect to find a corresponding arrangement of structural strength. Such, indeed, is the case. In the transverse framing the frames and floors maintain their maximum size for three-fifths of the vessel's length amidships, and are reduced in thickness for the remaining one-fifth of the length at each end. The floors are carried up the bilge to a height of twice their midship depth above the top of the keel, for one-fourth of the midship length of the vessel. Fore and aft of this distance, the ends are gradually lowered until the tops of the floors are level. At the extreme ends, however, the floors are increased in depth, as subsequently shown.

The height to which the reverse frames are carried varies according to the transverse dimensions of the vessel. When Lloyd's first number is below 45, the reverse frames are carried across every floor plate, and up the frame to the upper part of the bilges; when 45 and below 57 they extend alternately to the gunwale, and high enough to enable the double angle stringer above the bilges to be securely connected; or, if hold beams are fitted, high enough to get a good connection to the beam stringer angle. When the number is 57 and above, the reverse frames extend alternately to the gunwale and the stringer next below. When the number for sailing vessels reaches or exceeds 75, the reverse frames extend to the gunwale on every frame.

Except in spar- and awning-decked vessels, and in poops and forecastles, the beams, exclusive of hold beams, where less than three-fourths the length of the midship beam, are somewhat reduced—in many cases—in both depth and thickness.

In the vessels of which figs. 45, 46, and 47 are midship sections, only one tier of beams is required; but immediately the depth reaches and exceeds 15 feet 6 inches (Lloyd's depth), a tier of widely-spaced hold beams is fitted (figs. 48 and 49), with a continuous stringer plate on the ends. This, together with the gradual growth of the other framing, provides the additional transverse and longitudinal strength demanded by greater depth and length. When 24 feet is reached and exceeded, another tier of beams is required (fig. 50).

When the depth reaches 32 feet 6 inches, although another tier of beams is not required, a stringer plate supported on alternate frames by large bracket knees has to be fitted, and on its inner edge is riveted a large angle of the size of the centre keelson angles, converting it still further into an efficient girder.

When 36 feet depth is reached, the additions shown by the dotted lines in fig. 51 are required. Here we have a fourth tier of beams of extra strength fitted to every tenth frame. These are known as *orlop beams*.

Turning to the longitudinal framework, we find similar reductions in the scantlings taking place. The centre keelson plate standing upon the top of the floors, together with its rider plate, maintains the midship thickness for one-half the vessel's length amidships. Beyond this length considerable reduction takes place in the thickness, and the rider plate disappears altogether before and abaft of the three-fourths length amidships. Stringer plates at the ends of beams retain their midship dimension for one-half the vessel's length. Throughout the remaining one-fourth length at each end they gradually diminish in both breadth and thickness. The number of hold stringers and keelsons to below the bilge is regulated entirely by the depth of the vessel. The sizes of all the large angles for keelsons and stringers in the hold are the same as the centre keelson angles.

A glance at the outside shell plating covering the frames also shows a reduction, and this takes place on the one-fourth length at each end.

Where a steel deck of seven-twentieths of an inch or over is fitted, it is reduced in thickness fore and aft of the half length amidships. A little consideration will soon make it obvious that it would be absurd to deal out to every vessel of similar "2nd numerals" exactly similar longitudinal strengthening, for then a long, shallow vessel would receive actually less longitudinal stiffening than a shorter, deep one, since the introduction of hold stringers and keelsons is regulated almost entirely by the depth.

We have already seen that long, shallow vessels with small depth of girder possess less resistance to bending than shorter ones, not to mention shorter and deeper vessels. It is, therefore, the custom with classifying associations to fix upon a standard vessel of a certain number of depths to the length. Lloyd's Committee adopt a length of eleven depths as the standard worked upon. Vessels exceeding these proportions are, therefore, subject to the introduction of additional longitudinal strengthening over the middle of the length. In figs. 45 to 51, the number of depths in length is 12·5, excepting fig. 50, which is 11·6, and the additional strengthening therefore required is shown, as far as possible, on their respective sketches, as well as in the table of additions on each diagram.

When in three-decked vessels the length is more than eleven times the depth taken from the top of the keel to the top of the *middle deck beams*, special additional strength has to be introduced at the bilge and bottom. Such strength has been introduced into the foregoing midship sections. All vessels having a length of thirteen or more times the depth from the top of the keel to the top of the upper deck beams are to have a substantial erection, such as a bridge, extending over the half length of the vessel amidships.

A perusal of figs. 45 to 51, with the scantlings accompanying them, will, it is believed, fully verify the foregoing remarks upon the arrangement and growth of structural strength.

Local Strengthening—Space occupied by Machinery.— *Engine and Boiler Space.*—Perhaps one of the first things which strikes one in studying the arrangement of a steam vessel, from a structural point of view, is the concentration of weight, and also the vibration, especially in vessels of high speed, in the machinery space. In addition to this, there is a break in the arrangement of transverse strength owing to the omission of beams, and even at the upper deck the beams are cut in the way of the engine and boiler casings. The result of all this is to produce a tendency to vertical elongation, and to cause the upper breadth of the vessel to contract.

Moreover, if the vessel did not possess sufficient rigidity and stiffness to resist working under the vibration of the engines, the evil would rapidly increase, and assume serious dimensions. But, happily, all this can be provided against, and although the methods adopted may vary somewhat for different types of vessels, yet a few general hints may be given.

First of all, a good foundation must be secured for both engines and boilers. Lloyd's require that floor plates under engines be one-twentieth of an inch thicker, and under boilers two-twentieths of an inch thicker than are otherwise required, and by this means greater stiffness is given to the bottom of the vessel. No doubt the extra one-twentieth put upon boiler floors is on account of the fact that the damp heat created there has the effect of making corrosion more rapid than elsewhere. At the same time, it should be noted that a great deal of the corrosion in this locality is due to neglect, for if the bottoms of vessels were carefully watched, and frequently cement washed, there would be less cause for complaint regarding the condition of the floors of comparatively new vessels, and it would be found that a little care is cheaper in the end than new floors. As a rule, it is found, in vessels with ordinary floors, that the engine foundation or *seat*, as it is more commonly called, has to be built

Engine seat.

Elevation.

SHELL

INTERCOSTAL GIRDER

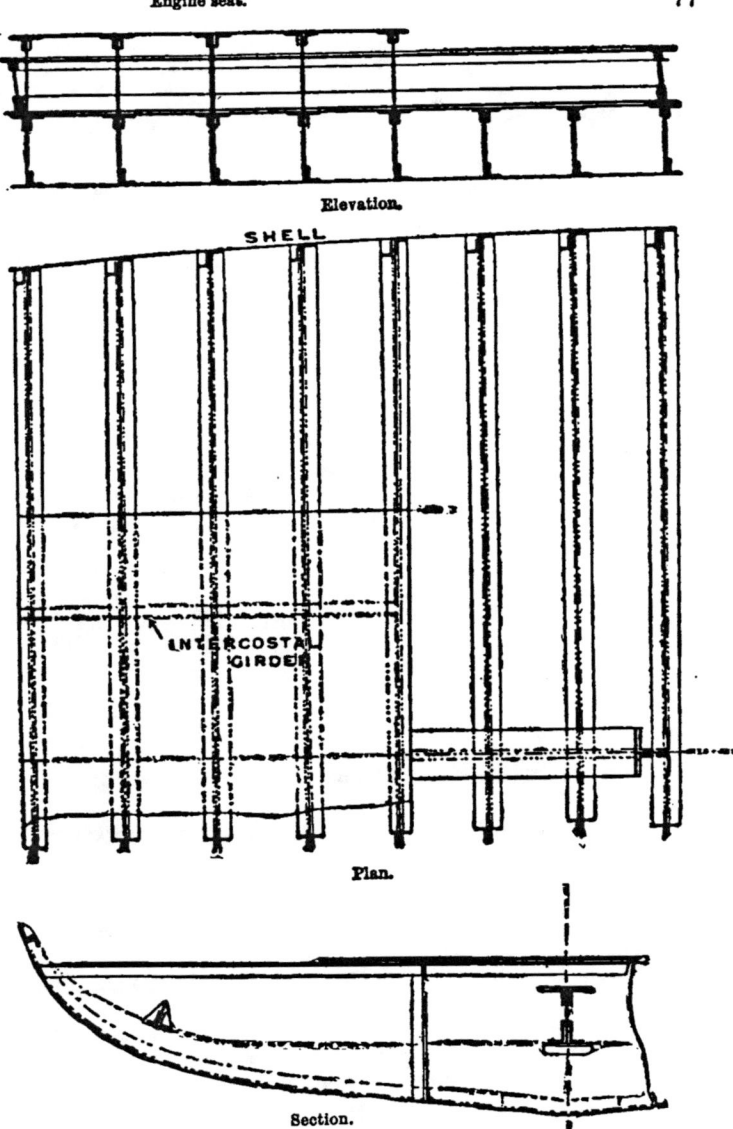

Plan.

Section.

FIG. 52.—ENGINE SEAT.

to a height considerably above the floors. When this is the case, and it is practicable, a splendid seat is constructed by making the floors deep enough to reach to the engine seat, and to extend with horizontal edges from side to side of the vessel. Across the top of all floors in engine and boiler space, Lloyd's require double reverse angles to be fitted at least from bilge to bilge. This not only stiffens the floor, but forms a good means of connection for the thick plating to which the bed plate of the engine is bolted. Intercostal plates, fitted between floors, on either side of the

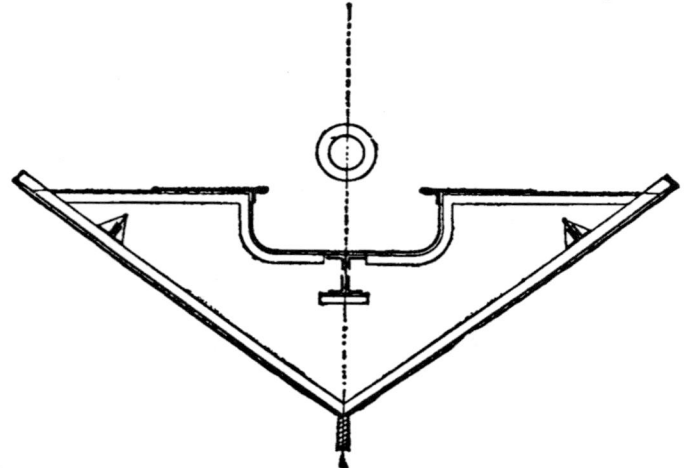

FIG. 53.—TROUGH UNDER ENGINES TO CATCH GREASE DRIP.

centre line, give further stiffness to the floors, and support for the condenser, etc. (see fig. 52).

Sometimes, especially in the case of yachts, a watertight trough is fitted in the engine floors immediately under the shafting, which prevents the grease drip from the engine finding its way into the bilges (fig. 53).

Where the floors are not made continuous from keel to engine seat, and from bilge to bilge, a seat has to be built simply in the way of the engine on the top of the ordinary floors; this, however, lacks in efficiency as compared with the previous method.

The boilers, too, require to be fitted on foundations which are firm and rigid, and securely connected with the bottom of the vessel (fig. 54). This is done by fixing two or three thick plates called *stools*, to the top of the floor plates under each boiler. The

STRUCTURE. 79

top edges of these plates are cut to shape so as to receive the boiler. They are attached to the double reverse angles on the top of the floors by means of double angles, and round their edges are fitted large double angles. These provide a good surface upon which the boiler can rest. Then, to hold the stools in place and prevent them tripping—that is, inclining one way or another—tie plates are fitted fore and aft, and connected to the double angles. In the case of vessels constructed with double bottoms, for water ballast, as shown in fig. 55, the engines generally stand upon the tank top, this usually being of sufficient height. Sometimes, however, it is necessary to build a girder seat upon the inner bottom plating, in order to raise the engines to the desired height. In any case, the inner bottom plating is increased

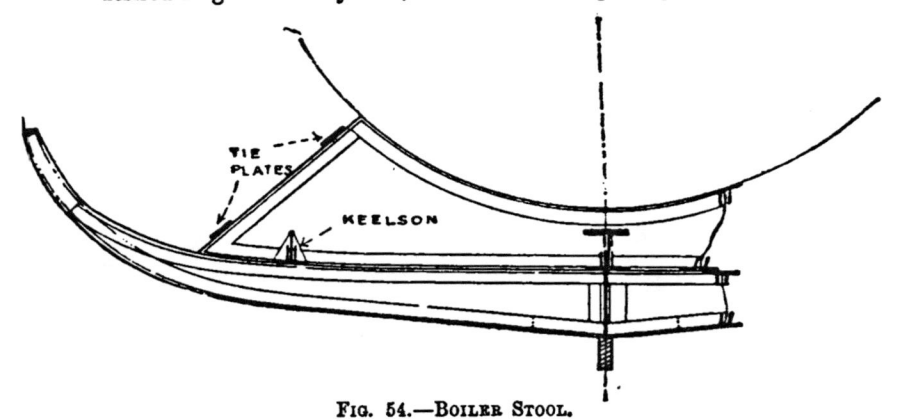

FIG. 54.—BOILER STOOL.

in thickness in way of the engine and boiler space, and immediately under the engines a thick foundation plate is firmly riveted. This plate may form a part of the inner bottom plating, or may be riveted on the top of the inner bottom plating when the engines stand immediately upon the tank top, or it may be riveted to the top of the girder seating when such an arrangement exists. To this the bed plate of the engine is bolted. The boiler stools are riveted to the inner bottom plating. As in the case of ordinary floors, extra stiffening is required at the bottom of the vessel under the engines and boilers, and this is done by fitting intercostal girders.

Having stiffened the bottom of the ship, it is necessary to provide for the loss of the beams, and also some means of keeping out the sides of the vessel. This is done in exactly the same

80   KNOW YOUR OWN SHIP.

manner as when hold beams are dispensed with—viz., web frames are fitted. In high speed vessels especially, care should be taken that sufficient of these are introduced. Valuable as web stringers are, in conjunction with web frames, it is found better in the

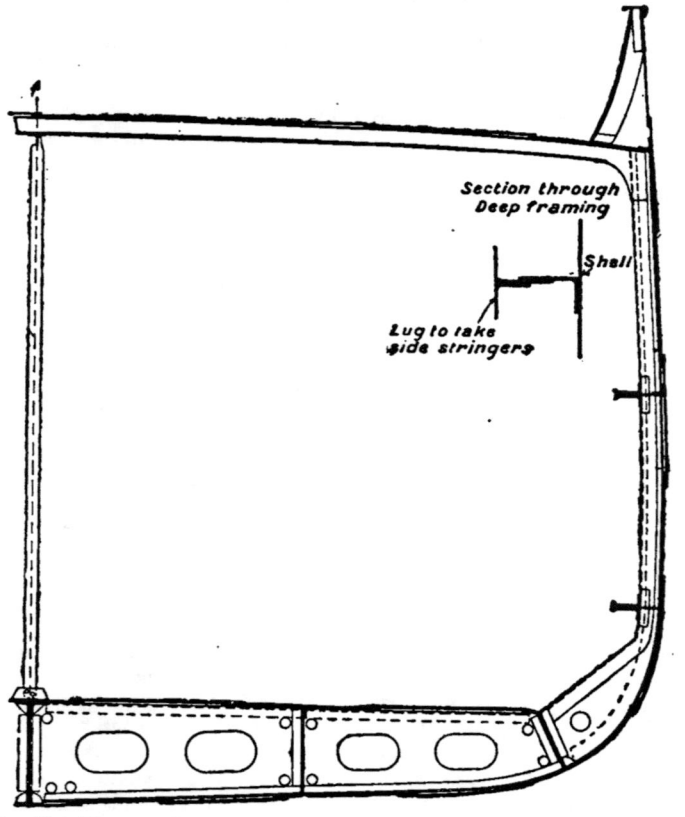

FIG. 55.—MIDSHIP SECTION OF A VESSEL BUILT ON THE "DEEP FRAME" SYSTEM WITH A CELLULAR DOUBLE BOTTOM.

machinery space to fit extra strong beams wherever practicable. Several forms may be adopted, according to the size of the vessel (see B, C, D, E and F, fig. 32).

It will often be found of advantage to further stiffen the frames in the engine and boiler space by fitting all the reverse angles to

STRUCTURE. 81

the upper deck, and in some cases the double reverse bars also.

Sometimes in vessels with the centre keelson fitted on the top of the ordinary floors (fig. 56), the height of the keelson throws

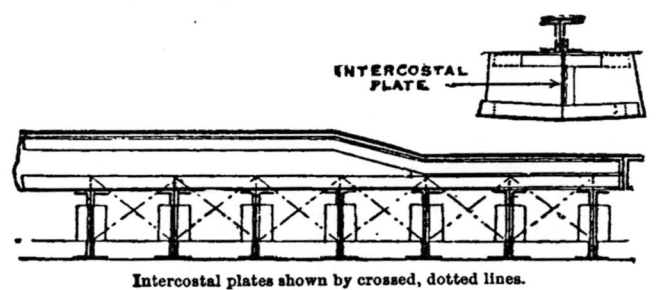

Intercostal plates shown by crossed, dotted lines.

Fig. 56.—Intercostal Compensation for Reduction in Depth of Centre Keelson under Boilers.

the boiler or boilers too high, and it is necessary to cut the keelson down. Were no compensation made for this reduction in the height of the girder, the weakness at this part would be a most serious defect. Several means, however, may be adopted to recover the strength.

In the first place, it will be seen that, owing to the depth of the keelson being reduced, the sectional area of the keelson is reduced also. This may be recovered by making the central plate thicker at this part until the sectional area has been regained. But even this does not fully recover the loss of strength, for though the sectional area is obtained, yet the reduced depth of the girder proves a loss of strength in resisting longitudinal bending. It, therefore, becomes necessary to increase the sectional area to beyond what it was in its original condition. A good method of compensation is to fit intercostal plates between the floors, and attached to the centre plate by means of its lower angles, as shown in fig. 56.

Again, it sometimes happens that, in the way of the boilers, it is necessary to reduce the width of the hold stringer plates. Compensation may be made, as in keelsons, by increasing the thickness of the plate in the way of the reduced width, and also by fitting strong angles on the inner edge of the stringer plate.

Mode of Strengthening Ship at Aft End of Shafting.—Leaving the engine and boiler space, and travelling aft, great vibration and strain is thrown upon the vessel, especially when in a seaway, by the action of the propeller. At one moment it is

F

totally submerged, and revolves with regular precision, but immediately after the wave has passed, it is totally or partly out of the water, and races at a high speed. Provision must, therefore, be made for this. To stiffen the sides of the vessel, the floor plates at the aft end should be made considerably deeper, and towards the stern post they should be carried above the shaft, so that the shaft passes through them. Then, again, it is most important that a sound connection be made between the outside plating and the stern post. By the midship sections we see that considerable reduction takes place in the thickness of the end outside plating. To connect thin plating to a massive iron stern post with large rivets widely spaced would, it is likely, result in the thin plate being unable to hold the stern post rigid under the strains mentioned, and leakage would ensue. So what is done is to increase the thickness of these endmost plates to at least the thickness of the plating at midships, and thereby get a more evenly balanced connection.

But a further means of rigidly fixing the stern frame is required, and this is done by fitting a deep plate, called a *transome plate*, against the upper part of the stern post, extending across the counter from side to side, and riveted to the frames. The connection between the stern post and the transome plate is effected by means of strong angles (see fig. 57, section A B).

It is well known to seamen that considerable leakage often takes place, notwithstanding these precautions to hold the post fixed. An additional means to secure this end is to have another post carried up at the fore side of the stern frame (Y in sketch), and this also is attached to a stout plate carried from side to side of the vessel, and connected with the frames as well as to a bracket plate at the lower deck. By this method, with sound workmanship, and by filling in the space above the propeller aperture between the two parts with cement, leakage is very improbable.

In the case of steamers with twin screws, a different arrangement of supporting the end shafting has to be adopted. One method, after leaving the skin of the ship, is by projecting struts, as shown in fig. 58.

The importance of having these struts securely attached to the hull of the vessel will readily be observed, for should any accident happen here, the vessel is entirely crippled. When the screws somewhat overlap each other, and revolve partly in an aperture in the stern frame, a sound and reliable connection of the struts with the stern frame may be made, as shown by fig. 58, *a* and *b*, the ends of the struts being welded out into broad palms capable of taking a sufficient number of rivets. If there is a weak point

STRUCTURE. 83

in this method, it is in the fact that both the upper palms of the struts, and also the lower palms, are connected by the same rivets, so that on each side of the main post, owing to the vibration of the shafts, there is this tendency to work or shear the rivets. Each rivet, therefore, bears a double strain. But where no aperture is needed for the propellers, the shell plating is carried out to the after stern post, and another means has to be provided for the connection of the struts. The commonest way, perhaps, is that shown in fig. 59.

Here the upper palms are simply fixed on to the outside shell

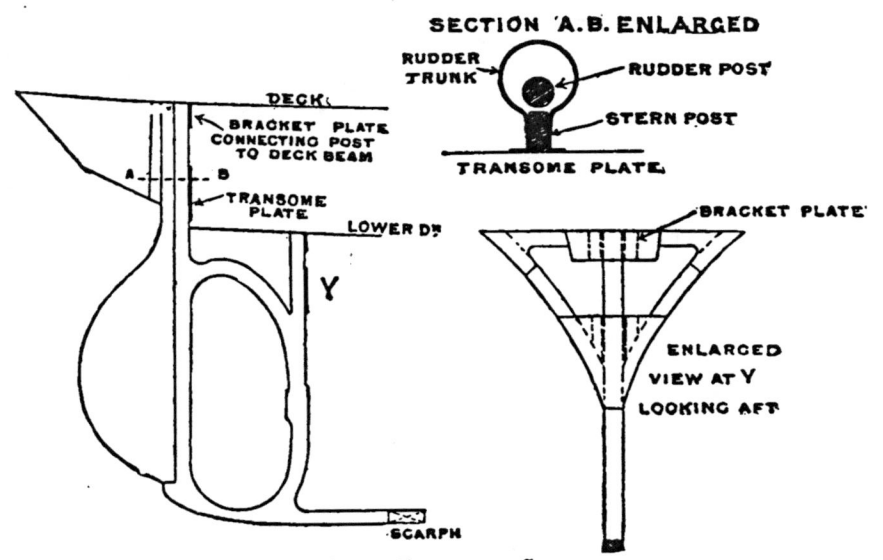

FIG. 57.—STERN FRAME AND CONNECTION.

plating. If this were all the connection, it would not take a keen observer to see that the immense vibration of the shafting, especially in high-speed steamers, would simply tear away the plating. It therefore becomes absolutely essential that the most effectual means be adopted to secure the best and surest connection. Two things have to be aimed at in accomplishing this. First, that at the point where the shafts emerge from the hull, the vessel be well stiffened and bound together, so as to reduce the vibration to a minimum; and second, that in the way of the strut connections the ship be well strengthened. A variety of means may be used

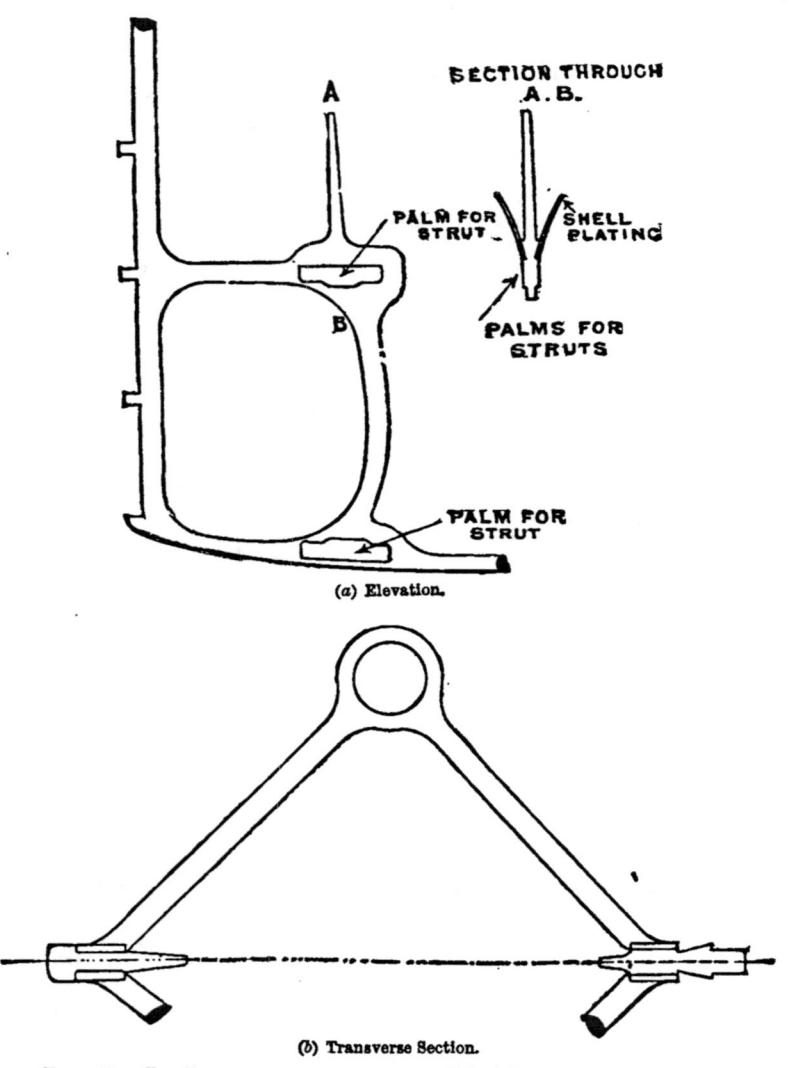

FIG. 58.—CONNECTION OF STRUTS WITH STERN FRAME IN TWIN-SCREW STEAMERS—*a*, ELEVATION; *b*, TRANSVERSE SECTION.

STRUCTURE. 85

to attain the former of these. For example, in some vessels the method shown in fig. 59 might be carried out.

Here a very thick plate or web is placed across the ship from side to side, and securely riveted to the main frames (A in fig. 59). Through this plate the shafts pass. If the vessel be not too wide at this part, this alone will form an efficient tie to the two sides, and give the required stiffness. However, in larger vessels, where the screws are further apart, it would be necessary

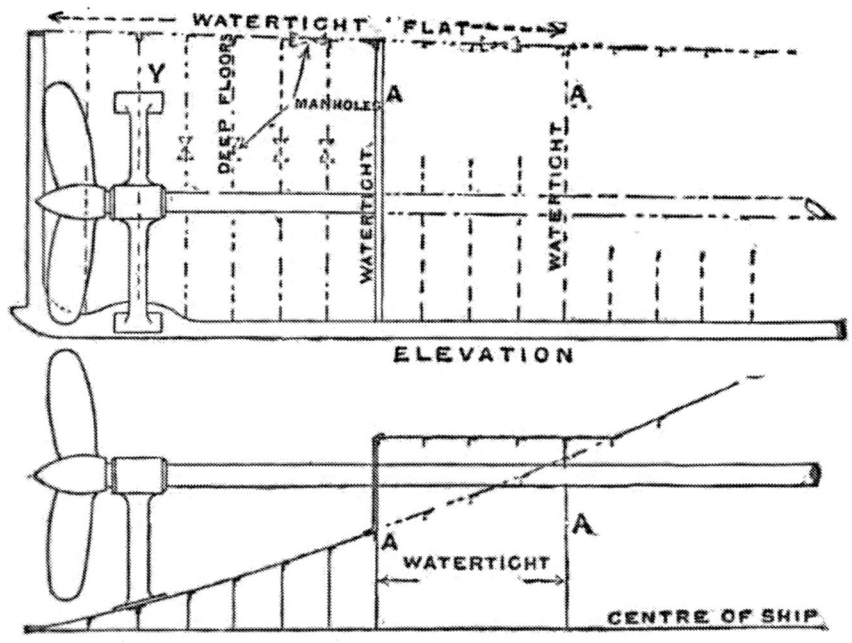

FIG. 59.—MODE OF STRENGTHENING SHIP AT THE AFTER END AND ATTACHING STRUTS TO SHELL PLATING.

to further stiffen this web by means of angle bars or bulb angles across its face, or even introduce more of these web plates on adjacent frames.

In way of the struts, the following system might be employed :—

1. Carry up all the floor plates from the stern post, to a few frames forward of the struts, above the height of the upper palms (see fig. 59).

86　　　KNOW YOUR OWN SHIP.

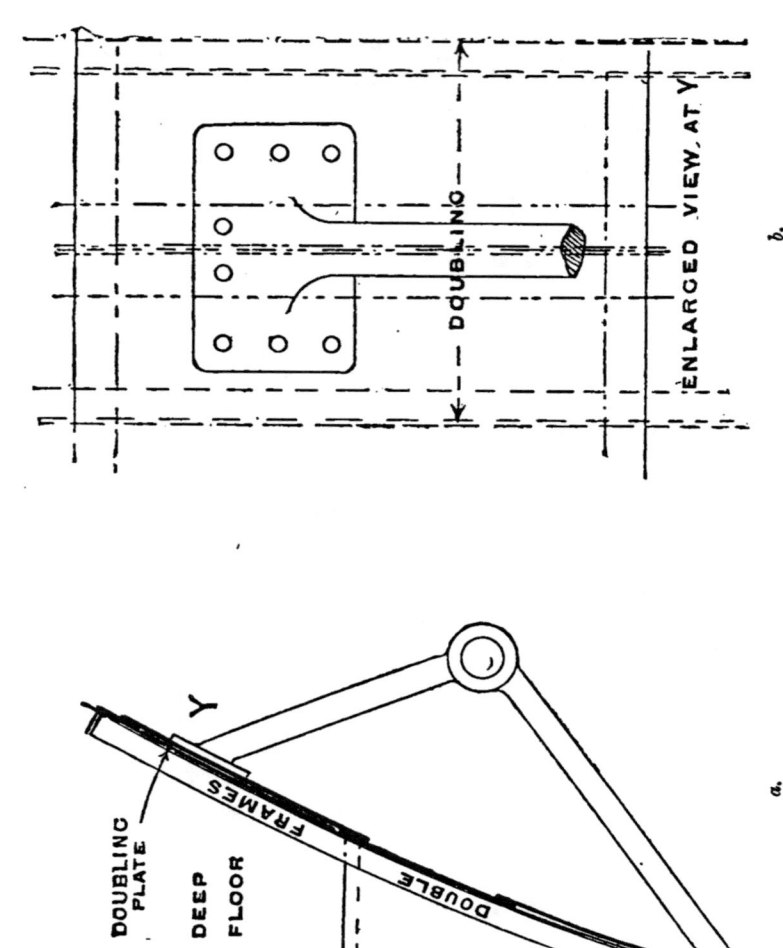

Fig. 59.—*a* and *b*.—Mode of Strengthening Ship at the After End and Attaching Struts to Shell Plating.

2. If possible, arrange for a double frame on each side of the vessel to come in the centre of the palms, having thick, broad flanges against the outside plating.

3. If the strake of plating, upon which the palms of the struts rest, be an *outside* one, fit a doubling plate between the edges of the two adjacent strakes, extending fore and aft from the after side of the frame abaft to the fore side of the frame before the double frame angles.

If the strake of plating be an *inside* one, the doubling plate may be fitted on the outside of the shell plating. With this additional strengthening, the palms can now be securely attached to the plating with rivets, through the shell plating, doubling plate, and large double-frame angles. Athwartship, the vessel is strengthened by the deep floor plates already mentioned.

In any case, in steam vessels there would be a watertight bulkhead at some little distance forward of the struts. But where in such vessels as these there is the greater possibility of leakage, it would be well to construct a watertight flat at a small height above the upper palms of the struts, so that, if leakage did occur, only this compartment would be flooded. From this bulkhead, it would also be well to continue the watertight flat just above the shafting, and extend it far enough forward to include the point where the shafts emerge from the shell, since this is another place where leakage might take place (A in fig. 59). This method of strut connection has the advantage of having the strain of the vibration of only one of the shafts thrown upon the rivets connecting the palms with the hull, each upper palm being a separate connection.

Another method of strut connection is that shown in fig. 60, *a* and *b*.

In this case, the struts are carried through the outside plating, and attached to a very thick intercostal plate, fitted between thick transverse plates. To insure watertightness where the struts pierce the shell, collars would have to be wrought and carefully caulked; and the chambers into which the struts enter should be made watertight also. This method has the same disadvantage mentioned in fig. 58—viz., the same rivets connect both palms.

A system has been adopted in recent years of supporting the end shafting in large twin-screw vessels, on first leaving the main body of the hull, by carrying out the framing and shell plating round the shafting. Further aft, however, where the shaft is considerably out from the hull, the main frame is carried down in the usual way (fig. 61), and a piece of frame bar is scarphed on to the main frame, which, together with the shell-plating, is worked round the shaft, binding the projection to, and, indeed, making it an integral part of the structure. By this method the struts are dispensed with, and a strong and efficient means of supporting the shafting is obtained.

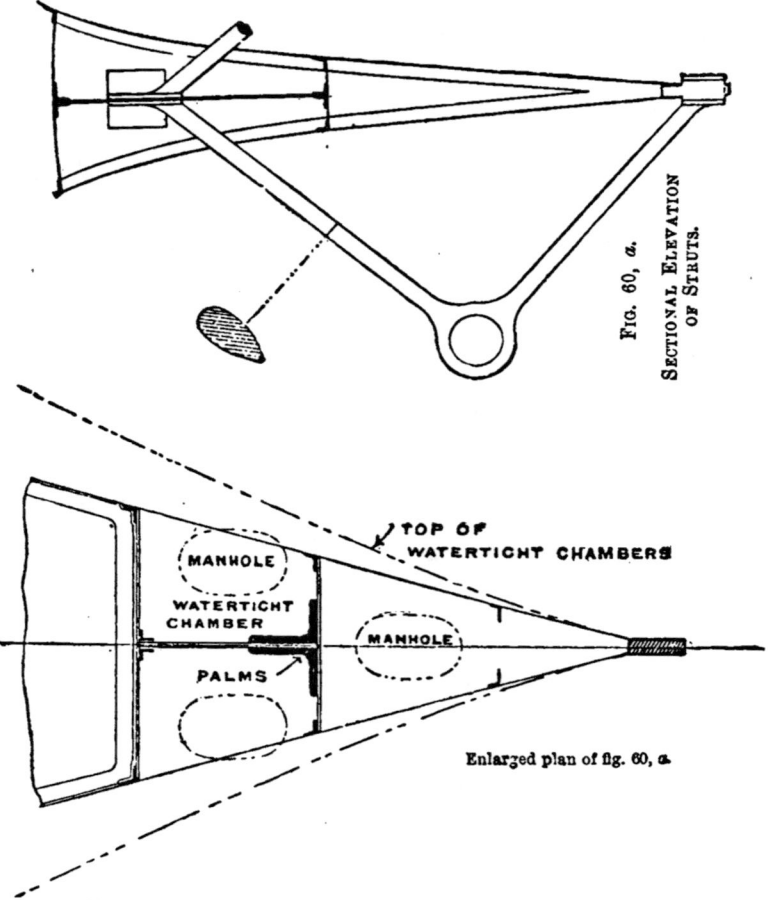

Fig. 60, a.—Sectional Elevation of Struts.

Enlarged plan of fig. 60, a.

Fig. 60, b.—Struts carried through Shell Plating into Watertight Chamber (Plan).

Panting.—Several means may be adopted to prevent panting, a few of which we shall notice. The common method is to fit plate stringers, called *panting stringers*, in addition to the ordinary stringers of the vessel, extending from abaft the collision bulkhead continuously into the stem, where they are joined by means of a plate called a *breast hook*, uniting, as it were, the two breasts of the vessel. These stringer plates are supported on beams, as shown in fig. 62.

An objection raised against this method of stiffening is that it forms an isolated rigid girder and the vessel is inclined to fall hollow on each side of it, especially if subject to encountering ice, as in the Baltic at certain seasons. What is wanted is a more even distribution of strength. This may be obtained by a slight modification in the arrangement of the transverse framing. The usual practice is to space the frames of a vessel throughout the length at equal distances apart, and measured in a fore and aft direction on the top of the keel. Let the frame spacing, as required by Lloyd's, be 24 inches. While the distance from heel to heel of the respective frames may measure exactly 24 inches as they stand upon the keel, yet, especially in very bluff-ended vessels, at waterplanes above the keel, where the shell rapidly curves in to the stem and stern, it may be found that the frame spacing measures as much as 26 inches—actually further apart than at amidships. Figures 62 and 63, which represent a somewhat fine-ended vessel, will, however, illustrate this. To give stiffness to the fore end, the frames should be spaced rather more closely instead of more widely, so as to produce, say 2 or 3 inches less than Lloyd's requirements when measured on the shell. It will also be found of advantage to increase the thickness of the shell plating. Whatever method be adopted, it is advisable to considerably increase the depth of the fore end floors.

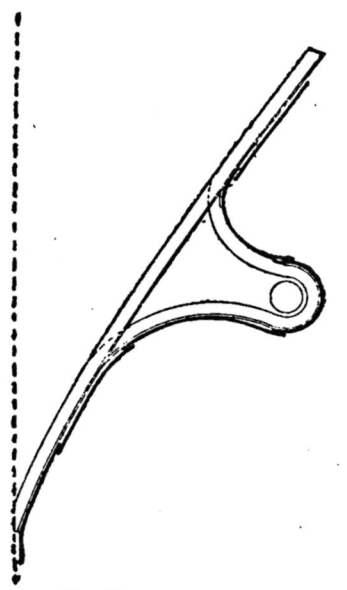

FIG. 61.—METHOD OF DISPENSING WITH STRUTS IN TWIN-SCREW STEAMERS.

The introduction of a few extra beams with stringer plates on their ends, where the ordinary stringers are widely spaced, will generally give sufficient stiffness.

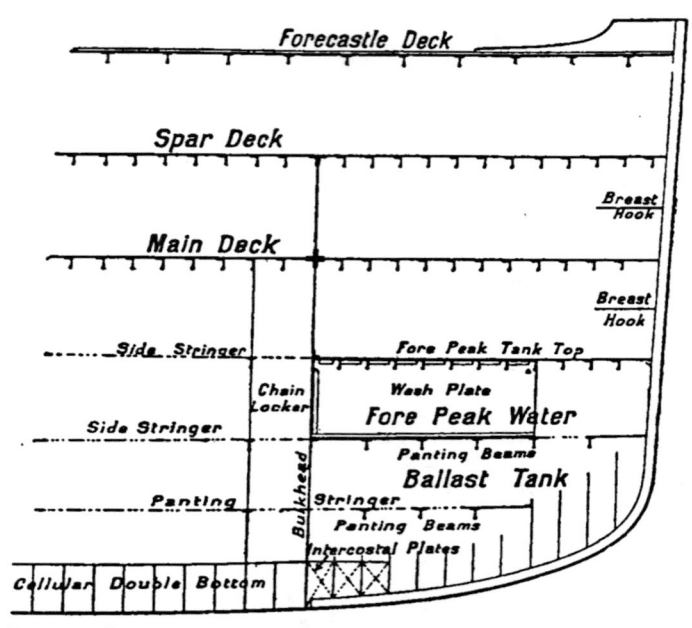

FIG. 62.—FORE AND AFT SECTION OF THE STEM OF A SCREW STEAMER, SHOWING SPECIAL STRENGTHENING AND PANTING ARRANGEMENT.

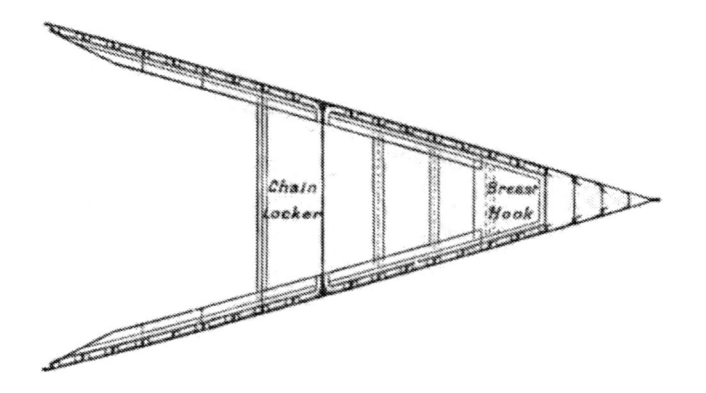

FIG. 63.—PLAN OF PANTING STRINGER IN FIG. 62.

Deck Cargoes and Permanent Deck Weights.—To provide against the strain caused by the weight of heavy timber or other deck cargoes, the important point to be observed is that the beams are held rigidly in place at their centres; it is only when the deck sinks at the centre that any damage can be wrought upon the beam knees. Therefore, to provide efficiently against these strains, extra strong stanchions should be fitted under the beams, with well-formed heads, and spaced not more than two frame spaces apart. Where heavy permanent weights are carried on the deck, such as winches, windlass, etc., local strength may be obtained by fitting extra strong beams supported by additional pillars.

Strains from Masts due to Propulsion by Wind.—The important point to be aimed at in this case is to transmit, especially in the case of vessels propelled wholly by wind, the immense strain which is thrown on to the hull, so as to make it as general as possible, though it is classed as a local one. This is accomplished by making good three conditions—

1. That the heel or lowest extremity of the mast be firmly secured and rendered immovable. If the mast is stepped on the top of a centre keelson, the heel may be secured as shown in fig. 64, a thick plate being firmly attached to the top of the centre keelson, with a circular, welded bulb angle well riveted on its upper face, into which the heel of the mast is stepped and firmly wedged. The mast is prevented from working at its lower extremity by means of a piece of T-bar riveted to the centre keelson, and notched into the mast heel.

2. That the mast be thoroughly secured at the upper deck. Since a great part of the strain is encountered here, it follows that the means to counteract it should be most efficient. If the deck is plated with iron or steel, and the plate in way of the mast is of reasonable thickness, a stout, circular, welded angle bulb to receive the wedging round the mast at the deck will provide a satisfactory means of transmitting the strain to the deck as a whole. Should, however, the deck not be entirely plated, a stout plate, called a *mast partner*, should be attached to the beams round the mast, which in its turn should be efficiently attached to the neighbouring beams and stringer plates by tie plates, as shown in fig. 65. In this way, again, the strain is transmitted to the deck as a whole.

3. That the upper reaches of the mast be held firm, and prevented from working. This is done by having a sufficient number of widely spaced shrouds.

Types of Vessels.—It is scarcely necessary to remind any reader that all vessels are not built for the same purposes, nor to engage in the same trades, or for the same class of harbours. Naturally, therefore, one vessel cannot fulfil all requirements, and be adapted for all traffic, and the result is a considerable variety in the types of vessels built. It is not uncommon to hear the complaint lodged against a vessel that she is ill adapted for her

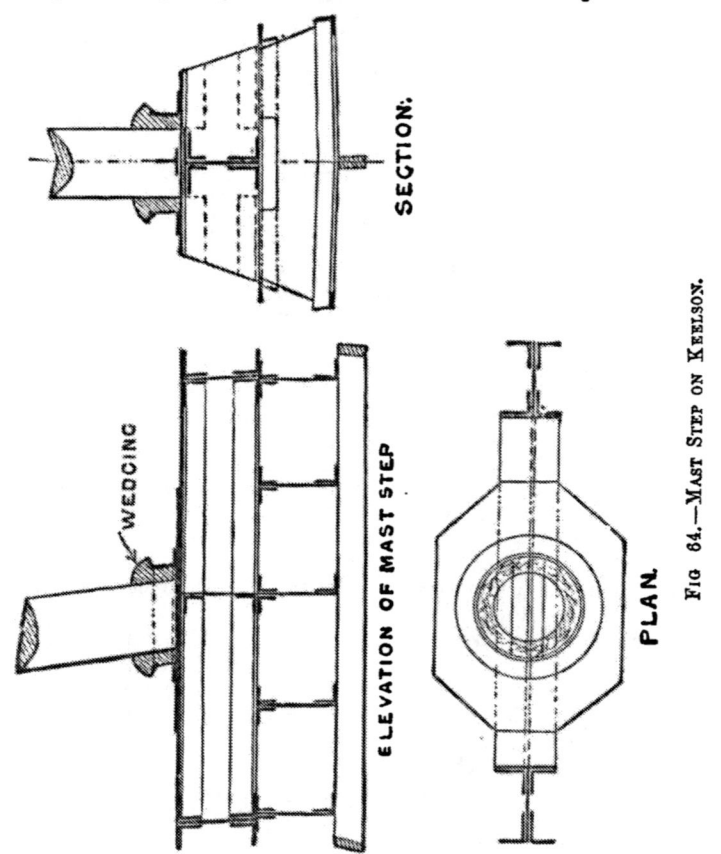

Fig 64.—Mast Step on Keelson.

special trade, and the reason undoubtedly is, in many cases, attributable to the fact that the wrong type has been selected to comply with the necessary requirements, neither the shipbuilder nor the naval architect with their wider experience, having been consulted on the matter. Type, in the majority of cases, depends entirely upon deadweight and internal capacity. The heaviest cargoes, in comparison with their bulk, require the least hold space. With the majority of cargoes it is quite possible to design and fix upon a type of vessel, so that, when

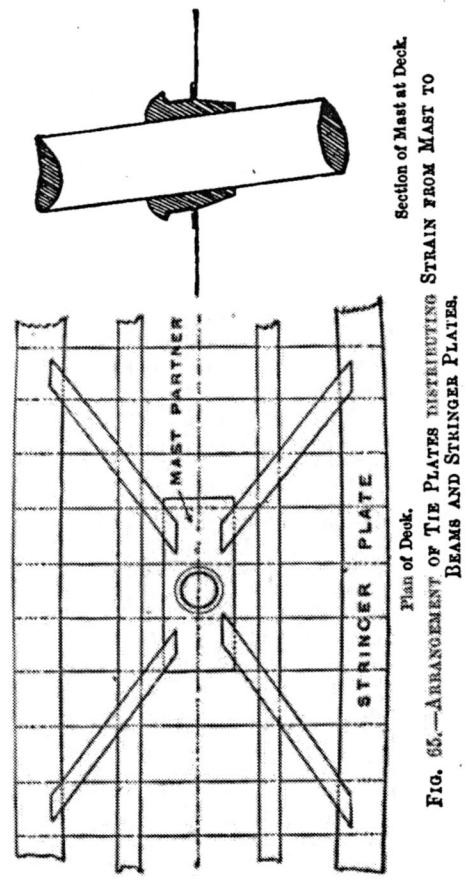

Fig. 65.—Arrangement of Tie Plates distributing Strain from Mast to Beams and Stringer Plates.

the holds are filled, she will float at her load water mark. But it must also be obvious that there are some cargoes, such as metal and ores, with which it would be impossible to fill the hold space, and not submerge the disc or extreme load line at the load water mark. However, for all purposes, there is a type of vessel most adapted for a particular kind of cargo. With light cargoes of small density which fill the holds and still leave large freeboard, the shipowner does not mind so long as the freights are satisfactory; but naturally, when he gets cargoes of greater density, which bring the vessel down to the waterline obtained by the lighter cargo before her holds are filled, he desires to decrease the freeboard, and to continue to decrease it as the cargoes increase in density. Unfortunately, experience has shown in some cases that this decrease would scarcely cease until the freeboard had approached vanishing point; and thus, for the sake of the safety of ships, cargoes, and lives, it has been absolutely necessary to fix upon standard types of vessels adapted to certain freeboards, which, as a rule, are most willingly accepted by shipowners, and, indeed, in some cases, more freeboard is given to vessels by the owners themselves than even the Board of Trade rules demand.

The *heavier* the cargo, especially in ships of great length to depth, the *greater* the bending and twisting strains experienced when among waves, and hence the greater necessity of *increased structural strength*.

Vessels, according to their structural strength, etc., are classified under different names or types, which are often heard and used, but less often understood. There are, again, degrees in each particular type; for example, by 100 A1 at Lloyd's is meant vessels built to the highest class of their particular type, and therefore fulfilling all the requirements for such class; 95, 90, and 80 A are lower classes, and do not therefore fulfil the requirements of the one above them. Such vessels have greater freeboard, and therefore carry less deadweight.

First.—The most important of these types is that known as the *Three Deck*. This is the best deadweight carrier, and therefore ranks as the strongest type of vessel built. This is the type of vessel used in illustrating the growth of structural strength for size of vessel in figs. 45 to 51, and includes the smaller vessels requiring only one or two decks.

Second.—A difficulty sometimes experienced in the three-deck type was, that partly owing to the loss of cargo space occupied by the shafting and enclosed by the tunnel, it was found impossible with some cargoes to trim the vessel by the stern, as desired, and hence, for part of the after length, the ship was increased about 4 feet in depth, and became known as the *Raised Quarter Decker* (see fig. 66). Such vessels are really parts of two different vessels joined together, and each built as near as possible to the previous rule for one-, two-, and three-deck vessels.

The sizes of all floors, frames, and reverses are governed by

Lloyd's first number to the main deck. The frames run up to the quarter deck, and the alternate reverse frames also. The number and arrangement of hold beams, beam stringers, and stringers in hold in the way of the raised quarter deck, are in accordance with the increased depth of the vessel at that part, and the height of the reverse frames is regulated by the Lloyd's first number, which the increased depth would give. The aim is to make the raised quarter deck an integral part of the whole, and the result is that a great amount of extra strengthening has to be introduced at the weakest part, which is in way of the break, by overlapping the decks and stringers, doubling the sheerstrake, and fitting webs between the overlapping part of the two decks, so as to better bind the structure together (see A, fig. 66). As the three-deck and the raised quarter-deck vessels are types which secure least freeboard, it is usual and

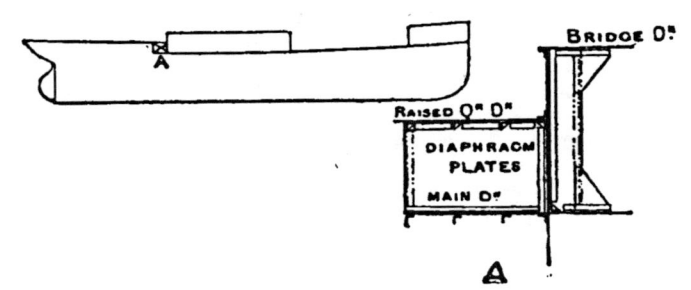

FIG. 66.—RAISED QUARTER DECKER.

certainly of great value to erect a bridge over the engines and boilers, and thus protect this most important part from the inroad of the sea. In addition to this, for the accommodation of the crew and passengers, a forecastle or poop is often erected, which, according to their respective values of affording additional effective buoyancy, may further be a means of reducing the freeboard (see chapter on "Freeboard").

Third.—When a vessel is required to carry cargoes of lighter density, it is usual to adopt the *Spar-Deck type*, as these vessels, while possessing the same interior cubical capacity as the three deck, are of lighter construction, and have greater freeboard.

Fourth.—Where the 'tween decks is required for passengers, or to carry very light cargoes of great bulk, the *Awning-Deck type* is usually chosen, as the awning deck is simply a light, entirely-closed superstructure over the main deck. Being of lighter construction, she is less capable than either the three-decker or the spar-decker for carrying deadweight, and therefore has the greatest freeboard.

Perhaps the simplest way of comparing the structure of these types of vessels will be by referring to a midship section of each of them.

Let figs. 67, 68, and 69 be midship sections of three vessels identical in their exterior appearance, each being 260 feet long, 35 feet beam, and 24 feet depth to the uppermost deck at the side from the top of the keel. Fig. 67 is the midship section of the three-decked type.

Fig. 63 is the midship section of the spar-deck type, the spar deck being the uppermost deck, and 7 feet above the main deck.

Fig. 69 is the midship section of the awning-deck type, the awning deck also being the uppermost deck, and 7 feet above the main deck.

We have already observed that Lloyd's first number governs the sizes of all the frames, reverse frames, floors, bulkheads, and pillars, and that the depth of one-, two-, and three-decked vessels is taken to the uppermost deck. In the case of spar-and awning-decked vessels, the depth and girth are only taken to the main deck, with the result that the first number for these vessels is considerably reduced, hence the difference in the sizes or *scantlings* of the frames, reverses, floors, bulkheads, and pillars (compare respective sections). Then, with a reduced first number, follows a reduced second number, governing the sizes of the outside plating, keelsons, stringers, etc.

Thus far the spar- and awning-decked vessels are identical, so we must note the distinction.

Spar-Decked vessels, according to Lloyd's, are supposed to have three tiers of beams, and to be not less than 17 feet depth from the top of the keel to the main deck. Should the depth be less than this, a modification must be made in the freeboard assigned. In all cases the frames extend from the keel to the gunwale; the reverse frames extend to the main and spar decks alternately; a thick sheerstrake is fitted to both the main and spar decks; the side plating above the main-deck sheerstrake is less in thickness than that below; the main and spar decks, when of wood, are each $3\frac{1}{2}$ inches thick, laid and caulked; spar deck beams, stringer and tie plates are lighter than those required for the upper deck of three-decked vessels. Since the depth is only taken to the main deck, and the strength above the main deck is considerable, greatly increasing the efficiency of the ship girder in resisting longitudinal strains, they do not require any additional strengthening for extreme proportions until over 13 depths to length, while in the three-deck type over 11 depths to upper deck in length is considered an extreme proportion, thereby requiring additional strengthening.

In *Awning-Decked* vessels all frames extend to the awning deck; the reverse frames extend to the main deck; the side plating above the main deck is greatly reduced, and no sheer-strake is fitted to the awning-deck; awning-deck beams are lighter than those required for spar-decked vessels. Since the awning-deck is simply a light superstructure, it is necessary that these vessels have a complete main deck laid and caulked. These vessels are considered of extreme proportions when over 11 depths to length, the depth being taken to the main deck.

The reader will be greatly aided in grasping these differences by comparing the sections.

Bulkheads.—Watertight bulkheads are iron or steel divisions arranged either transversely or longitudinally, dividing the vessel into watertight compartments. They also give strength, but their chief function is to afford safety, so that, should any compartment by any chance or accident be flooded, it is intended that the vessel will still float in comparative safety. Most of the large liners and warships built in these days are divided and sub-divided into numerous watertight compartments, having bulkheads far in excess of the requirements of Lloyd's, or any other classification society. Thus it is possible in many cases for several of these compartments to be damaged and flooded before endangering the safety of the vessel.

Screw steamers classed at Lloyd's have a bulkhead at each end of the engine and boiler space, and another near each end of the vessel. The necessity of having the engines and boilers encased in a watertight compartment is obvious. The foremost or collision bulkhead should be situated at not less than about half the midship beam of the vessel from the stem. The fore end of the vessel being the part most likely to be damaged in case of collision, and the strain which comes from panting tending to make leakage, explains the necessity of having this end of the vessel watertight. At the after end of the vessel, however, though free to some extent from the danger at the fore end, there is severe strain, especially in high-speed vessels, due to the vibration of the shafting. As the danger is that the aft end shell plate rivets may be worked loose and leakage occur, this part also is made watertight.

Though in short vessels it may be quite possible for one of these five watertight compartments to be flooded, and the vessel to remain afloat, it will be evident that in long vessels the lengths of the fore- and after-holds would be so great that were one of them damaged, and the sea to enter, the loss of buoyancy or floating power would be so great that sinking would be inevitable.

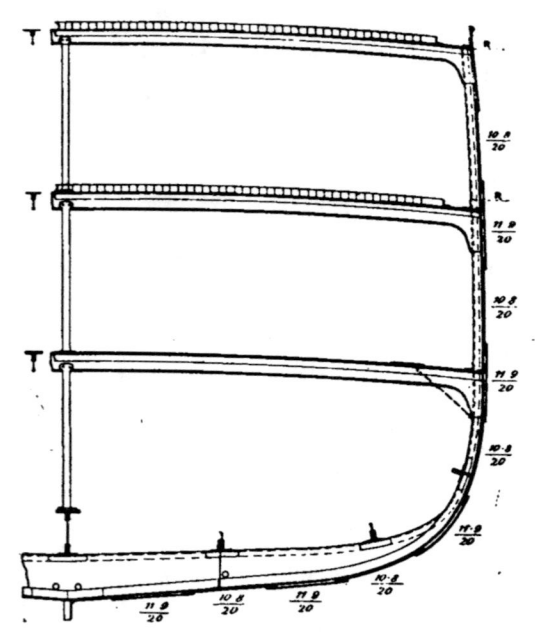

Three-Deck Type—10·4 depths to upper deck in length.
14·5   ,,   middle   ,,

FIG. 67.—MIDSHIP SECTION OF A THREE-DECKED TYPE OF VESSEL.

Lloyd's Numerals.

$\frac{1}{2}$ Girth,	39·1
$\frac{1}{2}$ Breadth,	17·5
Depth,	24·73
	81·33
	7
1st No.,	74·33
Length,	258
2nd No.,	19177

Frames, $5 \times 3 \times \frac{8-7}{20}$, spaced 24 inches.
Reverse frames, $3 \times 3 \times \frac{7}{20}$.
Floors, $23\frac{1}{2} \times \frac{9-7}{20}$.
Centre keelson, $22\frac{1}{2} \times \frac{13-11}{20}$.
Keelson continuous angles, $5\frac{1}{2} \times 4 \times \frac{9}{20}$.
Intercostal keelson plate, $\frac{8}{20}$.
Upper deck beams, $7\frac{1}{2} \times \frac{7}{20}$, bulb plate with } on every 2nd frame.
Double angles, $3 \times 3 \times \frac{9}{20}$
Middle deck beams, $8\frac{1}{2} \times \frac{8}{20}$, bulb plate with } on every 2nd frame.
Double angles, $3 \times 3 \times \frac{7}{20}$
Hold beams, $9\frac{1}{2} \times \frac{7}{20}$ } on every 10th frame.
Double angles, $4 \times 4 \times \frac{8}{20}$
Upper deck stringer plate, $52 \times \frac{13}{20} - 31 \times \frac{8}{20}$.
Middle deck stringer plate, $52 \times \frac{13}{20} - 31 \times \frac{7}{20}$.
Hold stringer plate, $34 \times \frac{7}{20} - 26 \times \frac{8}{20}$.
Gunwale angle bar, $4\frac{1}{2} \times 4\frac{1}{2} \times \frac{9}{20}$.
Sheer strake, $42 \times \frac{13-10}{20}$.
Garboard strake, $36 \times \frac{12-11}{20}$.
Keel, $9\frac{1}{2} \times 2\frac{1}{2}$.
Hold pillars, $3\frac{3}{4}$.
Upper deck, 4 inches thick, laid and caulked.
Middle deck, $3\frac{1}{2}$ inches thick, laid and caulked.

Additions for Extreme Length.

Centre keelson increased in depth.
To side keelson, a bulb is added for $\frac{1}{2}$ L amidships.
To bilge keelson, a bulb is added for $\frac{3}{4}$ L amidships.
To bilge stringer, an intercostal plate is added for $\frac{1}{2}$ L amidships.

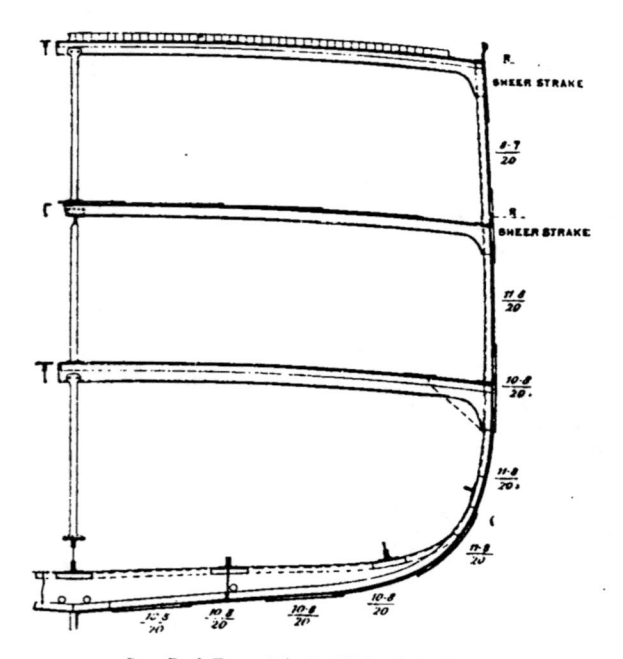

Spar-Deck Type—14½ - 2 = 12½ depths in length.

Fig. 68.—Midship Section of a Spar-Decked Vessel.

STRUCTURE.

Lloyd's Numerals.

½ Girth,	32·1
½ Breadth,	17·5
Depth,	17·73
1st No.,	67·33
Length,	258
2nd No.,	17371

Frames, $4 \times 3 \times \frac{7-6}{20}$, spaced 23 inches.
Reverse frames, $3 \times 3 \times \frac{7}{20}$.
Floors, $20\frac{1}{2} \times \frac{8-7}{20}$.
Centre keelson, $17 \times \frac{12-10}{20}$.
Keelson continuous angles, $5 \times 4 \times \frac{9}{20}$.
Intercostal keelson plate, $\frac{7}{20}$.
* Complete steel deck, $\frac{7}{20}$.
Spar-deck beams, $7 \times \frac{7}{20}$, bulb plate with $\Big\}$ on every 2nd frame.
Double angles, $3 \times 3 \times \frac{4}{20}$
Main deck beams, $6 \times 3 \times \frac{8-7}{20}$, bulb angle on every frame.
Hold beams, $9\frac{1}{2} \times \frac{9}{20}$, bulb plate with $\Big\}$ on every 10th frame.
Double angles, $4 \times 4 \times \frac{8}{20}$
Spar-deck stringer plate, $44 \times \frac{9}{20} - 29 \times \frac{8}{20}$.
Main deck stringer plate, $37 \times \frac{18}{20} - 29 \times \frac{8}{20}$.
Hold stringer plate, $32 \times \frac{9}{20} - 25 \times \frac{7}{20}$.
Gunwale angle bar, $4 \times 4 \times \frac{7}{20}$.
Garboard strake, $36 \times \frac{11-10}{20}$.
Keel, $9 \times 2\frac{3}{8}$.
Hold pillars, 3.
Spar-deck, $3\frac{1}{2}$ inches thick, laid and caulked.

Additions for Extreme Length.

To thickness of sheer strake, $\frac{2}{20}$ is added for $\frac{3}{4}$ L amidships.
To strake below sheer strake, $\frac{1}{20}$ is added for $\frac{1}{2}$ L amidships.
To bilge keelson, a bulb plate is added for $\frac{3}{4}$ L amidships.
To thickness of 2 strakes at bilge, $\frac{1}{20}$ is added for $\frac{1}{3}$ L amidships.

---

* Taking out the scantlings of this spar-decked vessel according to her numerals and proportions, a steel deck is required, as shown upon the Midship Section. But Lloyd's rules say : "In no case will the material at the upper part and the number and thickness of steel or iron decks be required to be greater than that of the three-decked vessel of the same dimensions." However, as a spar-deck vessel must have a complete main deck laid and caulked, though it may be of wood $3\frac{1}{2}$ inches in thickness, a common practice among shipowners of choosing a steel deck has been followed in this Midship Section.

Awning-Deck Type—14·5 depths in length.

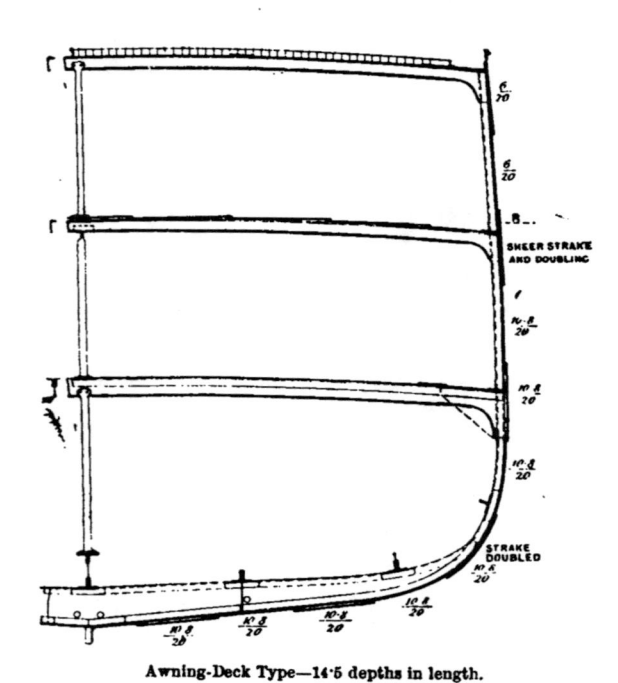

Fig. 69.—Midship Section of an Awning-Decked Vessel

STRUCTURE.

Lloyd's Numerals.

½ Girth,	32·1
½ Breadth,	17·5
Depth,	17·73
1st No.,	67·33
Length,	258
2nd No.,	17371

Frames, $4 \times 3 \times \frac{7-6}{20}$, spaced 23 inches.

Reverse frames, $3 \times 3 \times \frac{7}{20}$.

Floors, $20\frac{1}{2} \times \frac{8-7}{20}$.

Centre keelson, $17 \times \frac{12-10}{20}$.

Keelson continuous angles, $5 \times 4 \times \frac{9}{20}$.

Intercostal keelson plate, $\frac{8}{20}$.

Complete steel deck, $\frac{7}{20}$.

Awning-deck beams, $6\frac{1}{2} \times 3 \times \frac{8}{20}$, bulb angle on alternate frames.

Main deck beams, $6 \times 3 \times \frac{8-7}{20}$, bulb angle on every frame.

Hold beams, $9\frac{1}{2} \times \frac{9}{20}$, with } on every 10th frame.
Double angles, $4 \times 4 \times \frac{8}{20}$ }

Awning-deck stringer plate, $32 \times \frac{7}{20}$.

Main deck stringer plate, $37 \times 1\frac{1}{8} - 29 \times \frac{8}{20}$.

Hold stringer plate, $32 \times \frac{9}{20} - 25 \times \frac{8}{20}$.

Gunwale angle bar, $4 \times 4 \times \frac{9}{20}$.

Garboard strake, $36 \times \frac{11-10}{20}$.

Keel, $9 \times 2\frac{3}{4}$.

Hold pillars, 3.

Awning-deck, 3 inches thick, laid and caulked.

Additions for Extreme Length.

Sheer strake, doubled whole width below stringer plate, for ¾ L amidships.
To upper deck stringer, $\frac{1}{20}$ is added for ½ L amidships.
One strake at bilge, doubled for ½ L amidships.

104   KNOW YOUR OWN SHIP.

Lloyd's, therefore, require that in vessels 280 feet and over in length, an additional bulkhead be fitted in the main hold; and when over 330 feet in length, another bulkhead be placed in the after-hold.

All bulkheads should extend sufficiently high, so that in the event of any compartment being flooded, there would not be the danger of the water pouring over the top of any bulkhead. The collision bulkhead should extend to the height of the uppermost

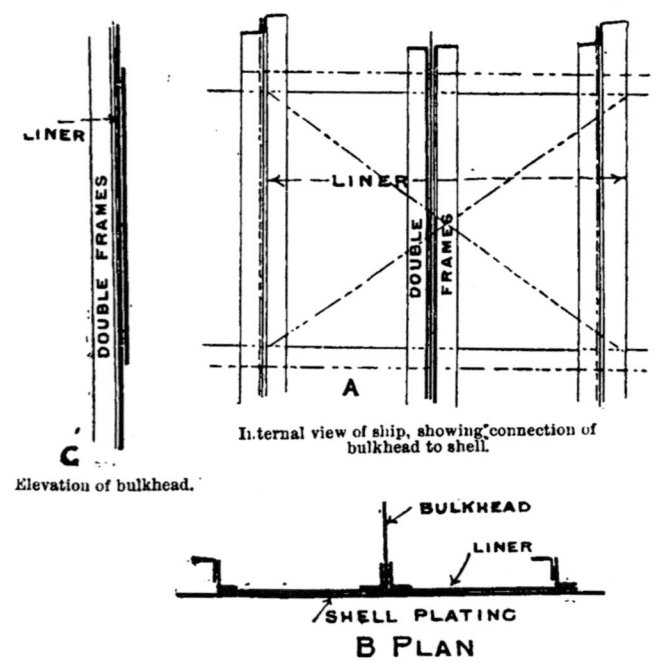

FIG. 70.—CONNECTION OF BULKHEAD TO SHELL.

deck, and its watertightness tested by filling the foremost compartment with water to the height of the load waterline.

The bulkhead bounding the engine and boiler space should extend to the upper deck in one-, two-, and three-deck vessels; to the spar-deck in spar-decked vessels; to the main deck in awning-decked vessels.

The aftermost bulkhead should extend to the upper deck, or to the spar-deck, and should also be tested to ensure watertightness by filling the after peak to the load waterline.

Should the vessel be long enough to require other bulkheads, they should extend to the upper deck in one-, two-, and three-deck vessels; to the spar-deck in spar-decked vessels; and to the main deck in awning-decked vessels.

The bulkhead plating is attached to the shell of the vessel by double frames (fig. 70).

It often occurs that it is not convenient to carry a bulkhead continuously from the keel to its required height, but it is recessed or stepped in the form of a plated flat at an intermediate part, and then continued to its prescribed height. (See A, fig. 71.) However, the watertightness must be maintained. This may be done by cutting the reverse frames and fitting angle collars round the frames in way of the flat, as shown by B, in fig. 71, or cast-iron chocks between the frames (C, in fig. 71), or else the frames may be cut, and the flat connected to the shell by a continuous angle, with brackets above and below joining the flat to the frames (D, in fig. 71). When a bulkhead is stepped on a water ballast tank it should be connected by double angles, or in the case where it terminates at a deck, or is fitted in a 'tween decks, it should be attached to the decks by double angles. Where a bulkhead extends above an iron deck, the longitudinal strength is preserved by keeping the deck continuous, and stopping the bulkhead at the under side, then continuing it again above the deck, making the connection by double angles. In fitting iron collars in way of a watertight flat, the reverse frame is cut and the main frame doubled for about three feet to compensate for the break in the reverse frame. Being very expensive, this method is not often adopted. When cast-metal chocks are fitted they are first bedded in cement, and a space of about three-quarters of an inch left all round. This space is filled with metal filings—the waste from drilling machines—which, when rusted and caulked, forms very satisfactory watertight work. The edges and butt laps in all bulkheads must be caulked. This is necessary on one side only. It is better to caulk the collision bulkhead on the after side, and the aftermost bulkhead on the fore side. One reason for this is that it is more easily done; but the chief one is, that when these peaks are tested by a head of water, should leakage take place, the exact spot is easily perceived and caulked. For the other bulkheads, it is of little consequence on which side they are caulked.

It is always better to avoid any abruptness in longitudinal strength, and thus all keelsons and stringers should be continuous through bulkheads. This necessitates making the bulkheads watertight at these places by fitting angle collars on one side of the bulkhead, and often plate collars on the other. The angle

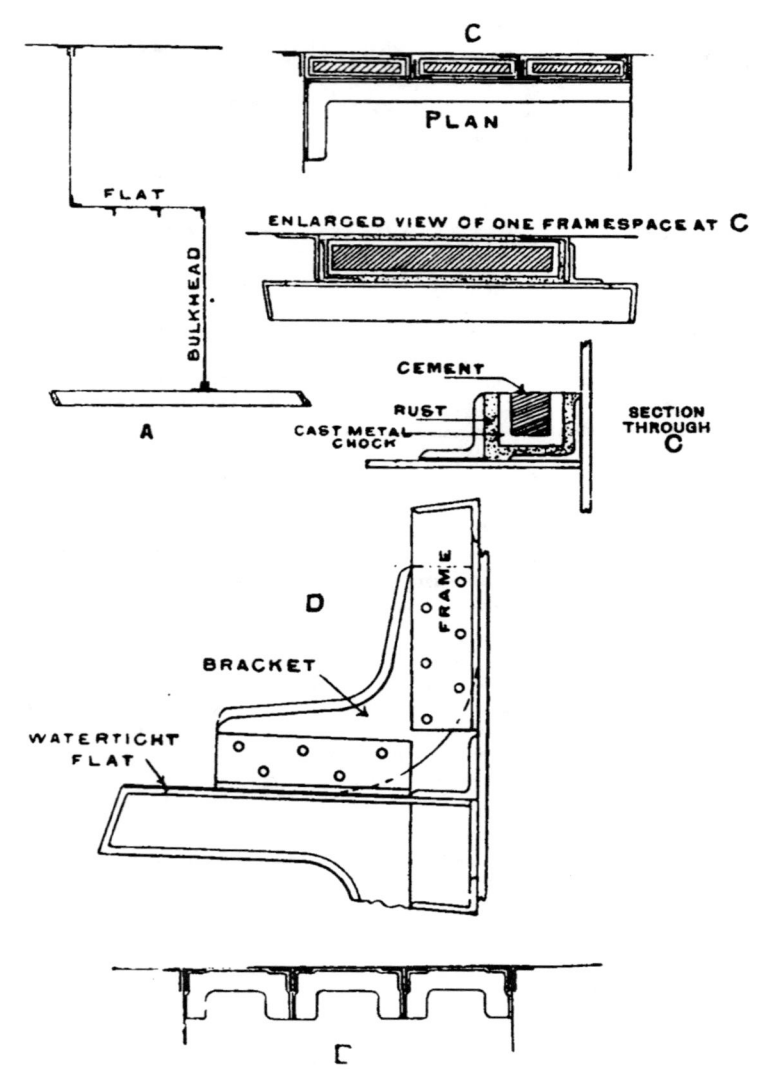

FIG. 71.—RECESSED BULKHEAD AND WATERTIGHT FLAT.

collar is fitted on the side on which the bulkhead is caulked, and is itself caulked (see fig. 72).

As has already been pointed out, if any part of the vessel is subject to any weakening, compensation must be made in some form or other to recover the strength. It will be noticed that in attaching bulkheads to the shell plating, a double row of holes has to be punched, one of which is spaced 4 to $4\frac{1}{2}$ diameters apart. The result is that the shell plating round the whole girth of the vessel has been greatly weakened. Compensation is made for this by fitting what are called *liners* in the way of all outside strakes, where practicable. These are plates extending, as shown in A, fig. 70, for at least two frame spaces from the toe of one frame to the heel of another, and from the edges of the two inside strakes adjacent. The riveting in bulkhead liners is arranged where convenient.

Longitudinal bulkheads continuous all fore and aft in twin-screw vessels also provide a means of subdivision, but at the same time they may afford great longitudinal strength if well constructed and stiffened, especially in very long vessels, and more particularly so in long, shallow vessels of the light draught type.

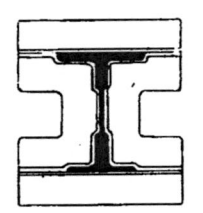

FIG. 72. — ANGLE COLLARS MAKING BULKHEAD WATERTIGHT ROUND KEELSON.

Their value in stiffening ships has been proved in cases where, before they were fitted, the vessels suffered greatly from vibration from the machinery, but after they were fitted the vibration was greatly reduced. The reason of this is explained by our girder illustration. The ship, being a hollow girder, has its upper and lower flanges (decks and bottom) the more effectively united, and by this means the whole structure acts more in unison in resisting either local or general strain.

All bulkheads must be thoroughly stiffened if they are to be of any service. To simply fit the sheet of comparatively thin plating of which bulkheads are made, from side to side of the vessel, would be useless in resisting any severe pressure. Bulkhead stiffening is composed of angle bars of the size of the main frames of the vessel placed vertically on one side, 30 inches apart, and on the other side, horizontally, 48 inches apart. According to the dimensions of the bulkhead, additional stiffening may be fitted in the form of semi-box beams, web plates, and bulb angles. In all cases the collision bulkhead should be additionally stiffened, since this is most liable to danger.

Especially in the case of passenger vessels and yachts, it is often found necessary to have a means of passage through a watertight bulkhead, in order to get from one part of the accommodation to another. This is done by fitting iron doors, which can be closed and secured so as to make the bulkhead perfectly water-

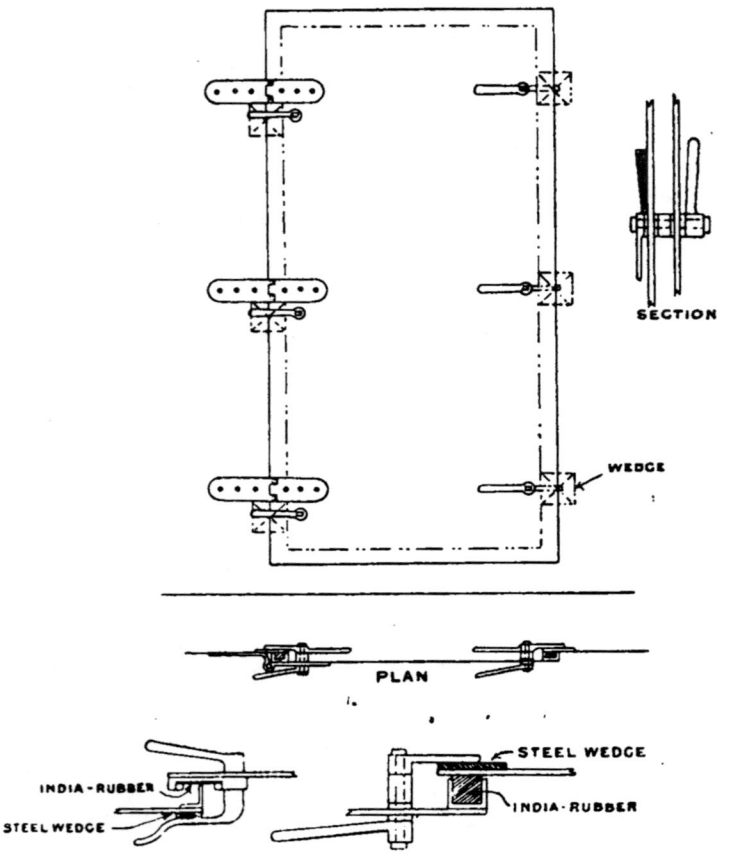

Enlarged sketches showing two different methods of making doors watertight.

FIG. 73.—BULKHEAD WATERTIGHT DOOR.

tight (fig. 73). When such a door is absolutely necessary, it must be provided, but whenever possible it should be dispensed with, for in case of accident it is very often found that the numerous watertight compartments into which the vessel has probably been divided in order to secure safety under the most exceptional circumstances, are rendered worthless, and sometimes a source of danger by these doors being left open, and either forgotten or unapproachable in the hour of need.

Riveting.—A most important consideration in the production of an efficient vessel is good workmanship, and the greatest attention should be paid to sound riveting and accurate fitting of butt straps, etc. (fig. 74). Blind or unfair holes are often the result of carelessness on the part of workmen in marking off the spacing, or in punching the rivet holes. Where such occur, however, the drift punch should be strictly forbidden, as its use not only tears and weakens the plate by the severity of its action, but to get thoroughly watertight work is rendered most difficult, as the rivet cannot fill up the cavities made in so damaging the plate. The drift punch is a tool, as shown in fig. 75, circular in section, which is driven into the hole, and a space for the rivet torn through.

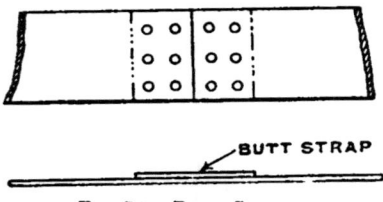

FIG. 74.—BUTT STRAP.

The proper method, when such holes do occur, is to rime them with a tool (fig. 75), which cuts away any projecting material and leaves a clean hole for the rivet.

After a rivet has cooled down, and is found to be slack, or that the head has been badly laid up, or that, on testing, leakage takes place, the rivet or rivets should not be caulked to ensure watertightness, as is sometimes done, but taken out and re-riveted.

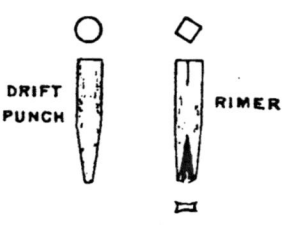

FIG. 75.—DRIFT PUNCH AND RIMER.

Were it not for the additional cost, it would be better to rime out punched holes before riveting, as this would not only clear the rivet holes, but to some extent restore the strength of the plate, which is weakened by the severity of the action of punching. Though much more costly, by far better work is obtained by drilling rivet holes, as this process not only gives fairer holes, but does not by its action weaken the plate.

In punching rivet holes in plates or angle bars, the holes are always punched from the *faying surfaces*—that is, from the surfaces which are to lie against each other. By this means the plates fit hard up to each other, as a rag edge is made on the outer surface of the plate in punching.

Thus in the plates in A, fig. 76, the holes are punched in the direction shown by the arrows. In machine punching, the rivet holes gradually increase in diameter in the direction in which the punch passes, as shown.

A few forms of riveting may here be taken with advantage. For watertight work, where one surface of the plating has to be flush, the pan head rivet is undoubtedly one of the best. Its form is as shown in B, fig. 76.

It will be observed that the neck of the rivet is expanded under

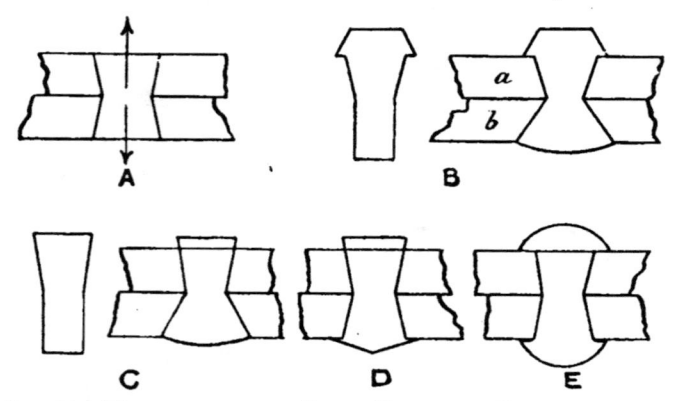

FIG. 76.—MODE OF PUNCHING RIVET HOLES, AND FORMS OF RIVETS.

the pan head; this, when heated and hammered, completely fills up the hole in plate a. In plate b, after the rivet hole has been punched, it is drilled by a machine into a more conical form, as shown. This is called *countersinking*. When the rivet head is thoroughly beaten up, it is cut off in a rounded form, leaving it rather full on the flush surface of the plate. In cooling, the rivet contracts, and this further tends to effectually close up the hole. Another form of rivet is shown in C, fig. 76. The further this rivet is driven into the hole the tighter it wedges itself, and, if well hammered and laid up, it produces good results. It is still further improved if the work is not prominently visible, by laying up a point as shown in D, fig. 76, producing greater holding power. This also applies to the pan head rivet, which, as has been proved by experience and experiment, is then the most efficient rivet.

For the sake of their better appearance, the snap head form of rivet has been considerably adopted in many parts of the structure exposed to view, especially in bulkheads. But hand-riveted snap head rivets have not produced satisfactory results. In clenching up the point of the rivet with the tool called the *snap cup*, the effect is to bring the edges of the rivet close, in the form of a point, as shown in E, fig. 76, leaving a hollow all round under the head of the rivet.

So long as the rivet is kept dry it works well enough, but when water gets to it the pointed edge rusts, and corrosion is set up in the plate also, with the result that, after a time, the rivet works loose, and, in many cases, they can be twisted round by the fingers.

Where the work has to be strictly watertight, a far better result is secured by simply hammering the point down hard on the plate, though the appearance is certainly not so taking. Nevertheless, appearance should be sacrificed, rather than efficiency and strength.

Where snap riveting is performed by a machine, however, the results are very satisfactory, since the machine has the power of more effectually clenching up the point of the rivet. But it is only in some parts of the structure where machine riveting can be carried out, although in every case it is superior to hand riveting.

CHAPTER VI. (Section I.)

STABILITY.

Contents.—Definition—The Righting Lever—The Metacentre—Righting Moment of Stability — Conditions of Equilibrium — "Stiff" and "Tender"—Metacentric Stability—Moment of Inertia—Agents in Design influencing Metacentric Height—How to obtain Stiffness—Changes in Metacentric Height during the Operation of Loading—Stability of Objects of Cylindrical Form—A Curve of Stability—Metacentre Curves—How the Ship's Officer can determine the Metacentric Height and then the position of the Centre of Gravity in any condition of Loading—Effect of Beam, Freeboard, Height of Centre of Gravity above top of Keel, and Metacentric Height, upon Stability—Wedges of Immersion and Emersion—Effect of Tumble Home upon Stability—Stability in Different Types of Vessels.

Definition.—By the term *stability* is meant the moment of force (usually measured in foot-tons, or in inch-tons), by means of which a vessel, when inclined out of the upright position through some external force, immediately endeavours to right herself. Stability is dependent upon design and loading. Many ocean-going craft have not sufficient stability to stand upright when light, and loll over, and some will not stand at all in this condition without ballast. But ships are not built to sail upon the open sea light, and these same vessels may be so regulated in the operation of loading as to be changed into splendid sea boats.

The Righting Lever.—In Chapter II. we explained the meaning of a foot-ton—viz., that 1 ton of force multiplied by a leverage of 1 foot equals 1 foot-ton. Now, if we say that a vessel of 1000 tons displacement has a righting moment of stability of 2000 foot-tons when inclined to an angle of, say, 30°, we mean that the weight, 1000 tons, is acting at a leverage of 2 feet, since 1000 × 2 = 2000 foot-tons; and this is the *righting tendency* possessed by this ship when inclined to an angle of 30°. The two factors in the moment of stability are *weight* and *leverage*. The weight is always the total displacement.

The leverage we shall now endeavour to explain.

STABILITY. 113

Fig. 77 represents two midship sections of the same vessel, one floating upright and the other inclined to an angle of about 14°. When she floated in the upright condition, W L was the waterline; the point B is, at the same time, the centre of buoyancy or centre of gravity of the water displaced by the ship, and G is, for that kind of loading, the centre of gravity of the ship and her loading. But when the vessel became inclined, as in the sketch, she floated at the waterline W' L'. Observe what has happened. The displacement has certainly changed in form, but not in total volume or weight; that remains exactly the same, because the weight of the ship and the cargo remain exactly the same. But the centre of buoyancy has moved towards the right to B', which is the new centre of displacement (that is to say, the new centre of the water displaced by the heeled ship), the old B being no

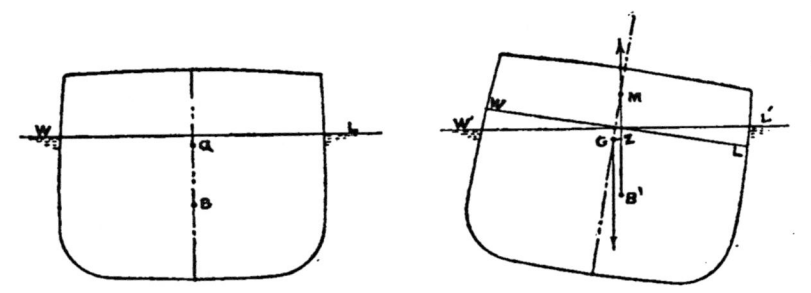

FIG. 77.—VESSEL FLOATING UPRIGHT, AND INCLINED 14°.

longer that centre. The centre of gravity of the ship and its loading, G, remains in the same position in the ship as before, whatever be the angle of heel, so long as we do not alter the loading. It must always be the centre of weight, and will never move, as just stated, so long as the weights on board remain stationary. The weight of the ship, like all weights, acts vertically downward through its centre of gravity, G, and the pressures of buoyancy act vertically upward through the centre of buoyancy, B or B', as the case may be. When the ship was upright, the two points were in the same vertical line, but after being inclined, their forces acted through their centres in the direction of the arrows on their vertical lines of action, leaving a horizontal distance between them, because, now they are not in the same vertical line. This perpendicular—that is to say, horizontal—distance between the two vertical lines of action, gives the lever, G Z; and this very important distance, G Z, is

H

the lever we want, and which we mentioned previously as one of the factors in the measure of a vessel's stability.

The Metacentre.—It will be noticed that the vertical line through the centre of buoyancy intersects the centre line of the section of the vessel at the point M. This point is called the *metacentre*, and is always at the intersection of these two lines, which is approximately a fixed point up to angles of about 12° or 15° of heel for ship-shaped objects. Usually, at greater angles, the vertical line through the new B' no longer intersects the centre line of the ship at the point M.

Righting Moment of Stability.—When M is above G, the lever is properly termed a *righting lever*, for then the action of the buoyancy is to push the vessel again into the upright position, as can easily be seen in fig. 77. When M is below G, the tendency is to add to the inclining force, and in this case the lever is an *upsetting* one. In the sketch, it is evident that the lever is a righting one, and therefore the displacement of the vessel multiplied by this leverage measured in feet, say, at any particular angle of heel, gives the righting moment of stability in foot-tons for that angle.

Conditions of Equilibrium.—So long as there is a leverage, the vessel, if left free to move, will not remain at rest, or, as it is generally termed, in a state of equilibrium; and the longer the leverage, the greater is the moment tending to bring her back to the upright condition, or to capsize her, as the case may be; and the longer the righting lever up to angles of 12° or 15° in ordinary vessels, the higher must the metacentre be above the centre of gravity. Thus, for a vessel to float upright in a state of equilibrium or rest, it is evident that the centre of buoyancy and the centre of gravity must be in the same vertical line, the force of gravity of the ship and buoyancy of the water neutralising each other. But it does not follow that in this condition the equilibrium will be stable. For if, under the effect of some external force (such as wind), the vessel heel, and the metacentre, before heeling, be below the centre of gravity, the result will be to push the vessel further over, not necessarily by any means to capsize her, but (it may be) to a position such as to cause the upward vertical line of action of the buoyancy to coincide with the downward line of action of the weight, when the vessel will again remain at rest. When the vessel floated upright before the inclination took place, she might have been compared to a child's play top carefully balanced upon the point, a condition it would not be likely to remain in for a long period. This equilibrium is called *unstable equilibrium*. When the metacentre is above the centre of gravity, and thus, under the least inclination, the vessel immediately endeavours to regain the upright condition, the equilibrium is termed *stable equilibrium*.

"**Stiff**" **and** "**Tender**."—When the metacentre is high (equivalent to large righting leverage), and the righting moment, when the vessel is inclined, is therefore great, the vessel is said to be "*stiff*," and when the metacentre is closer to the centre of gravity, and the vessel naturally possesses small righting leverage for small angles of inclination, she is said to be "*crank*," or, as sailors more commonly say, "*tender*."

When the centre of gravity and the metacentre coincide—that is, when both points occupy the same position,—the equilibrium is then *neutral*, neither stable nor unstable.

Metacentric Stability.—Having found the distance between the metacentre and the centre of gravity, the actual lever of stability may be found for small angles of heel not exceeding 15°, by multiplying this distance by the sine of the angle of heel (using natural sines*). Then leverage in feet multiplied by displacement in tons equals righting moment in foot-tons. This stability, which is deduced from the *metacentric height*, or, distance from M to G in fig. 77, is termed *metacentric stability*.

Our next endeavour must be to discover the position of these points, and to ascertain how they are influenced. The centre of buoyancy we have already studied in Chapter III., and have found its position to be always in the centre of displacement, and also, when the vessel is floating upright, that it is always a fixed point at any particular waterline.

The metacentre also is a fixed point for each successive draught when the vessel floats upright, and thus, by calculating a few of these points, a curve may be constructed, thereby enabling the position of the metacentre to be ascertained at any draught.

But, first of all, let us see how the design or dimensions of a vessel affect the position of the metacentre. The formula for finding the metacentre is:—

$$\frac{\text{Moment of inertia of waterplane}}{\text{Displacement in cubic feet}} = \left\{ \begin{array}{l} \text{Height of metacentre above centre of} \\ \text{buoyancy.} \end{array} \right.$$

To give the usual definition of *Moment of Inertia* would probably be to sound the keynote of despair to many a seaman. A simple, though perhaps not very scientific method may be used, reducing this to terms sufficiently plain to be understood by seamen with the most limited mathematical knowledge.

* See table of natural sines at end of book. Such a table is printed in many books on *Navigation*.

Moment of inertia may be understood as the measure of the tendency possessed by the superficial area of the waterplane of any floating object, to remain inert, dead still, or motionless. It must be clearly understood that this moment of inertia applies *only* to the waterplane *area* at which the vessel is floating, whether the vessel be a ship, a box, or a log of wood. Thus, the fact that there may be one part of the floating object out of the water, and another part in it, is left entirely out of consideration. Simply *area at the waterline* is dealt with. The formula for moment of inertia of a rectangular-shaped waterplane is as follows:—

Cube the side at right angles to the longitudinal axis, multiply this by the length of the waterline, and divide the result by 12.

Note.—All measurements should be in feet to match the displacement.

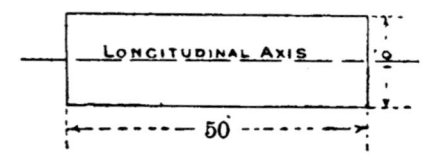

FIG. 78.—PLAN OF WATERPLANE.

For example, take fig. 78, which is a plan of the waterplane of a box-shaped vessel 50 feet long, 10 feet broad, and 8 feet draught. The moment of inertia would be:—

$$\frac{10^3 \times 50}{12} = \frac{10 \times 10 \times 10 \times 50}{12} = 4166 \cdot 66 \text{ moment of inertia.}$$

The student should note that since the dimensions and area of all the waterplanes of a box-shaped vessel, floating with the bottom parallel to the waterline, are equal to one another, the moment of inertia of each waterplane is the same.

But let us proceed to find the height of the metacentre. This is done by dividing 4166·66 by the displacement in cubic feet; the result is the height of the metacentre above the centre of buoyancy. The draught is 8 feet. It is, therefore, evident that the object, being of box form, the centre of buoyancy must be at half the draught, which is 4 feet above the bottom. The displacement in cubic feet will be the volume of the part of the vessel immersed, which equals:—

50 feet length × 10 feet breadth × 8 feet draught = 4000 cubic feet displacement.

Then $\frac{4166 \cdot 66}{4000} = 1 \cdot 04$ feet height of metacentre above centre of buoyancy.

Now, the question may be well asked at this stage—Of what value is this result? The only answer is, that these points, by themselves, are of no practical use, and give no idea whatever of a vessel's stability, until we get the position of the centre of gravity. Let us suppose the height of the centre of gravity from the bottom of the box to be 3 feet (the box having, of course, a load in its lower part, so as to keep the centre of gravity of the loaded box down to 3 feet).

Since the metacentre is 1·04 + 4 = 5·04 feet from the bottom of the box, therefore the distance of the centre of gravity below the metacentre is 5·04 − 3 = 2·04 feet, proving that the vessel is floating in a condition of stable equilibrium. While floating at this draught, imagine a weight of 20 tons already on board to be raised 13 feet. Observe distinctly the new state of affairs, supposing that after the alteration the vessel floats upright. The displacement, the positions of the centre of buoyancy and metacentre are all just the same as previously, no alteration having occurred in the draught. There has, however, been a change in the position of the centre of gravity, thereby affecting the distance between the changed centre of gravity and the unchanged metacentre. Any vertical movement of weight in the box must either raise or lower the centre of gravity, consequently, in the case before us, since the weight of 20 tons was raised, it follows that the centre of gravity must have travelled in the direction of the shifted weight. But how far? We have already discovered how to find this in the chapter on "Moments," viz. :—

Multiply the weight by the distance it has been moved, and divide by the total displacement (4000 *cubic feet* = 114·2 *tons*).

$$\frac{20 \times 13}{114·2} = 2·2 \text{ feet} = \begin{cases} \text{Distance centre of gravity} \\ \text{has moved upwards.} \end{cases}$$

We now see that the centre of gravity, instead of being below the metacentre as previously, has risen to (2·2 − 2·04) = 0·16 of a foot above it. If the box remain upright, it is in a condition of unstable equilibrium, and any exterior force upon it will readily cause it to heel. As the vessel has not sufficient stability to float upright, the vital question is—Will she capsize? In the example before us, the vessel having a good freeboard of about 5 feet, the answer is "no," for, after inclining to an angle of about 18°, she would remain at rest with this permanent list. Let us again investigate the circumstances. The centre of gravity has remained stationary under the inclination. We observe a slight movement of the centre of buoyancy from its old position to the new centre of displacement. In the earlier stage of this inclining movement, the vertical line through the centre of buoyancy intersected the centre line of the vessel below the centre of gravity, in fact, as always for very small heels, at the metacentre, making a leverage between the vertical lines through the centre of buoyancy and the centre of gravity. This point of intersection being below the centre of gravity, the heeling continues, for the moment is an upsetting one, the buoyancy expending itself in pushing the vessel further over. Reference to figs. 16 and 77 will assist in tracing the movements of these points. However, after inclining to an angle of about 18°, it would be found that the centre of buoyancy had travelled so far out towards the inclining side as to bring it vertically beneath the centre of gravity, the vertical lines of action through these two points coinciding. The upsetting lever having vanished, the box floats in a condition of equilibrium, having neither righting nor capsizing moment, and will move neither towards the port nor the starboard except by the application of sheer exterior force. For vessels of cylindrical form, and, therefore, circular in section, the metacentre is the point of intersection of the vertical line through the centre of buoyancy with the centre line of the cylinder *for all angles* of inclination, and, knowing this, we shall soon see how the whole range of such an object's stability may be readily determined. As we have observed, vessels of ship form differ from cylinders for large angles of inclination, and another method has to be adopted in order to trace out the whole range of stability. See Stability Calculation, Chap. X.

Agents in Design Influencing Metacentric Height.—The two important factors in the design of a ship influencing the height of the metacentre are *beam* and *displacement*. The formula for the moment of inertia of the waterplane of a box vessel we have already stated as:—

$$\frac{\text{Length} \times \text{Breadth}^3}{12},$$

so that, as the breadth is cubed, any addition to this dimension must have greater effect in increasing the moment of inertia of any waterplane than a similar addition to the length.

The formula for height of metacentre above centre of buoyancy is:—

$$\frac{\text{Moment of inertia of load waterplane}}{\text{Displacement}}.$$

It must, therefore, be evident that the smaller the displacement, the greater will be this height. Had the displacement of

STABILITY.

the box vessel, referred to previously, been reduced by cutting off the bottom corners and making it more triangular in shape, preserving the same area of waterplane, a much higher metacentre would have been obtained. Thus, fine vessels with good beam produce the highest metacentres. These points may be more vividly illustrated by means of a few simple examples:—

Let 2 feet be added to the beam of the box vessel with which we have just been dealing, the draught remaining the same. The dimensions will now be —Length, 50 feet; breadth, 12 feet; draught, 8 feet.

The moment of inertia $= \dfrac{50 \times 12^3}{12} = 7200$.

The displacement $= 50 \times 12 \times 8 = 4800$ cubic feet.

$\dfrac{7200}{4800} = 1\cdot 5$ feet $=$ metacentre above centre of buoyancy.

In the original condition, the metacentre was 1·04 feet above the centre of buoyancy, and 2·04 feet above the centre of gravity.

The metacentre, being raised (1·5 − 1·04 =) 0·46 of a foot, by the addition of 2 feet to the beam, in its turn increases the metacentric height to (2·04 + 0·46 =) 2·5 feet, greatly adding to the stiffness of the vessel. Had we added 2 feet to the length of the vessel, certainly the moment of inertia would have been increased, but so would the displacement, entirely neutralising what might have been imagined a means of raising the metacentre.

$\dfrac{52 \times 10^3}{12} = 4333\cdot 33$ moment of inertia.

$52 \times 10 \times 8 = 4160$ cubic feet displacement.

$\dfrac{4333\cdot 33}{4160} = 1\cdot 04$ feet metacentre above centre of buoyancy, and the same as the original height.

If 2 feet had been added to the draught, the moment of inertia would have remained the same: $\dfrac{50 \times 10^3}{12} = 4166\cdot 66$, and the displacement would naturally have been enlarged, $50 \times 10 \times 10 = 5000$ cubic feet. The evident result must be that in relation to the centre of buoyancy, the metacentre is now lowered, for $\dfrac{4166\cdot 66}{5000} = 0\cdot 83$ of a foot metacentre above centre of buoyancy, which is less than the original 1·04.

By these simple illustrations it is clear *that beam is the most important factor* in the dimensions of a vessel by means of which a high metacentre is obtained, simply because the breadth of the vessel is used in the third power, while the other dimensions are used only in the first power, in the process of calculation. It will also be noticed that the position of the weights carried governs the position of the centre of gravity of the loaded vessel.

How to Obtain Stiffness.—We have now discovered two means of obtaining stiffness—first, by placing heavy weights as low as possible, thus drawing the centre of gravity down from the metacentre; and second, by adding to the beam to raise the metacentre.

Changes in Metacentric Height when Loading.—We can also the better understand how it is that some vessels, especially when loaded with homogeneous cargoes, get tender when the last part of the cargo is being put on board, and the load waterline approached. For some distance below the load waterline, little or no increase has occurred in the area of the waterplane. In fact, where the vessel possesses great *tumble home*,* it may even happen that the width of the waterplane at the load line is less than at 1 or 2 feet below it, and we have seen what effect any reduction in the beam has upon the moment of inertia. Now, while with increasing load, no increase may have been occurring in the moment of inertia while nearing the load waterline, the displacement has certainly continued to increase with the increasing load. The metacentre and the centre of buoyancy have, therefore, come nearer each other. But at the same time, the centre of buoyancy has certainly risen above the keel some distance, being now the centre of a greater displacement, and may have risen more than the metacentre has sunk. The result is that the metacentre may actually be higher from the keel than before. But the continuation of the loading producing this rise in the centre of buoyancy above the keel has also raised the centre of gravity above the keel, with the effect of reducing the distance between the metacentre and the centre of gravity— that is to say, the metacentric height is reduced, or, the vessel has become more tender. Before any cargo was placed on board, the metacentre and the centre of gravity had perhaps only 5 or 6 inches between them. In this position, the vessel would be tender, not at all uncommon when light. Now, let the centre of gravity of the unloaded ship be at half the depth of the hold. It is clear that when the operation of loading commences, and as long as weight is being placed below the centre of gravity, the centre of gravity must be gradually lowering, and the vessel becomes very stiff in consequence, but as cargo continues to be loaded, and the holds are about filled, the centre of gravity rises again. It is also evident, that if more heavy cargo be placed above the original

---

* "Tumble home" is the difference between the amidship breadth at the uppermost deck edge and the moulded breadth.

centre of gravity than below it, the centre of gravity must be higher than it was previously. We must not forget, however, that the metacentre has varied with every change of draught. If we had a curve of metacentres for the vessel, we could readily ascertain its correct position. If it were found that we had so loaded our vessel as to bring the centre of gravity again into proximity to the metacentre, the result would naturally be a small metacentric height and a tender ship.

Stability of Cylindrical Objects.—At this stage, it will repay us to give a little attention to the stability of objects of cylindrical or cigar form, and from these simple shapes to deduce such principles as will help us in dealing with the more complicated ship forms.

Let fig. 79 be such an object 50 feet long, 10 feet diameter, and for the sake of example, a solid piece of timber, floating half

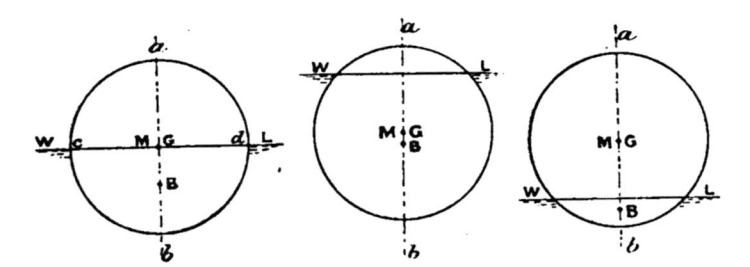

FIGS. 79, 80, 81.—STABILITY OF FLOATING CYLINDRICAL OBJECTS.

immersed—that is, at 5 feet draught. The centre of gravity will be in the centre of the log at G. Every reader knows from observation that such an object will float as readily in one position as another, and with any of the points, a, b, c, d, uppermost. Such a condition is, therefore, one of neutral equilibrium, and this being the case the metacentre and the centre of gravity must coincide. Let us endeavour to prove this. It is clear that G is the centre of weight. The centre of buoyancy will be in the centre of the half circle in the water at B. If a piece of cardboard were cut to this shape, and balanced upon a point, it would be found to be ·4244 of the half diameter of the cylinder down from the line C D = ·4244 of 5 feet = 2·122 feet. The moment of inertia of the waterplane will be

$$\frac{50 \times 10^3}{12} = 4166\cdot 66,$$

The displacement is

$$\frac{10^2 \times \cdot 7854 \times 50}{2} = 1963 \cdot 5.$$

Therefore, $\frac{4166 \cdot 66}{1963 \cdot 5} = 2 \cdot 122$ metacentre above centre of buoyancy, bringing it exactly up to G, thus proving the fact that the vessel floats in a condition of neutral equilibrium.

Had the above cylinder been made of heavier wood, so as to float deeper, as shown in fig. 80, its equilibrium would still have been neutral, and in like manner the metacentre and centre of gravity would have coincided.

Or again, had the object been made of lighter material, and floated as in fig. 81, the equilibrium would have been unchanged, for still the metacentre and centre of gravity would have occupied the one position.

The fact to be remembered from these examples is, that *the transverse metacentre is always the centre of the circular section whatever be the draught.*

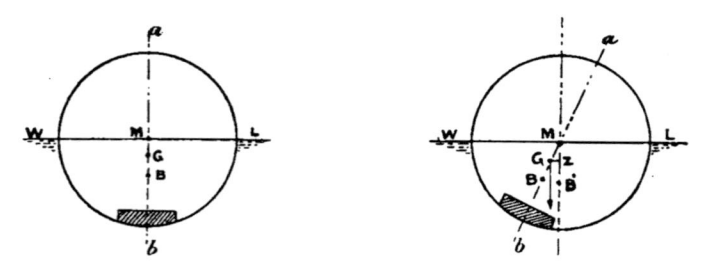

Fig. 82.—Vessel loaded with a Fixed Weight.

But suppose the vessel is hollow, and a weight is placed inside and firmly fixed, with the effect of lowering the centre of gravity, say, 1 foot below the metacentre. The vessel floats, say, at half the diameter draught, 5 feet. To whatever angle the vessel be now heeled, the centre of buoyancy must always be in the centre of the immersed semicircle, and the centre of gravity is immovable in its position.

When floating upright, as shown in fig. 82, the metacentre, centre of gravity, and centre of buoyancy are in the same vertical line; but if the vessel be heeled, the centre of gravity and centre of buoyancy will no longer be in the same vertical line. The distance between the two vertical lines through the points G and B' indicates the lever of stability, G Z.

STABILITY. 123

In heeling, it is clear that a part of the vessel of wedge shape, L M *l* (fig. 83), formerly out of the water, is now immersed, and another wedge-shaped part, W M *w*, formerly in the water, has become emerged. Whenever it happens, whatever be the type of floating object, that the immersed wedge is identical in shape with the emerged wedge, with each of their centres the same distance from the vertical line through the centre of buoyancy, this vertical line will intersect at the point M, in the line *a b*, thus keeping the distance from M to G (the metacentric height) the same.

The lever of stability can be found by *multiplying the distance*, M G, *by the sine of the particular angle of heel.* This is true for all floating objects of cylindrical form; and thus it matters not how great may be the angle of heel, it is always found that the immersed wedge, L M *l*, is equal both in shape and volume to W M *w*, and also that their centres, P and K (see figure), are at equal distances, *x*, from the vertical line through B'. Knowing this, we can proceed to ascertain the whole range* of stability for our cylindrical vessel. The calculations for the levers of stability will be made at every 10° of heel.

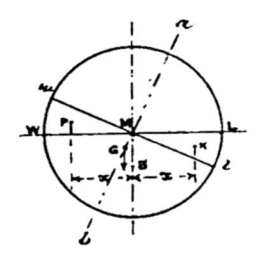

Fig. 83.

A very simple method of illustrating the levers of stability for cylindrical vessels (and one which the reader might well try for himself, thereby proving by measurement the accuracy of the calculation) is as follows:—Cut a piece of cardboard circular in shape, and mark upon it in black dots the positions of the metacentre, the centre of buoyancy, and the centre of gravity (1 foot below the metacentre) in the upright condition. Loosely attach the cardboard to a flat board, placed vertically by means of a screw through the point indicating the metacentre, so as to be free to revolve. Over the head of the screw loop a thin length of cord, and to the other end attach a button or round piece of lead, so as to exactly cover the centre of buoyancy. Knot another piece of cord, and pass it through the back of the cardboard at the point indicating the centre of gravity, and at an indefinite length hang another button or weight. When upright, the two cords will hang together, but immediately the cardboard diagram is inclined the cords will separate, and the perpendicular distance between them represents the lever of stability. By revolving

* By *range* is meant the extent of the inclination from the upright position to the angle at which a ship's righting force vanishes.

124   KNOW YOUR OWN SHIP.

the diagram at intervals it will be found that the levers measure

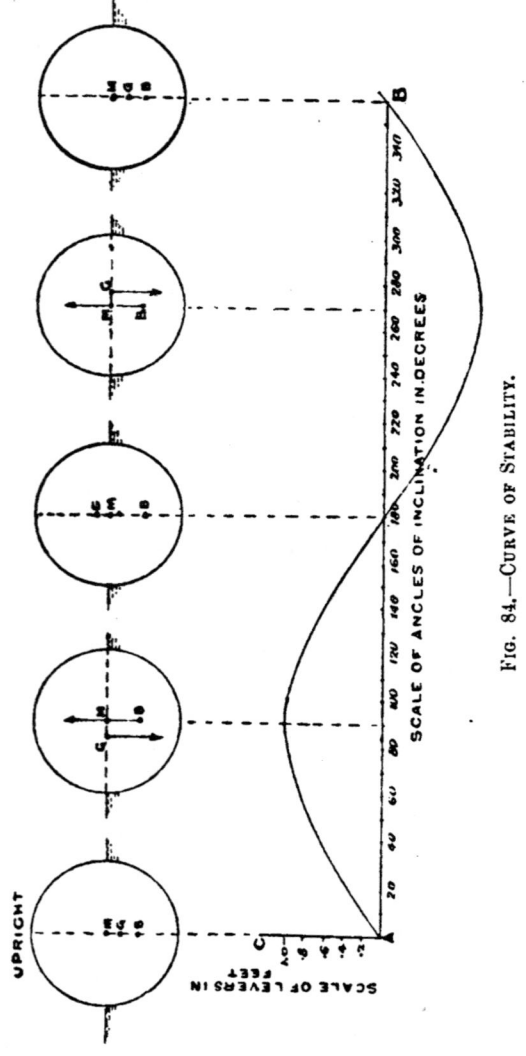

Fig. 84.—Curve of Stability.

according to the calculation G M × sine of angle of heel = lever of stability. [*Note.*—G M = 1 foot.]

STABILITY. 125

Lever at 10°	= 1 ×	sine of angle of 10°	·1736	=	·17
,, 20°	= 1 ×	,, 20°	·3420	=	·34
,, 30°	= 1 ×	,, 30°	·5000	=	·50
,, 40°	= 1 ×	,, 40°	·6427	=	·64
,, 50°	= 1 ×	,, 50°	·7660	=	·76
,, 60°	= 1 ×	,, 60°	·8660	=	·86
,, 70°	= 1 ×	,, 70°	·9396	=	·93
,, 80°	= 1 ×	,, 80°	·9848	=	·98
,, 90°	= 1 ×	,, 90°	1·000	=	1

To Construct the Curve of Stability (see fig. 84).—Draw the horizontal line A B, and upon it at regular intervals mark off spaces, each indicating 10° of heel. The 10° spaces may be further subdivided into tenths, each representing 1° interval. From the point A draw the vertical line A C, and upon this line construct a scale of levers of stability, each space representing ·1 of a foot. Using the scale A C set up at the 10°, 20°, 30°, etc., intervals, their corresponding leverages, ·17, ·34, ·50, etc., and through all these points run a curve. By means of this curve, leverage at any intermediate angle of heel can now be readily measured. In the figure before us, we see by the diagram that the righting lever of stability steadily increases up to 90°, where

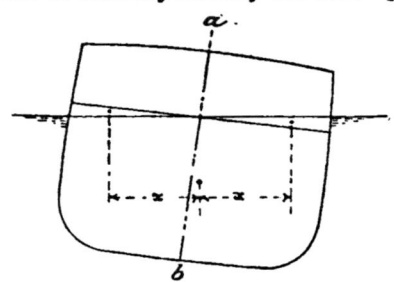

FIG. 85.—WEDGES OF IMMERSION AND EMERSION PRACTICALLY EQUAL SECTORS OF THE SAME CIRCLE FOR SMALL ANGLES OF INCLINATION.

it attains its maximum length; after that it gradually decreases, exactly opposite to the way in which it increased, until at 180° it vanishes altogether. The lever then begins to grow again, no longer a righting one, but an upsetting one, and it continues to increase up to 270°, where it is longest. After that, it again diminishes, until, when a whole revolution has been made at 360°, the vessel once more becomes stable.

But the question may be asked—How comes it that for vessels of ship form the vertical line through the centre of buoyancy only intersects the centre line of the ship at the metacentre for small angles of heel up to 12° or 15°, and at larger heels usually does not? The reason is simply this:—

The wedges of immersion and emersion of an actual ship for *small angles* of heel are practically sectors of a circle, and thus resemble the wedges with which we have just been dealing, in the vessel of cylindrical form, the wedges being exactly equal in volume, and practically identical in shape, with their respective centres at practically equal distances (x) from the vertical line through the centre of buoyancy (fig. 85). So long as these conditions remain unaltered, the vertical line through the centre of buoyancy, at whatever angle the vessel may be inclined, will always intersect the line $a\,b$ at the metacentre, and when such is the case, the distance M G multiplied by sine of angle of heel will give the lever of stability, G Z. But in vessels of ordinary ship form when inclined to large angles of heel, the wedges of immersion and emersion, although exactly equal to each other in volume, are dissimilar in shape, with their centres at quite unequal distances from the centre of buoyancy.

Metacentre Curves.—When this is the case, the vertical line through the centre of buoyancy does not intersect the line $a\,b$, fig. 85, at the metacentre.* For large heels the position of the point M (originally the metacentre for the upright condition) is more difficult to ascertain, and is, therefore, discarded in present practice in determining the range of stability. Now, although it is advisable for a seaman to thoroughly understand displacement, buoyancy, the metacentre, and the principles governing and affecting the same, yet it is not necessary for him to enter into the mathematical method of calculating the same, for when these matters are understood, every necessary information about their values may be supplied to him in the form of curves, by the shipbuilder or naval architect who designs his vessel. Thus we have shown in figs. 3, 5, 20, and 21 curves of displacement, curves of "tons per inch," and also curves of centres of buoyancy. Since these quantities are always the same at particular draughts, all that the seaman requires are the curves themselves, the knowledge of their value, and how to read them. The same applies to the metacentres. These also are fixed points for each draught for the vessel in the upright position, and, as we have shown, for small angles of heel. We shall now proceed to give an illustration of such a curve, and show how it is constructed and read.

The principle of the calculation † for the metacentre of a vessel at a particular draught is the same as for a box, except that on account of the varying shape of the waterplane, a slight modification has to be made, to find the moment of inertia of the water-

---

* The metacentre is a term which ought only to be applied to the point M so long as it is constant in position, which is only for small angles of inclination.   † See Chap. X. for metacentre and other ship calculations.

STABILITY. 127

plane, *which, divided by the displacement at the particular draught, gives the height of the metacentre above the centre of buoyancy.* So that, first of all, we require the positions of the centres of buoyancy. In fig. 20 a curve of centres of buoyancy is given for a certain vessel. This we shall transfer to fig. 86, and set off the metacentres for the same vessel.

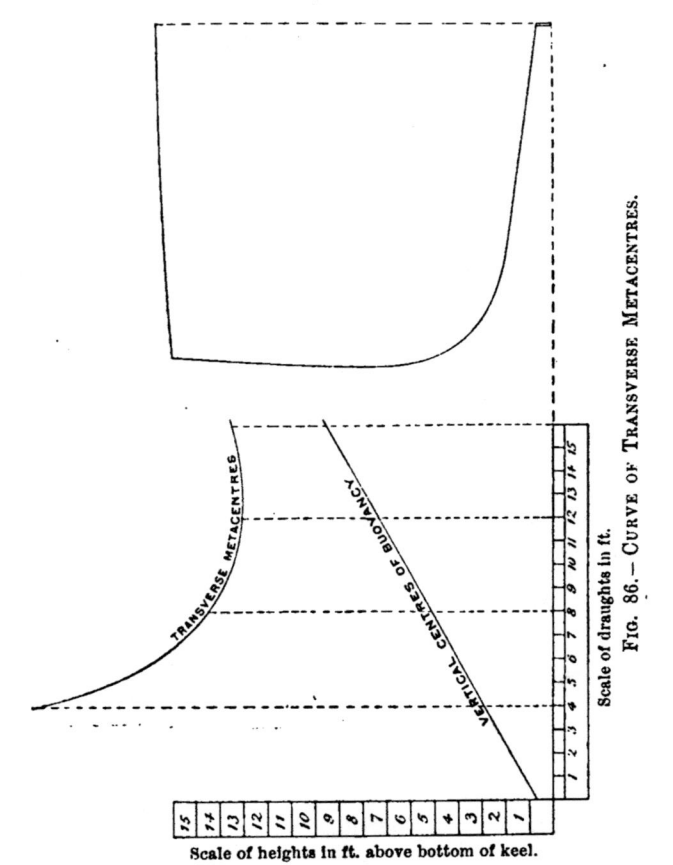

FIG. 86.—CURVE OF TRANSVERSE METACENTRES.

The metacentres, as calculated, were as follows:—

At 4 feet draught, 19·0 feet above centre of buoyancy.
,, 8 ,, 9·2 ,, ,,
,, 12 ,, 5·75 ,, ,,
,, 16 ,, 4 ,, ,,

At the point in the horizontal scale of draughts, representing 4 feet draught, draw a vertical line intersecting the curve of centres of buoyancy, and extending above it. Using the vertical scale of feet at the side, set up the height of the metacentre above the centre of buoyancy (19 feet), which shows the position of the metacentre to be 22 feet above the bottom of the keel. The same process is performed at the other waterlines, and when all the points representing the metacentres have been set off, a curve is run through them, which is that required, and enables us to read off heights of metacentres at any draught.

Unlike the centres of buoyancy and metacentres, the centre of gravity is not a fixed point except in certain conditions, and, therefore, cannot be supplied to the ship's officer. Every variation in the arrangement of weight or cargo, whether it be a yacht or cargo steamer, will affect the position of the centre of gravity, since, as we have already observed, the centre of gravity is the centre of weight. So that about the only conditions in which the centre of gravity may be relied upon as occupying a constant or fixed position, are, when—(1) the vessel is light, with bunkers empty and no stores on board, (2) bunkers full, boilers full, and all stores on board, and (3) in the case of cargo vessels— the same as 2, with the holds *filled* with homogeneous cargoes which *exactly* bring them down to the load draught. To know the position of the centre of gravity in conditions 1 and 2 is practically all that is required for yachts, whether they be sailing or steam, as these are about the only conditions in which they float. But for vessels carrying miscellaneous cargoes, perhaps wheat on one voyage, cotton on another, coal on another, and so on, the centre of gravity may possibly after loading seldom occupy the same position twice in succession, so that it becomes advisable to ascertain the metacentric height under certain conditions of loading. We shall, therefore, now endeavour to show how the ship's officer may determine the metacentric height himself.

How to find the Metacentric Height and the Position of the Centre of Gravity.—Perhaps some reader is imagining that the method will be that described in the chapter on "Moments," and it is quite true that the centre of gravity could be found by striking a horizontal line at the bottom of the keel, and multiplying each individual weight constituting the ship and the cargo (shell plating, frames, floors, decks, beams, masts, stores, cargo, engines, boilers, winches, windlass, and the host of other items) by its height above the horizontal line mentioned, and dividing the sum of the moments by the sum of all the weights, the result being the height of the centre of gravity above the horizontal line. If carefully done, the method would be all very well, but the immense labour entailed must be evident to every reader. Happily, an accurate, as well as a very simple method may be adopted, by means of which the centre of gravity can be determined in a very short time by experiment. The day chosen should be as calm as possible. The vessel, lying either in dock or river, should be moored only over stem and stern; no ropes abeam, and, if possible, with what breeze there may be blowing directly fore and aft, so as to lend no assistance in heeling the vessel. Place a known weight with its centre over the centre line of the vessel, as near as possible to midships, and capable, when afterwards moved to the port or starboard, of inclining her 5° or 6°.

This weight may consist of anything heavy enough in its nature, and the centre of which and its weight can be accurately determined. Pig iron* may be conveniently used, or blocks of ballast iron, etc. These should be carefully ranged over as little space, according to the size of the vessel, as possible, and may weigh from 1 ton or less, to perhaps 15 or more tons.

This part of the operation having been carried out, the next thing to do is to carefully note the draught at which the vessel is floating. Let it be, say,

13 feet 6 inches forward } a mean draught of 14 feet.
14   ,,   aft

This, on the displacement scale belonging to the vessel, reads 1400 tons (see fig. 3). At the centre line of the vessel suspend two plumb lines, one forward, and the other aft. Let both lines hang freely, and mark clearly a definite length on each of them, measuring from the point from which each one hangs, in our case, say, 8 feet. Great care should be taken to see that the lines hang perfectly plumb on the centre line of the ship. Having carefully arranged all this, we can now proceed with the experiment.

Let the weight of (say) 12 tons be moved from the centre of the vessel, first to starboard, as far as possible, in our example, say, 14 feet. This distance of 14 feet is measured from the centre of the weight when it was on the centre of the vessel to the centre of the weight when afterwards moved to starboard. Having done this, on going to the plumb lines it is found that by the fore one a deviation of (say) $5\frac{1}{2}$ inches has occurred on the length of 8 feet, but on going to the aft one the deviation is found to be 6 inches. The weights should now be carried to the port side, and placed at the same distance as on the starboard side, 14 feet from the centre line of the ship. On going again

* A reliable and most convenient method is to fit a large fresh-water tank on each side of the ship, and use the weight of water in this (if sufficient) for inclining purposes.

to the plumb lines it is found that by the fore one 6¼ inches deviation has occurred, and 5¾ inches by the after.

If all these be added together and divided by 4, we shall have the mean deviation :—

```
Port forward       = 6¼
   ,,   aft        = 5¾
Starboard forward  = 5½
   ,,   aft        = 6
                   ─────
                4)23½
```

5·875 inches. Mean deviation.

Having obtained this result, the practical part of the experiment is now finished, the remainder being a matter of simple calculation. Three results have to be found.

1. *The distance the actual centre of gravity has moved to one side.* Perhaps some reader is saying, "but the centre of gravity has not been found." True, nevertheless we can find how far it has moved in the direction of the shifted weight. According to our study of moments in Chapter II., the rule is :—*Multiply the weight moved by the distance it is moved, and divide the result by the total weight.*

Weight moved = 12 tons.
Distance moved = 14 feet, and the total weight equals the displacement, which was found to be 1400 tons.

$$\frac{12 \times 14}{1400 \text{ tons displacement}} = 0\cdot12 \text{ of a foot} = \text{distance the centre of gravity has}$$
moved in a line, parallel to the line joining the centres of the weight, in its original and in its new position.

2. The next thing to be done is to find the *cotangent of the angle to which the vessel has been inclined.* This is arrived at by *dividing the length of the plumb line in inches by the mean deflection of the plumb line at that length in inches.*

$$\frac{8 \times 12}{5\cdot875} = 16\cdot3 = \text{natural cotangent of angle of inclination.}$$

Although it is not needed in this calculation, still by referring to the table of cotangents at the end of the book, it is seen that the vessel has inclined to a mean angle of 3½°.

3. The last part of the operation is to find *the metacentric height,* or the distance from the centre of gravity to the metacentre. This is done by *multiplying the shift of the centre of gravity by the cotangent of the angle of heel.*

0·12 × 16·3 = 1·95 feet, distance of metacentre above centre of gravity.

This result is the metacentric height of the vessel in its present condition—that is, with the weight used for heeling upon the upper deck. Now it is not likely that this weight will be carried in this position when the ship goes to sea. Should the weight be placed on board simply for the experiment, with the intention of placing it ashore afterwards, a correction must be made in the metacentric height, for it is evident that the weight being as we now find it, say, 8 feet above the centre of gravity, its effect is to raise the centre of gravity higher than it would be were the weight not there. Therefore, by taking the weight away, the centre of gravity must be lower. Then *weight removed, multiplied by its distance from the centre of gravity, and the result divided by the total weight*, which must be reduced by the removal of the weight, will give us the *distance the centre of gravity has lowered.*

$$\frac{12 \times 8}{1400 - 12} = \frac{96}{1388} = 0\cdot06 \text{ of a foot} = \text{how much centre of gravity is lowered.}$$

Thus after the removal of the weight, the centre of gravity is $1\cdot95 + 0\cdot06 = 2\cdot01$ feet below the metacentre.

But suppose the weight used for heeling is one which is intended to be kept on board, being perhaps part of the ballast iron in the case of a yacht. Then, by lowering the weight into the hold again, the effect must be to lower the centre of gravity, and we proceed as in the previous case. (Weight × distance moved ÷ displacement = distance centre of gravity has lowered.) Measure the distance from the centre of the weight on deck to its centre in the new position it will occupy in the hold, say, 15 feet. Then,

$$\frac{12 \times 15}{1400} = 0\cdot12 \text{ of a foot} = \text{distance centre of gravity is lowered.}$$

Therefore the corrected metacentric height is

$$1\cdot95 + 0\cdot12 = 2\cdot07 \text{ feet.}$$

Having become acquainted with the points known as the centre of buoyancy, the metacentre, and the centre of gravity, and to some extent the causes affecting them, and having given some attention to the wedges of immersion and emersion when the vessel is inclined, we are now more capable of pursuing our study of the subject, and of endeavouring to understand how the levers and range of a vessel's stability are affected under greater angles of heel.

Valuable as a knowledge of the metacentric stability (that is, stability at very small angles of heel) of a vessel may be under certain circumstances, yet *it alone is no safe criterion of a vessel's resource of safety*, when exposed to severe weather and subject to **excessive** heeling forces.

For example, in the loaded condition one vessel might have a metacentric height of, say, 1 foot, which for small angles of heel would generally give good righting force, but for greater angles of inclination the righting lever might rapidly decrease and soon vanish altogether; while another vessel, with perhaps only 6 inches metacentric height, and possessing small righting force for small angles of heel, might yet have a very long range of stability and good righting force at greater angles of inclination.

It thus becomes evident that further investigation is necessary, and we have yet to discover those features in a vessel's design, condition, etc., which so powerfully influence her stability under all angles of heel.

No.	Dimensions.* L. B. D. in feet.	Freeboard in Feet and Inches.	Centre of Gravity from bottom in Ft.	Metacentric Height in Feet.	Remarks.
1	100 × 12 × 10	5 0	6·3	−·1	Showing effect of increase in beam, other dimensions remaining unchanged.
2	100 × 15 × 10	5 0	6·3	·57	
3	100 × 20 × 10	5 0	6·3	2	
4	100 × 25 × 10	5 0	6·3	3·9	
5	100 × 30 × 10	5 0	6·3	6·2	
6	100 × 20 × 10	1 0	6·3	2	Showing effect of increase in freeboard, other dimensions remaining unchanged.
7	100 × 20 × 10	1 6	6·3	2	
8	100 × 20 × 10	2 0	6·3	2	
9	100 × 20 × 10	4 0	6·3	2	
10	100 × 20 × 10	8 0	6·3	2	
11	100 × 20 × 10	1 6	8·05	·25	Showing effect of variations in position of centre of gravity, dimensions remaining unchanged.
12	100 × 20 × 10	1 6	7·3	1	
13	100 × 20 × 10	1 6	4·3	4	
14	100 × 20 × 10	1 6	2·3	6	
15	100 × 12 × 10	1 6	6·3	−·1	
16	100 × 12 × 10	1 6	5·3	·9	
17	100 × 30 × 10	1 6	6·3	6·2	
18	100 × 30 × 4	16 0	14·5	6·2	

* L = Length; B = Breadth; D = Draught.

To treat of this effectively by means of actual ship stability data, would necessitate such a graduated variety of vessels as to make such a task a very laborious one. However, this difficulty is easily surmounted. Our purpose will be served by using vessels of box shape, which, while simpler in form than actual ships, are nevertheless capable of lending themselves to the main features we wish to illustrate, and of proving the principles it is desired to make prominent. The three great factors upon which the stability of any floating object depends, whether it be

133

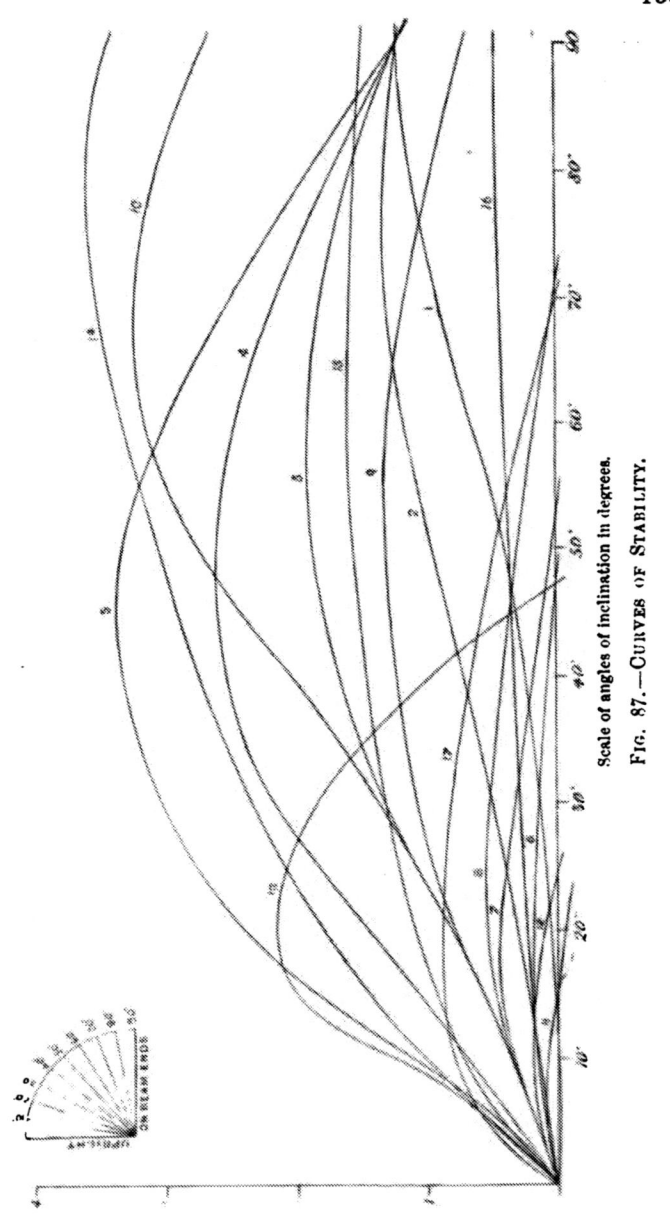

Fig. 87.—Curves of Stability.

of ship or box form, are *beam, freeboard*, and *position of centre of gravity*. By means of a series of box vessels, the particulars of which are given in the table (p. 132), and the curves of stability in fig. 87, the endeavour will be made to reveal the importance of each.

1. **Effect of Beam.**—Vessels, numbers 1, 2, 3, 4, and 5, with the curves of corresponding numbers, will serve for reference and examples in this case. Here we have five vessels, each 100 feet long, 10 feet draught, 5 feet freeboard, and with beams of 12, 15, 20, 25, and 30 feet respectively. In each case we have imagined the centre of gravity to remain stationary at 6·3 feet from the bottom of the box, for the depth and draughts being assumed to remain the same in each vessel, it is quite reasonable to take it for granted that no change would occur in respect to the position of the centre of gravity.

It has already been shown, earlier in this chapter, that increase of beam raises the metacentre. This has, therefore, happened in the examples with which we are dealing, the result being a metacentric height of $-0·1$, ·57, 2, 3·9, and 6·2 feet respectively.

Let us examine the curves of each of these vessels, and see what can be gathered from them.

Curve No. 1 commences with the metacentre 0·1 foot *below* the centre of gravity. In this condition the vessel is incapable of floating upright. Will she capsize? The curve answers the question most emphatically that she will not. If undisturbed she would take a slight list and then lie at rest. If forcibly inclined, the righting lever of stability would continue to grow in length, until, when on her beam ends (90° on curve), she has barely attained her maximum stability.

There is, however, nothing to elate one very much in the fact that a vessel has splendid stability at 90° of inclination, for every seaman knows that long before that angle is reached, it would be impossible to stand upon or work such a ship, and, moreover, weights on board which are considered as permanent and fixed would be on the move, and then most disastrous results would inevitably ensue. If the levers of stability are good up to 50° or 60°, and even then are decreasing and vanish altogether at 90°, not much fear need be entertained, for rolling to angles of even 30° or 40° on each side is considered very excessive.

Had we been guided in this particular case entirely by the metacentric height, the conclusion might have been come to that her condition is much more serious than it actually proves to be. Certainly the vessel is too tender; but what is needed is either ballast of some kind in the bottom, if the vessel is not down to the load waterline, or else a re-arrangement of the cargo so as to bring the heavy weights lower, and thus increase the metacentric height. Either of these methods would add immensely to the improvement of the vessel's condition. Further observations on this type of vessel are made at a later stage, in the remarks on "Rolling" and "Behaviour at Sea."

Curve No. 2.—The difference between this vessel and the previous one is, that owing to an increase of 3 feet in the beam the metacentre has been raised, and now the metacentric height is 0·57 foot. The curve shows longer levers of stability up to 90° of heel, where it crosses Curve No. 1. The vessel is therefore stiffer, and has greater righting force up to this point, but her maximum lever of stability is reached sooner than in the foregoing vessel.

Curve No. 3.—The beam is now 20 feet, giving a higher metacentre, and a metacentric height of 2 feet.

Curve No. 4.—Here the beam is 25 feet, with a higher metacentre, and a metacentric height of 3·9 feet.

Curve No. 5.—In this case the beam increasing to 30 feet, the metacentre continues to rise, and the metacentric height attains 6·2 feet.

Let us see what information can be gathered from these curves for similar vessels increasing in beam only.

First. With every increase in beam, and consequent increase in the metacentric height, the successive curves rise steeper and steeper, indicating greater stiffness and resistance to heeling.

Second. Each successive curve reaches a greater height than its predecessor, giving a longer maximum lever of stability, and, consequently, a greater righting moment.

Third. In each successive curve the maximum lever of stability is reached at a smaller angle of inclination.

Fourth. In each successive curve the lever of stability vanishes at a smaller angle of inclination.

Judging from these curves, we might at first be inclined to give the entire credit of the great growth in the lengths of the righting levers in each successive curve to the increased metacentric height resulting from the increased beam; but a further comparison of curves will assist us in arriving at a more correct conclusion.

2. **Effect of Freeboard.**—*Curve No. 6.*—The vessel represented by this curve has the same length, breadth, draught, metacentric height, and height of centre of gravity above the bottom of the keel as No. 3. Instead of 5 feet freeboard she has only 1 foot. Mark the effect in the curve. Instead of the splendid sweep of No. 3 reaching its maximum lever of 1·9

feet at about 60° of inclination, and vanishing at considerably past 90°, No. 6 has never more than a lever of 0·2 foot, and that at the angle of 20°, the curve vanishing altogether at an angle of less than 50°.

Curve No. 7 is for a similar vessel, with 1·5 feet freeboard.
,,   8   ,,   ,,   ,,   2   ,,   ,,
,,   9   ,,   ,,   ,,   4   ,,   ,,
,, 10   ,,   ,,   ,,   8   ,,   ,,

Let us gather again what these latter curves indicate.

First. Even good beam with good metacentric height, unless combined with suitable freeboard, is no guarantee for either a good range or good levers of stability. This is proved by a comparison of Curves Nos. 6 and 7 with Curve No. 3, and comparing also No. 17 with No. 5. These two latter are also identical vessels, with the exception of the freeboard, which is 1 foot 6 inches in the former and 5 feet in the latter. The great difference in the curves, which can only be attributed to the difference in the freeboard, is very apparent.

Second. Increase in freeboard with undiminished metacentric height, increases not only the length of the levers, but the maximum lever in each case is at a greater angle of inclination, and the range of stability is lengthened out also. Curves Nos. 6, 7, 8, 9, and 10, with their increasing freeboard, prove this, each of the first four in its turn approaching No. 3, until No. 10, with the great freeboard of 8 feet, far surpasses it in both maximum levers of stability and range.

It may be all very well to make the statement that freeboard has this and the other effect upon a vessel's stability; but some reader may be asking the question, "How is it that freeboard is capable of producing such an effect?" An endeavour to explain this will be made by the aid of the following graphic illustration (fig. 88).

Figs. i. and ii. are the vessels in the table, Nos. 8 and 3 respectively. They are similar to each other, except that i. has 2 feet, and ii. 5 feet freeboard. Floating in the upright condition, they have exactly the same metacentric height, 2 feet. Let fig. i. be heeled until the deck edge is down to the level of the water, as shown in fig. iii., the angle of inclination being 10°. To find the lever of stability we must turn our attention to the centre of buoyancy. Let fig. x. represent the buoyancy of the vessel, which corresponds in every respect to the immersed part of fig. iii. The centre point in fig. x. is B, and this is the centre of buoyancy. By transferring the position of B to fig. iii., we see its position in relation to the centre of gravity. The distance between the vertical lines

137

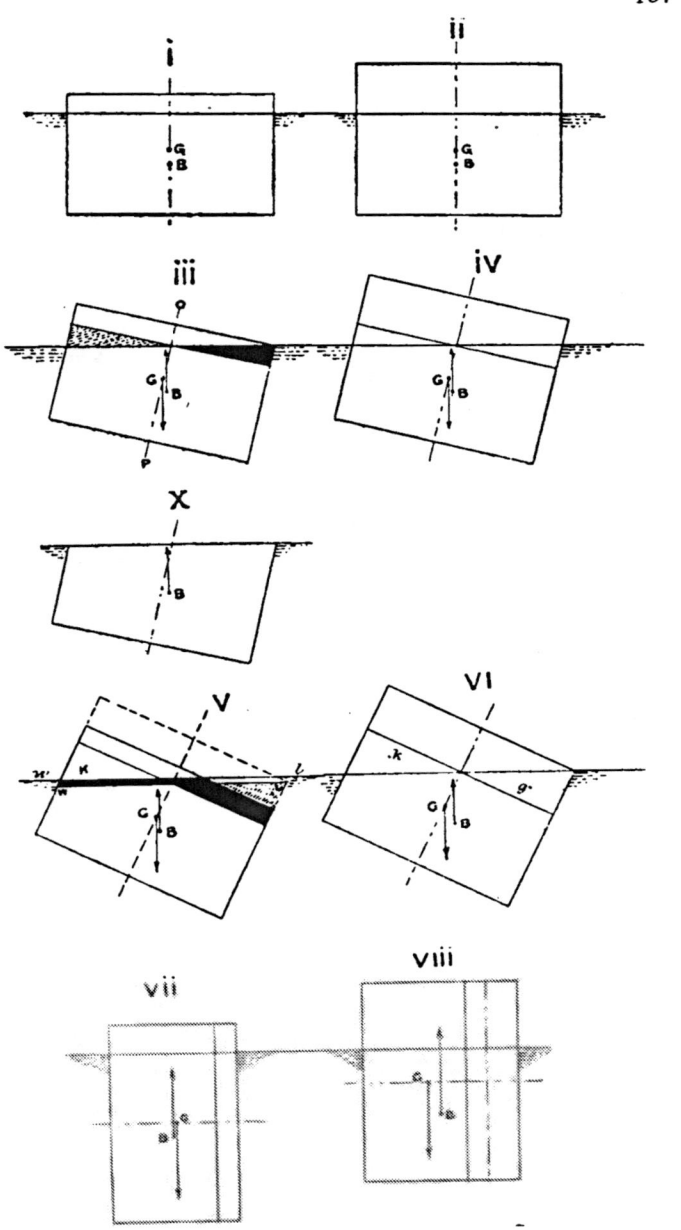

through these points is the lever of stability, and since it intersects the centre line, O P, above the centre of gravity, it is a righting lever. Although the shape of the buoyancy has altered in form under the inclination, owing to the transference of the wedge of buoyancy shown by the hatched lines, to the other side of the vessel, indicated by a black wedge, yet its volume is unchanged. The centre of buoyancy, therefore, travels in the direction in which the buoyancy is transferred, just in the same manner as shifting a weight upon a lever, the distance in the case of the wedges being reckoned from centre to centre.

Turning to fig. ii. with the greater freeboard, on inclining this vessel to the same angle 10°, her deck edge is still considerably out of the water (fig. iv.). However, in this condition her form of buoyancy is similar to fig. iii., simply because the wedges of immersion and emersion are identical in shape. The centre of buoyancy, therefore, occupies the same position, and it follows that the lever of stability must be similar to that in fig. iii.

Let fig. ii. be now heeled until its deck edge reaches the waterline, as shown in fig. vi. The reasoning applied to fig. iii. will apply in this case also:—A wedge of buoyancy has been transferred from one side to the other, giving the form of buoyancy shown below the waterline, the centre of which is the centre of buoyancy, B. This point again shows the relation of the centre of buoyancy to the centre of gravity, and the lever of stability is found to have increased simply because a larger wedge has been transferred a distance of k to g (the centres of the wedges from each other).

Fig. i. at this angle of inclination shows a different state of affairs (see fig. v.). There is certainly a wedge of emersion, but owing to the decreased freeboard, there cannot be a similar wedge of immersion. The dotted line indicates the boundary of the greater buoyancy of fig. vi. Now, supposing fig. v., when inclined, to float at the waterline W L with the wedge of emersion K, then there ought to be a wedge similar in volume immersed. But this cannot be, for the immersed wedge is deficient by the volume of the small wedge indicated by the hatched lines in the diagram. Now, the volume of the emerged part, of whatever shape, must equal the volume of the immersed part, and since the immersed wedge is deficient by the small wedge already referred to, this loss of buoyancy can only be regained by the vessel sinking to a deeper draught, thereby spreading the buoyancy of the lost wedge along the waterline, and thus causing the vessel to float at the new waterline $w\,l$. The black part in the figure shows the new immersed buoyancy. The important point is to find what effect this new immersed buoyancy will have upon the lever of stability. The more buoyancy that can be placed towards the side to which the centre of buoyancy has already begun to move, the further will this point be brought towards that side. Owing to the lower freeboard the loss of the hatched wedge of buoyancy has robbed the vessel of a most effective agent in bringing out the centre of buoyancy from the centre of gravity, and by placing this lost buoyancy along the waterline, a further check is made upon the outward movement of the centre of buoyancy, and the tendency is to draw it back again; hence the decreased lever of stability, as compared with fig. vi. Passing now to the conditions shown in figs. vii. and viii., we find these vessels heeled to an angle of 90°. The centre of buoyancy can easily be determined in these cases. For fig. vii. it will be *half the distance from the bottom* (not the bottom of the figure, but the bottom of the ship) *to the deck*, which is $\frac{10 + 2}{2} = 6$ feet, and the centre of gravity being 6·3 feet, the result is an upsetting or capsizing lever of 0·3 of a foot. Fig. viii., however, has the advantage of the buoyancy afforded by the additional freeboard shown beyond the dotted line. This naturally tends to draw out the centre of buoyancy, and its position now is $\frac{10 + 5}{2} = 7·5$. $7·5 - 6·3 = 1·2$ righting lever.

The foregoing remarks, with a little study of the figures themselves, will leave little doubt in the mind of the reader of the importance of freeboard as a factor in affecting stability.

3. **Effect of Position of Centre of Gravity.**—Let us take our final illustration, and note the effect of obtaining metacentric height, not by means of increasing the beam, and thereby raising the metacentre, but by retaining the same beam and same position of metacentre, and lowering the centre of gravity.

This may be done to a considerable extent in merchant steamers, in loading miscellaneous cargoes, by keeping the heavy weights low, but to a very much greater extent can it be done in yachts, where the centre of gravity can almost be placed wherever desired by means of permanent ballast.

Curves Nos. 11, 12, 7, 13, and 14 are for vessels 100 feet long, 20 feet broad, and 10 feet draught, with 1 foot 6 inches freeboard, the metacentre being, therefore, at the same height from the bottom of each vessel—viz., 8·3 feet. The only difference between them is in the height of the centre of gravity.

For No. 11, the centre of gravity is taken at ·25 foot below the metacentre.
,, 12, ,, ,, ,, 1 ,, ,,
,, 7, ,, ,, ,, 2 feet ,,
,, 13, ,, ,, ,, 4 ,, ,,
,, 14, ,, ,, ,, 6 ,, ,,

The effect of this alteration in the position of the centre of gravity is shown by the curves. *First, at a particular draught,* by every downward movement of the centre of gravity, thereby causing an increase in the metacentric height, both the levers and range of stability are lengthened. *Second,* each curve rises more steeply than its predecessor, indicating greater stiffness.

We have already seen the effect of beam upon stability. Suppose 10 feet be added to the beam of a vessel identical in every respect with No. 14. The effect is to raise the metacentre, and give much greater metacentric height. But let the centre of gravity also be raised from 2·3 feet from the bottom of the vessel to 6·3 feet, the metacentric height being now 6·2 feet, the result is to give a vessel identical with No. 17. Nos. 14 and 17 differ now practically in beam only.

An error is sometimes made in a case like this, it being imagined that the broad vessel, even with the same metacentric height, is better than the narrow one of the same depth and freeboard. If the curves of these two vessels be compared, their stability is seen to be widely different, the narrow vessel possessing more stability in every respect than the beamy one, except perhaps at the beginning of the curve.

Now how does this happen? In the first place, the advantage of the increased metacentric height, which was obtained by the increased beam, was robbed from the vessel by raising the centre of gravity, and making the metacentric height similar in both cases. Perhaps some reader is still saying—"Having made the metacentric height similar for both vessels, why are the curves not similar?" This is just where the error is often made in depending upon the metacentric stability, which, as has been previously stated, is no guide for a vessel's range of stability, but only for small angles of inclination. Up to 7° or 8° these curves are approximately the same. For greater angles of inclination, we must turn our attention to the centre of buoyancy, and trace its movements in relation to the position of the centre of gravity.

In A and B, fig. 89, no levers of stability are seen, since the vessels float upright and are perfectly stable, with the centre of gravity, centre of buoyancy, and metacentre (the last not indicated) in the same vertical line, *a b*. In figs. C and D the same vessels are inclined to angles of 45°. It is now seen that the levers of stability in this condition are very different, those for the smaller vessel being much the greater. Both the centre of buoyancy and the centre of gravity, though chiefly the latter, are accountable for this result. Turning our attention first to the centre of buoyancy, we find that, in both figs. C and D,

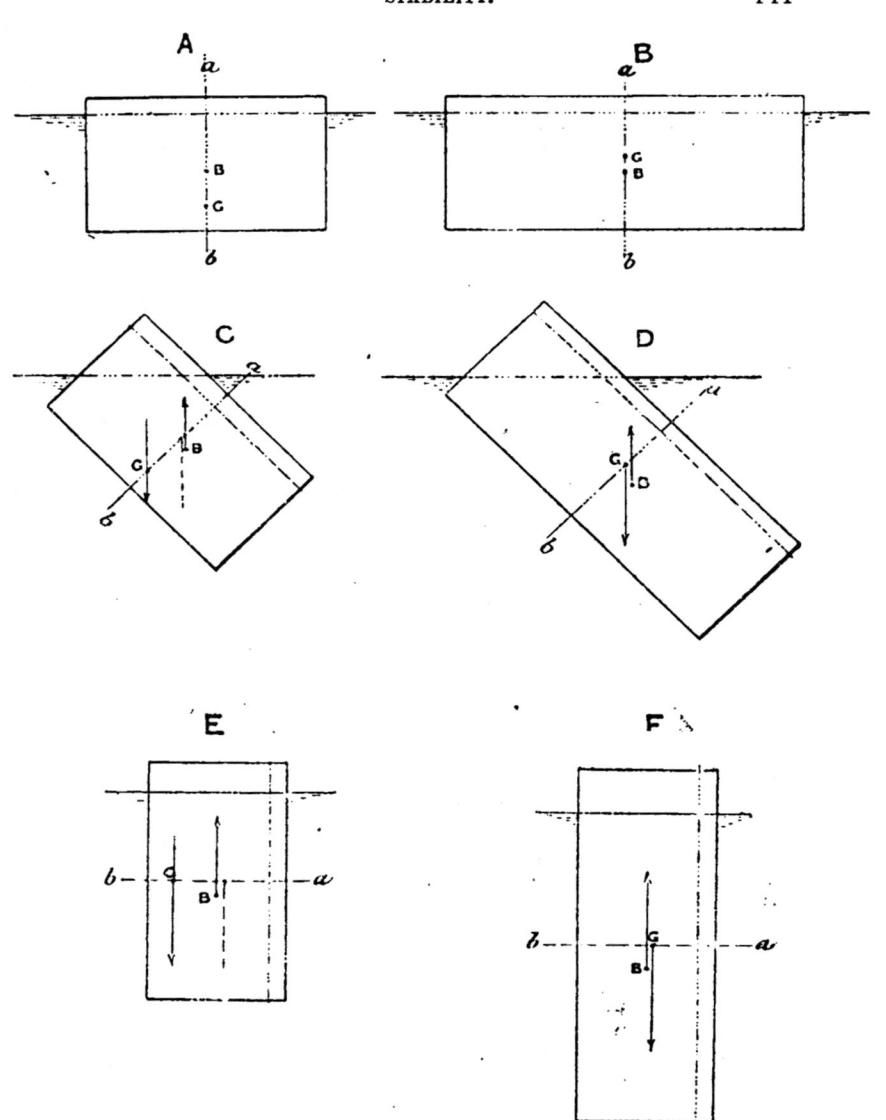

Fig. 89.—Comparison of Vessels with different Beams but similar Metacentric Heights.

owing to the small freeboard in each case, the deck edge has become immersed when inclined to a very small angle. This at once checks the outward movement of the centre of buoyancy though fig. C, owing to its smaller beam, has the advantage to some extent, since greater inclination would be needed to immerse its deck edge than would be required for fig. D.

Turning now to the centre of gravity in each case, we find here the chief factor in producing the great difference in the stability levers. It is first noticed that their difference in position from the bottom of each vessel on the line $a\,b$ is very great, although the metacentric heights in the upright condition are practically identical. In fig. B it was easy to get a good metacentric height, owing to the great beam, but in fig. A the greatly reduced beam made it necessary to very much lower the centre of gravity, in order to get the same metacentric height. Hence the difference in their positions. It will also be observed that the *lower the position of the centre of gravity* is on the line $a\,b$, the *greater must be the righting lever of stability* (a glance at the figures will show this clearly), and on the other hand, the *higher the position of the centre of gravity* on the same line, the *smaller the lever*. This, then, accounts to a great extent for the difference in the levers of stability. So that even on their beam ends, at angles of 90°, we still find the narrower vessel (fig. E) with a large righting lever, while the broader one, F, has actually an upsetting lever.

However, the case we have taken is certainly an extreme one for cargo or passenger vessels, for while in the broad vessel the position of the centre of gravity would very often be found high in comparison with the depth, in the narrow vessel, it would be impossible to load her and have the centre of gravity as low in comparison with the depth. Were the two vessels loaded in the same manner—that is, in relation to the vertical position of the weights of the cargo, the centre of gravity in both vessels occupying the same position from the bottom of the box —a vastly different result would arise. Let the centre of gravity of the narrow vessel be raised by the loading of cargo to the same position as in the broader one—viz., 6·3 feet from the bottom of the vessel. The dotted lines on the figures show the new vertical line through the centre of gravity of the narrow vessel, and curve No. 7 will show the whole range of stability. The tables are now turned against the narrow one, indicating much reduced stability. Hence the necessity of wisdom in loading, as, in the latter case, the heavy weight of the cargo would have to be placed much lower in order to get anything approaching similar levers of stability for the two vessels.

STABILITY. 143

But while the position of the centre of gravity at 2·3 feet from the bottom of the vessel for curve No. 14 is an exaggerated case for merchant vessels, it is by no means out of the way for sailing yachts, for in order to get great stiffness and long levers of stability, which are necessary to carry great sail area, especially with small beam, the method of bringing the centre of gravity very low by placing ballast either in the keel, or else as low as possible, has to be adopted.

Before leaving this part of the subject of stability, the reader is again warned against jumping to the conclusion that even a

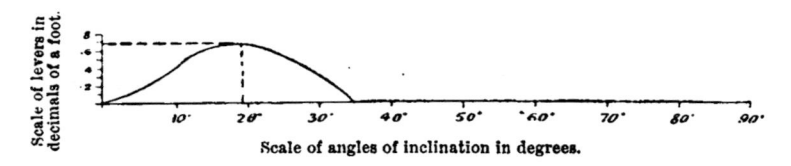

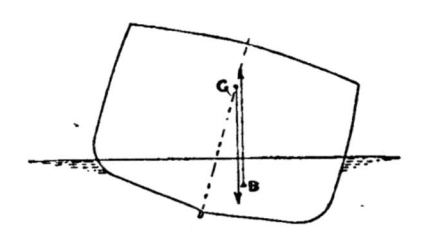

Inclination showing maximum lever of stability.

Curve of stability for a sailing ship, 270 feet long, 41 feet beam, and 26 feet 3 inches depth, in a light condition, with about 112 tons of ballast on board. Metacentric height, 2·9 feet. Centre of gravity, 20¼ feet above the top of the keel. Freeboard, 17¼ feet.

FIG. 90.

combination of good beam, good freeboard, and good metacentric height will always produce satisfactory stability. This has already been shown by the several box vessels and their curves, and will further be emphasised by a glance at the curve of stability (fig. 90), which is for a sailing ship in a light condition. She has 41 feet beam, 2·9 feet metacentric height, and 17½ feet freeboard, the last of which is extremely great, and yet the curve of stability in this condition represents both short levers and very short range, this being attributable to the fact that, in the light condition, the heavy top weight of masts, spars, etc., brings the centre of gravity very high, and it has already been pointed out that the higher the centre of gravity, the shorter are the levers of stability, and the sooner does the vertical line through the centre of buoyancy intersect the centre line of the ship below the centre of gravity, thereby creating a capsizing moment. The same vessel in her loaded condition, with only 5½ feet freeboard and 3 feet metacentric height, would have both immensely greater levers, and greater range of stability, because then the centre of gravity is much lower in its position.

In a box vessel, say, 20 feet deep, if the centre of gravity be at half the depth, the righting lever of stability must vanish at 90° of inclination, whatever be the freeboard, since in this condition the centre of buoyancy and the centre of gravity are in the same vertical line. If the centre of gravity be higher than half the depth, the levers will be shorter and the range less also, but if the centre of gravity be lower than half the depth, then the levers will be longer, and the range will extend beyond 90° of inclination. Now, all actual ships are not of box form, though it is granted that in some cases it is somewhat difficult to draw that distinction. Of two vessels of similar beam and depth, the one with most buoyancy in the upper half of her depth, being therefore most fined away at the bilge and bottom, can afford to have the centre of gravity the higher, and the nearer the box section is approached, the lower must be the centre of gravity.

The actual box ship is, therefore, the worst case, since it brings the centre of buoyancy into the lowest possible position; and, on the other hand, the vessel fullest at the waterline, and well fined away below, has its centre of buoyancy in the highest possible position, in which position the longest levers and the greatest range of stability are produced, other features in the design being favourable. It is thus impossible to stipulate a particular position for the centre of gravity applicable to all ships.

For a box-shaped vessel, if the centre of gravity is from about 0·5 to 0·6 of the depth from the top of the keel, with fair metacentric height, a fairly good range of stability may be expected, though the righting levers may be small under certain circumstances. For vessels of finer underwater form greater stability would be developed, and when the centre of gravity is less than 0·5 of the depth from the top of the keel, great stability may be anticipated.

Effect of Tumble Home.—In fig. 91, let G be the centre of gravity, B the centre of buoyancy in the upright position, C D E the immersed wedge, and K the centre of the immersed wedge.

The greater the distance from the original centre of buoyancy,

B, to the centre of buoyancy of the immersed wedge, K, as shown by the line P, the greater will be the effect in drawing the centre of buoyancy out from its original position. Let B' be the new centre of buoyancy when inclined, and G Z the righting lever of stability.

Now, supposing a piece of the shape of the black wedge be cut off from the vessel, let us observe the effect upon the stability. Owing to the loss of this buoyancy when inclined, compensation must be made by apparently drawing upon the reserve buoyancy, and taking a layer off all along the waterline to the dotted line, simply because the wedge of immersion is now less than the wedge of emersion; and they are equalised by adding a layer to the wedge of immersion at C D, and deducting a layer from the wedge of emersion at W C.

The centre of the immersed wedge, K, will have travelled

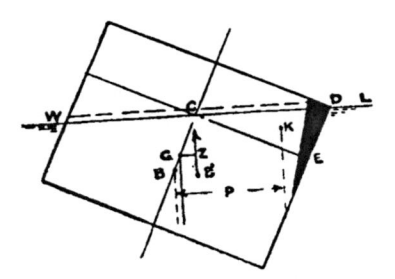

FIG. 91.—EFFECT OF TUMBLE HOME ON STABILITY.

towards the left of the figure, causing B' to move in the same direction, and the layer of buoyancy along the waterline, W to C, will also have aided to produce this effect.

G Z will now have become shortened, which means reduced righting moment of stability.

Taking the case of a vessel of ship form, we can easily see the application of the above illustration. Instead of carrying the sides up perpendicularly they are usually curved in, as shown in fig. 15. This is known as *tumble home*. A valuable piece of buoyancy is lost; in fact, the very part of the wedge which is most efficacious in drawing out the centre of buoyancy from the centre of gravity is cut away. Thus in vessels of low freeboard, and especially if at the same time possessing narrow beam, the effect of much tumble home may be to assist in causing deficient stability in certain conditions. However, as it is not usual to give an ordinary mercantile vessel more than a few inches of tumble home at the main deck, the effect is not serious in the ordinary types of modern broad beamed cargo steamers.

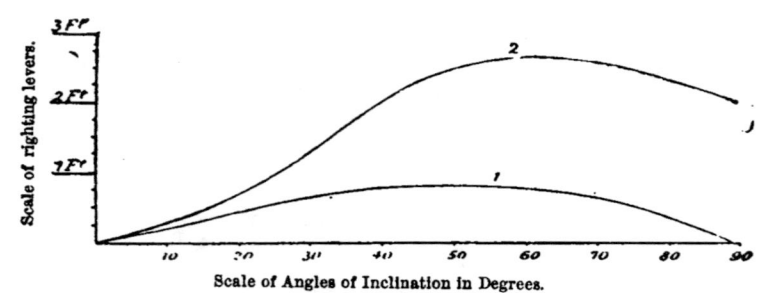

Curves of stability for a cargo steamer—Length, 480 ft.; beam, 57 ft.; depth, 40 ft.
Curve No. 1, Light condition, metacentric height, 2·68 ft.
  ,,   2, Load   ,,   ,,   3·67 ,,

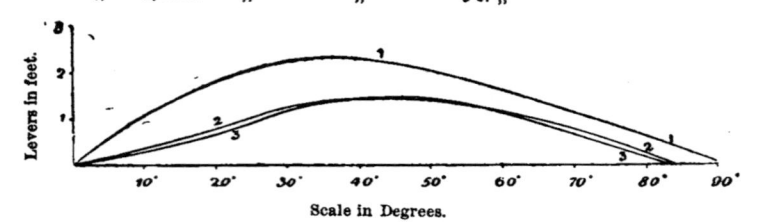

Curve for a steamer 410 ft. long, 50 ft. 6 ins. beam, 32 ft. depth.
No. 1, Light condition, metacentric height (G M) = 11·06 ft.
No. 2, Loaded with homogeneous cargo, 7 ft. 6 ins. freeboard, G M 1·85 ft.
No. 3, Same as No. 2, with coal consumed. G M 1·58 ft.
Note.—1 ft. in the scale of levers in this figure is equal to 2 ft. for *No. 1 Curve*. The levers are thus only half length on this diagram.

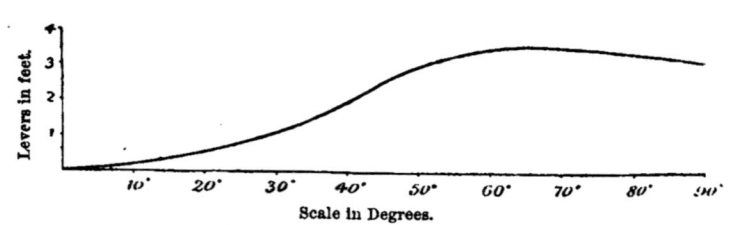

Curve for a vessel 360 ft. long, 45 ft. beam, 30 ft. 1 in. depth. Metacentric height, 1·48 ft.

FIGS. 92, 93, AND 94 ARE EXAMPLES OF ACTUAL SHIP CURVES.

A point which is sometimes overlooked on the part of the owner or his representative in the design of a new vessel, is the value of sheer. That it adds to the appearance, gives valuable rising power, and tends to prevent the shipping of water over the stem and stern, must be clear to everyone.

But one of its best features is that it produces increased freeboard, the use of which has already been discussed.

Stability in different Types of Vessels.—As regards types of vessels best adapted to produce good stability when of suitable dimensions, and the loading properly carried out, those with most freeboard must come first.

Thus we have the *awning-decker*, with its completely closed-in light superstructure between the main and awning decks, splendidly adapted for *carrying passengers* or light cargoes.

Next comes the *spar-decker* with a stronger superstructure, and adapted for *carrying cargoes* of greater density with smaller freeboard in comparison with the awning-decker.

Last, we have the strongest type of ship, the *two* or *three decker*. This is the *best deadweight carrier*, having least freeboard. Unfortunately, structural strength and stability are in no way related to each other, and thus, as statistics prove, especially the older types of these vessels, with their small beam to depth, and also small freeboard, have produced the most disastrous results, through lack of stability. A vessel which has found great favour among shipowners during late years, because of its special adaptation for certain trades, is the *raised quarter-decker*, which is simply a modification of the strong two or three deck type previously referred to, and whose comparatively greater freeboard assists in producing more favourable stability. (See also page 192 for further remarks upon types of vessels, etc.)

Note.—Awning- and spar-decked ships are equally as strong as three-deckers in relation to the deadweight they carry at their respective load lines.

CHAPTER VI. (Section II.)

ROLLING.

Contents.—Rolling in Still Water—Relation of Stiffness and Tenderness to Rapidity of Movements in Rolling—Resistances to Rolling—Danger of great Stiffness—Rolling among Waves—Lines of Action of Buoyancy and Gravity—A Raft, a Cylinder, and a Ship among Waves—Synchronism, how Produced and Destroyed—Effect of Loading upon Behaviour—Effect of Transverse Arrangement of Weights upon Rolling Motions—Alteration in Behaviour during a Voyage—The Metacentric Height—Fore and Aft Motions—Fore and Aft Arrangement of Weights.

Rolling.—After the consideration already given to the subject of stability, we are now able to proceed further, and observe the relation between stability and rolling at sea, and what means can be adopted to reduce the latter to a minimum.

Rolling is often spoken of as though it were a particular quality belonging to a ship. For instance, it is not uncommon to hear a ship described as a heavy roller; or another, as being very steady. A little investigation will show that it is not strictly correct to so characterise any vessel. At the same time, however, we shall see that the design of some vessels lends more encouragement to rolling than others; and, on the other hand, it is possible to a considerable extent to overrule even the influence of design, and make a vessel either steady, or specially inclined to heavy rolling, in spite of design.

Let us briefly enumerate the points we have already studied, which will help us.

First.—If a vessel rolls under the influence of some external force, the power she possesses which brings her back to the upright is her stability.

Second.—If a vessel has a great metacentric height, it follows that, at least for small angles of inclination, she possesses considerable righting moment, and the curve representing levers of stability will rise the more steeply the greater the metacentric height. Such a vessel is said to be stiff.

Third.—If the metacentric height is small, the reverse of the previous case will be the consequence, the righting levers will be small for small angles of inclination, and the curve of stability will rise slowly.

The effect of metacentric height in relation to rolling is exactly the opposite to what one would at first imagine. The stiff ship with great metacentric height offering great resistance to inclination, is the very one which generally rolls most in a seaway; and the tender vessel, with small metacentric height and small resistance to heeling, is usually the steady one. How comes this? We shall be better prepared to answer if we make a few mental experiments upon a vessel for ourselves.

Rolling in Still Water.—Let us imagine a ship with large metacentric height and a fair range of stability to be lying in the dock. By means of some external force let the vessel be heeled over to, say, 10° of inclination, and held there. We know that the centre of buoyancy will have shifted into the centre of the new shape of displacement, and there is now created a lever between the vertical lines passing through the centre of gravity and the centre of buoyancy. It is, therefore, evident that the vessel possesses an amount of righting force exactly equal to the external heeling force required to so heel her, and by means of which, when the latter is removed, she will come to the upright. In this position the available righting moment will have disappeared, since the centre of gravity and the centre of buoyancy are again in the same vertical line. Moreover, the greater the metacentric height the greater the amount of available righting moment, and thus the more rapidly will she reach the upright position.

Resistances to Rolling.—Let us now free our vessel lying at the angle of 10°. The result is, that in the space of a few seconds she has reached the upright. But does she remain there? Not at all, for just as in the case of a pendulum in travelling from an angle of inclination to the vertical, an amount of energy of motion (*kinetic energy*) is accumulated, which carries her over to the other side, where again a righting lever is created acting in opposition to the last roll. Were there no resistance of any kind this process of rolling would be endless; but experience shows us that after a series of rolls, the vessel will come to rest. This is brought about by the united action of several kinds of resistance.

First.—The *friction of the air* upon the exposed surface of the vessel.

Second.—The *friction of the water* upon the immersed surface.

Third.—*Head resistance*, caused by projections on the immersed surface.

Fourth.—*Wave resistance.*

As the great object is to get a safe and steady ship, let us see how far it lies in our power to modify these resistances.

Nothing can be done to increase the air friction, unless it be by means of sail, which will certainly tend to produce steadiness. It would be an easy matter to make a rough skin upon a vessel; but this would deduct so enormously from the speed, that it is preferred to get the smoothest surface possible. In the third case, however, a very great deal may be done to produce steadiness by fitting projections in the form of keels or bilge keels. The day of doubt as to the efficiency of this means is past. Not only are naval experts agreed, but the testimony of every seaman who has experienced the efficacy of bilge keels, especially when fitted upon vessels which had previously been without them, is unanimous as to their great value in reducing both the number and angles of roll, or oscillations.

By an *oscillation* is meant *a complete roll* from port to starboard, and the time occupied to perform such oscillation is termed the *period of oscillation*. An example, taken from the experiments of the late Mr. Froude, upon the model of the war vessel "Devastation," will serve as an illustration in passing :—

Number and Description of Bilge Keels.	Number of double Oscillations before Vessel was brought to rest.	Period of double Oscillation in seconds.
1. No bilge keels,	31½	1·77
2. One 21 inch bilge keel on each side,	12¼	1·9
3. ,, 36 ,, ,, ,,	8	1·9
4. Two 36 ,, ,, ,,	5¾	1·92
5. One 72 ,, ,, ,,	4	1·99

In speaking of wave resistance, we do not refer to sea waves—for, as was formerly stated, the vessel upon which we are experimenting is supposed to be lying in a dock—but to waves created by the vessel in her rolling movements in the water. Such might at first appear to be very trivial, but to create such waves, even though very small, means an immense expenditure of energy, and this, therefore, must be deducted from the total available energy, which incites the vessel to roll. The combined effect of these agencies is to diminish the angle of inclination, and, finally, to produce extinction. A noteworthy point to be observed, as shown by the above table, is that for moderate angles of inclination the period is approximately the same for the larger as for the smaller oscillations. Thus we see that it is great stability which conduces to rapidity of rolling motion, though not necessarily to great angles of inclination.

Danger of great Stiffness.—The danger of very stiff vessels with good range of stability is, not that they will capsize, but by the severity of their movements that they will damage themselves by straining the structure and causing leakage, or by shaking their masts overboard, not at all an unheard-of-occurrence, where broad-beamed sailing-ships, owing to pure ignorance, have been ballasted in a manner producing enormous stiffness.

On the other hand, the vessel with the small metacentric height, when forcibly inclined to the same angle of 10°, and then set free, returns to the upright much more slowly, having shorter righting levers, and, therefore, less stored energy. The energy of motion acquired in returning to the upright is less, and adding to this the resisting agents, it follows that the angles to which she rolls, and the number of oscillations before coming to rest, will be reduced.

The nearer the immersed portion of any object approaches the shape of a circle, and the nearer the metacentre and the centre of gravity are together, the less power to regain the upright will it possess until we reach the minimum in the actual cylindrical type with the centre of gravity and the metacentre coincident. Vessels of this latter type possess no righting force at all, and thus, when inclined to an angle, they remain there, even though entirely free. A very small external force, therefore, will heel them, and turn the underside uppermost altogether.

Rolling among Waves.—Now the question arises, since the motion of rolling is so governable, is it better to have the steady type of ship with small metacentric height, or the stiff one with great metacentric height? But this we shall better answer if we first briefly consider her more complicated motions among waves, as thus far our considerations have dealt exclusively with vessels in still water. Here, however, peculiarities arise, and although the principles deduced from forced rolling in still water still hold good to a great extent, we shall find our ship behaving very differently at times from what we should imagine if we depended solely upon our knowledge of rolling in still water. In the first place, it is scarcely necessary to inform any reader who has ever noticed a piece of wood floating in the sea among unbroken waves, that it is not the mass of water composing the waves which moves onward, but the form only. A slight forward and backward motion of the floating object shows that the only movement of the wave water is slightly forwards and backwards. At a comparatively small distance below the surface of the water there is apparently no motion whatever. An old, though not

152   KNOW YOUR OWN SHIP.

strictly correct, illustration is that of wind blowing over a field of corn, causing a waving motion as the heads incline with the gusts of wind, and then rise again.

Raft.—As a complex form like a ship is a form more difficult to deal with than that of a flat floating piece of wood, let us examine, first, the behaviour of a small raft. In smooth water we know that, owing to its great stiffness, its oscillations are exceedingly rapid and its period very short, and that a condition of rest is soon obtained.

We have also noticed that, when among wave water (figs. 95 and 100), its deck is always parallel, and the mast perpendicular to the surface of that part of the wave upon which it is floating. It is

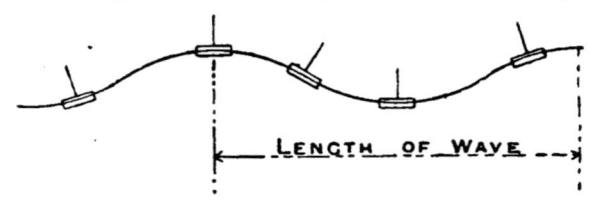

FIG. 95.—BEHAVIOUR OF A SMALL RAFT AMONG WAVES.

therefore upright on the summit and in the trough of each wave, and its greatest angle of inclination is at about half the height of the wave where the slope is greatest. In this case the raft, being very small, behaves practically as though it were actually a particle of the surface wave water. Such agreement becomes less and less as the beam increases relatively to the length of wave, as it can no longer lie flat on the surface, or have the greatest angle of inclination where the wave slope is greatest, until at last, where exceptionally large beams are reached, as in the Czar of Russia's yacht "Livadia" (153 feet), the vessel no

FIG. 96.—ILLUSTRATING A VERY BEAMY VESSEL AMONG WAVES.

longer takes of the motion of a small raft at all, but maintains a comparatively horizontal deck, as in fig. 96.

The longer the waves are in comparison with the breadth of such a vessel, the greater inclination she would reach in endeavouring to follow the angle of the wave surface. But in a short sea she would be practically steady.

Cylinder.—Let us take as another example a vessel of the cylindrical type.

ROLLING. 153

In fig. 97 we see the object as it would float in smooth water. Being of wood, and of equal density throughout, the centre of gravity is in the centre, as is also the metacentre.

The centre of buoyancy is in the same vertical line through this point, and the object floats at rest, as it will do at any angle of heel, since it never has any stability; a state which can only exist when there is no righting lever, the vertical lines through the centre of gravity and the centre of buoyancy always coinciding.

In fig. 98 we see the same object among waves, and on a wave-slope. Let us examine its condition now.

From observation every reader knows that no revolving or heeling motion occurs. The line ab remains vertical, and the

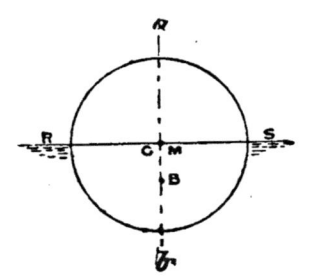

FIG. 97.—BEHAVIOUR OF A CYLINDRICAL VESSEL IN SMOOTH WATER.

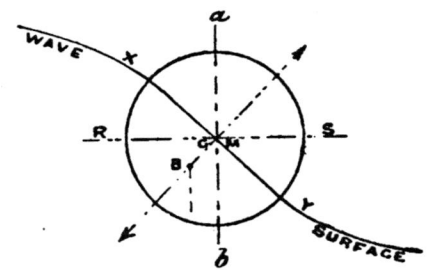

FIG. 98.—BEHAVIOUR OF A CYLINDRICAL VESSEL AMONG WAVES.

waterline varies from R S, when floating in still water, or on the crest, or in the trough of the wave, to X Y, the greatest wave-slope. But on examining the object on the wave-slope, it is found that the centre of buoyancy has shifted into the centre of the immersed part. If we drop a vertical line through the centre of gravity and through the centre of buoyancy, we see that these points are no longer in the same vertical line, but that a distance exists between them. If this distance represents the length of the lever of stability, the vessel cannot remain in this condition without making some effort to bring the centre of buoyancy and the centre of gravity into the same vertical line, which effort must incline the vessel more or less. But observation proves that such is not the case, for the object makes no movement to the one side or the other, the only interpretation of such behaviour being that no lever whatever exists, and that the downward force of the weight of the ship and the upward force of buoyancy are evidently being subject to other forces causing them to act differently from the manner in which we have been accustomed to consider them in still water. It is just on this point where many of those, whose knowledge of the subject of stability, etc., is very limited, are apt to come to a wrong conclusion regarding the behaviour of ships among waves.

It is this apparently contradictory behaviour of ships which has given rise to so many theories on the subject. But it was not until the late Mr. Froude brought forward the now generally accepted wave theory that so much light has been thrown upon the subject. To discuss at length the theory of deep sea waves would form a volume in itself, and therefore lies outside the aim of a book such as this. Those wishing to pursue this branch of the subject can find ample information in the volumes of the Institute of Naval Architects, and also in the admirable works mentioned in the preface. We can, however, make a few brief observations, borrowing from the theory mentioned, such principles as may be of assistance to men of practical experience at sea, the class of men which it is the chief aim of this work to assist.

A feather in the air would fall in a straight line to the earth if there were no wind, owing to gravitation. Such fall, however, is always more or less overruled by the force of the wind when wind is blowing. Again, an iron plumb ball suspended from a cord, would hang vertically, if undisturbed, owing to the downward attraction of gravitation. On approaching it with a magnet sufficiently close to produce induced magnetism, gravitation is interfered with, and the iron ball seeks to follow the magnet. These instances are related simply to show that under certain circumstances the power of gravitation (such as causes the weight of a ship to act through its centre of gravity in a vertical line) can be over-governed by the introduction of other forces. Thus in waves we have what is termed *centrifugal force*, which, acting along with the gravitation force, gives a resultant force approximately perpendicular to the wave surface.* Turning to fig. 98,

---

* If centrifugal force were approximately perpendicular to the wave surface, the fluid pressures in waves could not possibly be perpendicular to the wave surface also, for gravity would show itself by producing a resultant which would certainly not be perpendicular to the existing wave surface, but considerably deflected from it ; but this is impossible, since the wave surface at any point, at any moment, is perpendicular to the instantaneous resultant of several forces, of which the centrifugal force is one.

Referring again to fig. 98, though the cylinder is on a wave slope, gravity still acts vertically through its centre of gravity G, and were gravity the only force exercising any influence upon the cylinder, it would cause the cylinder to slide down the wave slope, but this does not happen, for here again we find that the natural vertical force of the weight of the cylinder is interfered with

ROLLING. 155

We can now better understand how it happens that no righting lever was set up, the reason simply being that the lines of action

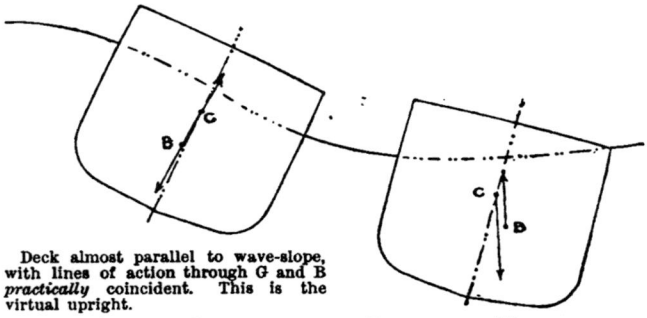

Deck almost parallel to wave-slope, with lines of action through G and B *practically* coincident. This is the virtual upright.

FIG. 99.—BEHAVIOUR OF A SHIP AMONG WAVES.

through the centres of buoyancy and gravity coincided, as shown by the arrowed line.

by the internal wave forces, producing a resultant *which is the virtual upright*, approximately perpendicular to the wave surface, and therefore parallel to the resultant buoyancy pressures in the wave. Thus an instantaneous position of equilibrium is set up without any tendency whatever for the cylinder to slide down the wave slope.

Where the water surface is horizontal, the buoyancy pressures act in upward vertical lines. This is as true for the smooth surface of the vast ocean as for the water in a bucket. And even when waves have been created, the upward pressures from the ocean depths are in nowise changed. But on coming to the actual waves themselves (which are only surface disturbances extending to a very small depth as compared with the depth of the ocean), we find that the buoyancy pressures are now exerted in lines of action approximately perpendicular to the wave surface.

It is not supposed that this theory, known as the "Trochoidal Wave Theory," covers the whole question of wave forms at sea. But it is at least a good working hypothesis for simple waves in very deep water, and has the advantage of covering all forms between the two "limits" of trochoids, viz., the cycloid and the straight line, the last being, of course, smooth water.

The centrifugal force is perpendicular to the wave surface at the crest, and acts directly in opposition to the universal gravity force, which we can never leave out of account. Whatever influence this may have in reducing the weight of the particles of wave water, it does not deflect them from the vertical line. At the wave hollow the centrifugal and gravity forces also act (this time together) in a vertical direction. In no other positions than wave crests and hollows does centrifugal force exert itself perpendicularly to the wave surface. In all other parts of the wave surface, its force acts more or less obliquely, and gravity, as it always does, acts vertically, the resultant of which two forces (together with any other less important yet possibly existing forces) is approximately square to the wave surface. And thus the original lines of buoyancy pressures, which were vertical in still water, are continually changing the direction of their lines of action in wave water. (See fig. 100.)

156 KNOW YOUR OWN SHIP.

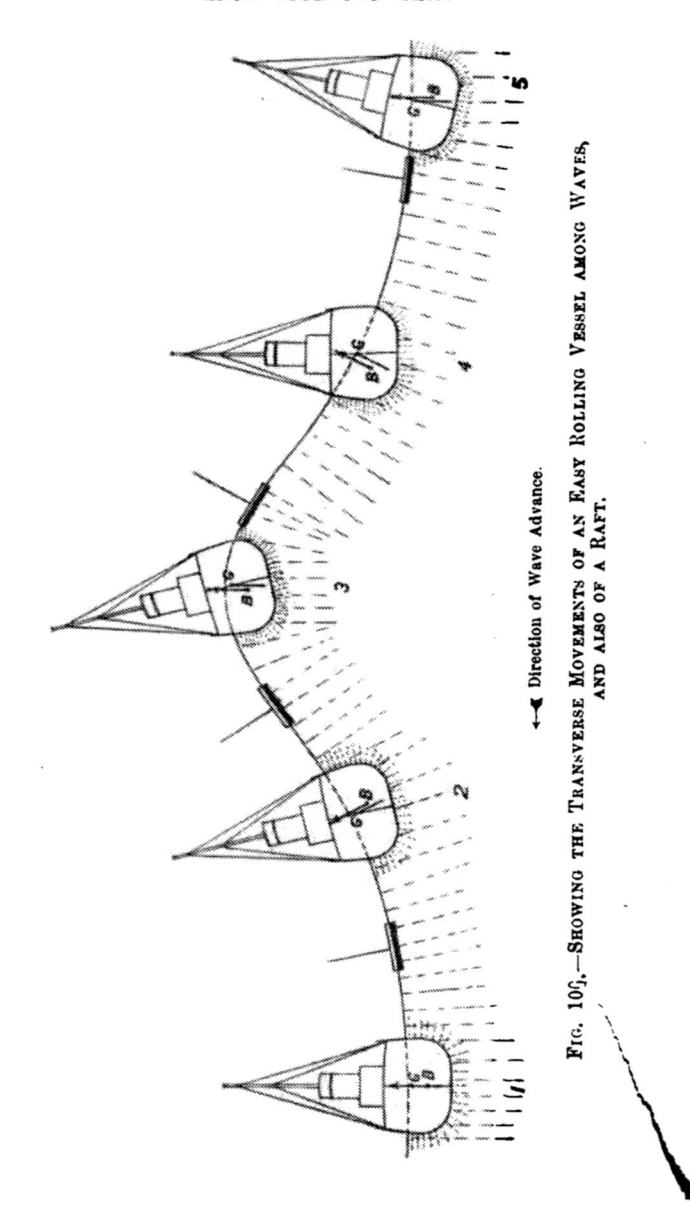

FIG. 106.—SHOWING THE TRANSVERSE MOVEMENTS OF AN EASY ROLLING VESSEL AMONG WAVES, AND ALSO OF A RAFT.

Direction of Wave Advance.

The short, dotted lines, perpendicular to the immersed surface of the vessel, indicate the water pressures. The longer dotted lines, approximately perpendicular to the wave surface, indicate the direction in which the action of buoyancy is exerted. These lines likewise indicate the *virtual* upright. When a vessel floats with its deck parallel to the wave surface, the line of action through the centre of buoyancy passes through the centre of gravity, and no righting lever exists under such circumstances (see first position in the figure, and also the first position in fig. 99).

Position 1 shows the vessel upright, when the hollow of the wave reaches her. The lines of action of gravity and buoyancy are vertical and coincide, and the vessel possesses no tendency whatever to incline to the one side or the other.

Position 2. Here the wave has advanced and the vessel is upon the slope. The direction of the buoyant action has changed, and the line through the centre of buoyancy does not pass through the centre of gravity. The righting lever now existing (the distance between the parallel lines passing through B and G) *tends* to bring the vessel's masts perpendicular to the wave surface.

Position 3. The vessel is now upon the wave crest. Here she is still lagging behind in her efforts to rear herself perpendicular to the wave surface, and instead of being upright she is considerably inclined. The distance between the parallel lines through B and G indicates the righting lever.

Position 4. Here the vessel is upon the other slope of the wave, having only succeeded in reaching the vertical position. Considerable righting lever exists, as shown, still tending to bring her perpendicular to the wave surface.

Position 5. The righting moment produced in position 4 creates a momentum, which, by the time the vessel reaches the wave hollow (position 5), has carried her beyond the vertical—in this case the perpendicular to the wave surface. The righting lever is again indicated.

The raft, being very stiff, and therefore much more rapid in its movements maintains a condition always perpendicular to the wave surface.

In dealing with a cylinder of no stability, we must not forget that the least external effect of wind or water washing over it might make it revolve; the only resistance offered to this would be the friction of the water on its immersed surface. In the instance we took as an example, we considered it as not affected by any external force, but simply under the influence of the unbroken wave water. Now a modern ship is neither like a raft nor a cylinder, yet it includes in some measure the qualities of both, and may approach either in behaviour.

Ship.—Let us continue our experiment, and, placing an actual ship among waves, watch her behaviour (figs. 99, 100). If she is very stiff indeed—that is, has great metacentric height, with her still-water rolling period less than half that of the waves she is among, she will act very similarly to the raft, which makes two complete rolls on a single wave. Supposing her to be floating in the upright position, immediately the base of the wave reaches her, she will at once seek to keep her masts perpendicular to the wave surface. As the wave passes under her, she will reach, or approximately reach, her greatest angle of inclination on the steepest part of the wave-slope; she will be upright at the summit, and again upright in the trough. She will, therefore, make two complete rolls in passing a complete wave (summit to summit). Her greatest angle will always occur approximately where the wave-slope is steepest. So that the danger in such a ship would lie not in capsizing, for she scarcely ever expends any of her stability, but owing to the rapidity of her movements, to shift the cargo, or strain her structure.

Synchronism.—But let us suppose that our vessel is tender, possessing a small metacentric height and long rolling period. When the wave reaches her and passes underneath, she will endeavour, as did the other ship, to rear herself perpendicularly to the wave surface. But we observed in our remarks upon rolling in still water that she moves slowly, and so she cannot keep up with the rapid motion of the wave, and falls behind. Thus, by the time the steepest part of the wave is under her, she is still at a considerable distance from that angle. Immediately that point is passed, the less inclination of the wave, as the summit is approached, checks the heeling influence, and at the summit the tendency is to bring her to the upright again. However, here she is yet lagging behind the wave, having still some inclination. When the other slope of the wave is reached she has possibly just reached the upright, and, before she can heel far under its inclining influence to the other side, the trough is reached, where the tendency is to bring her to the upright. We see then, that, by her slower movements, she lags behind the wave, and never reaches the angle of the greatest wave-slope, and at the summit and the trough is generally still inclined, not having reached the upright. Taking two such ships upon a single wave, the stiff vessel with the great metacentric height will always reach greater angles of inclination than the tender one with the small metacentric height, simply because, as we have shown, the stiff ship can better follow the angle of the wave; while the tender one of slower motion cannot reach the greatest angle of the wave on the one side of the slope, while, after the wave has passed beneath her, the other slope tends to push her back, and heel her to the opposite side. But, although the passage of the first wave may not have the effect of producing any great angle of inclination, owing to the usually slower movement of the ship in comparison with the speed of the waves, or more properly speaking, *the longer double roll period to the wave period*, it must be clear that a time may come when a ship may reach her greatest angle of inclination when the greatest angle of wave-slope reaches her. The result will then be, that a comparatively sudden additional impulse is given to the heeling of the vessel, and she will take an extraordinary, and what seamen have often called an unaccountable, lurch. Such a condition of waves and ship reaching their greatest angle of inclination at the same moment at regular intervals, is termed *synchronising*, or in other words, *keeping time*, and the effect is to produce considerably greater angles of inclination in the ship than the steepest wave-slope. The worst case is that where the period of a ship's single roll is half that of the wave period, as under such circumstances the impulse is given on each wave, and excessive rolling is naturally set up. This can be further illustrated by a simple pendulum (fig. 101).

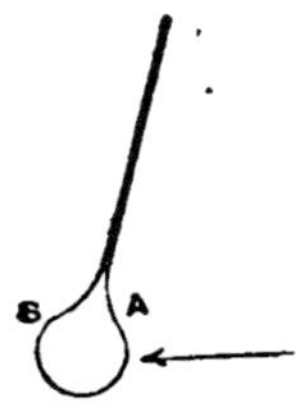

Let us imagine that the pendulum has just swung out to almost its greatest angle of inclination in the direction of the arrow. Suppose it receive a sudden impulse on the side B, it will naturally be checked, and commence its return to the vertical position. But suppose, on the other hand, the impulse had been given on the side A, at the moment the pendulum reaches its greatest angle, when there is neither return nor outward motion. The result is that a slight impulse will considerably increase the extent of its outward movement, and produce a greater angle from the vertical. This is

FIG. 101.—INFLUENCE OF EXTERNAL FORCES ON A SWINGING PENDULUM.

exactly what happens with a vessel whose period of roll synchronises with the wave period; if a sudden impulse be given near the extremity of her outward motion, a considerable augmenting of the angle of heel will result. The effect is bad enough when the synchronism occurs periodically—that is, with a series of waves—but when it happens on every oscillation, the effect is still more excessive, and the motion experienced by the vessel is rapid and jerky, with the greater probability of producing dangerous results. Stiff vessels with quick periods of about four to six seconds, would be the most likely to develop such behaviour. Vessels of longer or shorter periods may destroy synchronism altogether in most cases.

Were all sea waves of the same length, period, and height, it would be quite possible to design a warship or a yacht, whose equipped conditions are of an unvarying nature, to give a rolling period in still water which would produce great steadiness among waves.

But sea waves, at different times and places, vary greatly in length, period, height, and character. Atlantic storm waves reach 500 feet and over in length from crest to crest, with periods of 9, 10, or 11 seconds, and heights of 28 feet and over, while in other localities the length may not be more than 200 or 300 feet, with varying periods of 6 to 8 seconds, and height of about 12 feet.

Effect of Loading on Behaviour.—As waves, therefore, vary, according to the locality, the force of the wind, etc., it must be fairly clear that to design either a warship or a yacht to behave always in the same manner among waves is impossible, for although it is not likely that vessels with long rolling periods will be subject to heavy rolling, yet it is most probable that at some time they may fall in with waves which synchronise with their own period, and this inevitably produces heavy rolling. With merchant steamers the difficulty in producing steadiness is more marked than in any other case. In the first place, there is the difficulty, especially in coasting vessels, whose loading has to be rapidly conducted, with possibly part of the cargo arriving just before they sail, of obtaining a certain metacentric height which is known to have produced steadiness on a former occasion; or, if the metacentric height is the same, considerable difference may have taken place in the positions of the weights of the cargo, not vertically, but out on each side from the centre line of the ship.

This brings us to another very important point. While small metacentric height conduces to steadiness, the error must not be fallen into that this mode of procedure can always be carried out.

So long as the levers of stability at considerable angles be good, and the range satisfactory, such a method is all very well. But in tender vessels with short levers and short range, as seen by curves Nos. 11, 12, 6, and 7, fig. 87, such a method is extremely dangerous, for should synchronism be set up, they may take an excessive roll and capsize altogether, so that it is evident some vessels need more metacentric height than others, in order to ensure safety, even though it produces more lively motions among waves.*

Effect of Transverse Arrangement of Weight on Rolling Motion.—A safe method which can be adopted to assist in producing steadiness in such a case, is to wing out the heavy weights of the cargo, on each side of the vessel, without altering their position vertically. Such an arrangement of cargo will have a steadying effect upon lively transverse motions, and, on the other hand, concentrating the weights in the middle line of the vessel would tend to increase the rapidity of the transverse rolling.†

Alteration in Behaviour during a Voyage.—In ocean ships, whose loading is possibly not so hurried, or at any rate the nature of whose cargo is often understood beforehand, because it is all or nearly all alongside before commencing to load, it is certainly possible to so carry out this method when a knowledge of a ship's stability is understood, as to closely approximate to a particular metacentric height, and moreover to arrange the weights so as to be best fitted for steadiness.

Supposing we have secured a certain metacentric height which has produced great steadiness even in a heavy sea, it is sometimes found that this same vessel in a long, low ground swell of greater period, labours in a most extraordinary manner. Such is not an unknown experience to seamen, and the cause is simply due to the fact that the vessel has now fallen in with waves which synchronize with her own period. An instance bearing on this point was related to the author by a captain. Coming from the Mediterranean with a light cargo, he encountered heavy weather. His ship was naturally tender, and behaved splendidly for some time, but on approaching the Bay of Biscay,

---

* It may here be noted that it is utterly impossible to specify a metacentric height adapted to all vessels. For vessels in the Royal Navy, it varies from 1 foot to 12 feet. In steam yachts 2·5 feet is probably an average; in sailing ships 2 to 4 feet is common, and in passenger and cargo steamers it ranges from about 0·5 to about 3 feet. Shipbuilders with their wide experience of the various types of vessels are undoubtedly most capable of suggesting the best metacentric height for any particular condition of loading. The foregoing metacentric heights are *only* for loaded conditions.

† See *Appendix*.

a long heavy swell set in, and the ship began to roll so heavily, especially at periodic intervals, that he imagined she would capsize altogether. He immediately set to work and filled one of the water ballast tanks, with the result that the vessel completely altered in her behaviour, and again regained comparative steadiness. The ship had evidently fallen in with waves which synchronized with her own period, and caused the heavy rolling, but by filling a ballast tank the metacentric height was increased and the ship's period altered. Then, there being no longer synchronism, she steadied. Every seaman naturally learns from experience that where heavy rolling is suddenly set up, it can be modified by an alteration in the course or the speed. The reason is not far to seek. If the synchronism is produced by a beam sea, by changing the course more towards the waves, the apparent wave period is decreased, the crests being now passed more rapidly. By taking a course in an oblique direction, away from the direction at right angles to the waves, the wave period is increased, and in any other sea than one direct abeam, an increase in the speed without altering the course will decrease the wave period. Thus synchronism can be prevented either by altering the course or speed, and thereby altering the apparent period of the waves, or by altering the period of the ship through shifting weights in the ship.

Synchronism is not always produced by a beam sea, for the sea coming in an oblique direction may cause the vessel's period to synchronize with the wave period, when no such result would have happened with a beam sea, and therefore such large angles of roll could not have been experienced. How to obviate this has just been mentioned—by changing the course or speed. To attempt to alter the ship's period by filling the water ballast tanks when rolling heavily is by no means a safe experiment, for the moment of the free water dashing from side to side before the tank is filled, may add to the angle of heel, instead of reducing it. We must now be able to see that great stability is not the best condition for a ship, for it will either make her movements exceedingly rapid, following, as in the case of the raft, the wave-slope, or, if not so stiff, tending to produce synchronism, with consequent heavy rolling.

The Metacentric Height.—The *best vessel* is undoubtedly the one with *moderate metacentric height, good levers of stability at considerable angles of inclination,* and *good range.* She will probably thus be slow in her period, easy in her movements, and when not subject to synchronism (which she is less likely to be) will be comparatively steady among waves.

To secure steadiness at the cost of small metacentric height, with short levers and range of stability, would only make disaster more probable. This is the very reason why a ship's officer should possess stability curves of his vessel in the various conditions under which she is likely to proceed at sea, from which he will undoubtedly be more able to intelligently manœuvre the condition of his ship in order to produce seaworthiness.

Could we imagine a vessel rolling among waves unresistedly (that is, without being subject to resistance from wind, immersed surface, or keel resistance of any sort), whose own period synchronized with that of the waves, the effect would be that the continued impulses given by the synchronizing waves would eventually capsize her, whatever might be her stability, just as a child's swing pushed synchronically would at last overset. These resistances have the same effect among wave water as when rolling in still water. There is one important point, however, to be observed, and that is, the more rapid the motions of the vessel the more resistance is offered. And thus, upon a vessel whose period synchronizes with the wave period, when she begins to attain large angles of heel and great rapidity of motion, the various resistances grow in proportion until a point may be reached where the effect of these resistances is just sufficient to prevent greater oscillation being attained, and capsizing is also averted. Thus, where synchronism produces great angles of oscillation, it does not follow that the ship will capsize, except in unusual cases where the range of stability is very short.

The great and important value of bilge keels in offering resistance and reducing rolling, has already been shown from Mr Froude's experiments. These will produce good results upon large ships, but the effect is still more apparent upon small vessels of quick period.

Fore and Aft Motions.—Thus far our remarks have been confined entirely to transverse stability and behaviour relatively to transverse motions at sea, simply because that it is in these directions that danger is most likely to occur. Could we heel our ship in every possible way it would be found that she possessed least stiffness or stability when inclined transversely than in any other direction, and that her transverse metacentric height is the smallest possible. Thus, on heeling in any skew direction, more stability is developed, and most of all when inclined longitudinally (*pitching*). In ordinary types of vessels it is, therefore, only possible for them to capsize transversely, unless it happens that, through damage and the admission of water, the loss of buoyancy at either end is so great as to cause the vessel to go down by head or stern, as the case may be.

This explains why it is only necessary to be provided with

curves of transverse stability, and to be thoroughly aware of the vessel's condition in this respect. The greatness of longitudinal metacentric height will be obvious when it is observed that the moment of inertia of the waterplane about a transverse axis must be immensely increased beyond that for a longitudinal axis, simply because what was formerly considered as beam in the formula for the transverse metacentre becomes length, and the length becomes beam.

The principles which govern transverse behaviour apply in a similar manner when considering the longitudinal motions of a ship, though here again the design may exercise much influence in the production of objectionable qualities in behaviour. For example, take an ill-designed vessel considerably full on the load waterline aft, but fined away forward with sides almost vertical to the gunwale. Such a vessel will be admirably adapted for diving into the sea and shipping huge volumes of water on her deck, with her stern probably high and dry. However, it must not by any means be inferred that the load waterline forward should be bluff or even identical with the after end, but certainly where any degree of comfort is desired, there should be some reasonable approach to equality. A vessel, very fine under water, may be considerably improved by giving her reasonable *flam* * or flare above the waterline, the additional buoyancy produced by which forms a valuable check upon diving.

But we have also noticed that winging out the weights transversely from the centre line produces slower rolling motion, that concentrating them on the centre line creates greater liveliness, and, moreover, that the latter result is always produced by a large metacentric height, and steadiness by a moderate metacentric height.

Fore and Aft Arrangement of Weights.—With the enormous longitudinal metacentric height—that is to say, length metacentric height, not beam metacentric height—possessed by most vessels, it is impossible to make any visible effect upon the longitudinal motions through this agent, for even were it possible to reduce or increase it by a few feet, the comparative difference would be exceedingly slight. Moreover, such alteration in the position of the centre of gravity might seriously imperil the safety of the vessel transversely. Thus, the only alternative is to influence longitudinal motion by a proper adjustment of the heavy weights of the cargo in a fore and aft direction. If liveliness is required—that is, quick rising motion—they should

* By *flam* is meant exactly the opposite to "tumble home." It is most noticeable at the bow of a ship, where her sides slant outwards, greatly increasing her beam above the load waterline.

be stowed nearest to midships; if a slower movement is required they should be spread out longitudinally (that is to say, more fore and aft). But it must not be forgotten that placing heavy weights at the extremities of a vessel has the tendency to excessively strain the structure when subject to the varying support of wave water, and also to some extent when lying at rest in still water.

CHAPTER VI. (Section III.)

BALLASTING.

Contents.—Similar Metacentric Heights at Different Draughts—Wind Pressure—Amount and Arrangement of Ballast—Means to Prevent Shifting of Ballast—Water Ballast—Trimming Tanks—Inadaptability of Double Bottom Tanks alone to Provide an Efficient Means of Ballasting—Considerations upon the Height of the Transverse Metacentre between the Light and Load Draughts, and Effect upon Stability in Ballast—Unmanageableness in Ballast—Minimum Draught in Ballast—Arrangement of Ballast.

Ballasting.—The number of losses and disasters happening annually, not only to old, but often to fine new ships when in ballast, abundantly proves that something is wrong. This is all the more manifest from the random way in which ballast is often thrown into a ship. One man *considers* 400 tons sufficient, and another, 800 tons for the same ship, and all pitched into the hold. Both cannot be right, since both methods cannot produce similar results. One is either dangerously stiff, conducing to heavy rolling and tending to shift the ballast, or the other is too tender with too small righting moments.

Before ballasting can be intelligently carried out, it is necessary that a few important facts be kept in mind.

1. That *metacentric height alone is no guarantee* for a vessel's stability.
2. That *freeboard alone is no safeguard*.
3. That although a certain metacentric height on one occasion may be very good for a vessel at a particular draught, the same metacentric height would be unsafe at a different draught, and even *if* it were possible to get the same lengths of righting levers at a certain angle of inclination at light and load draughts, the righting moments in each case would be immensely different.

Similar Metacentric Heights at Different Draughts.—Reference to curves Nos. 5 and 18 (fig. 87) will considerably help in illustrating these points. Curve 5 is for a box vessel 100 feet long, 30 feet broad, 10 feet draught, and 5 feet freeboard in the load condition, with a metacentric height of 6·2 feet, the centre of gravity being 6·3 feet from the bottom of the box. Curve 18 is for a box vessel 100 feet long, 30 feet broad,

4 feet draught, and 16 feet freeboard in the light condition, with a metacentric height of 6·2 feet, the centre of gravity being 14·5 feet from the bottom of the box.

Taking metacentric height and freeboard as the only guides, the latter vessel should have by far the greatest stability. A comparison of the curves contradicts such a conclusion, and shows that the higher the centre of gravity is with a certain metacentric height and freeboard, the smaller will the angle be at which the vertical line through the centre of buoyancy intersects the centre line of the ship below the centre of gravity, hence the increased range of curve No. 5. Fig. 90 is the stability curve for a sailing barque in the light condition, with 112 tons of ballast aboard. The length is 270 feet, the breadth 41 feet, and the freeboard 17½ feet, with a metacentric height of 2·9 feet and a displacement of 1390 tons. The maximum lever of stability is 0·69 at 18° of inclination, and the righting moment 1390 × 0·69 = 959 foot-tons. Moreover, the stability vanishes altogether at the comparatively small angle of 34°.

Wind Pressure.—Both the maximum lever and range are exceedingly small for a heavily-rigged vessel with large sail area, and the effect of a sudden squall of wind with much sail set is easily perceived. In the loaded condition, however, with only 5½ feet freeboard, 4000 tons displacement, and a much lower centre of gravity with the same metacentric height, this same vessel would have longer levers and much greater moment, as well as greater range of stability.

Amount and Arrangement of Ballast.—Now let us take a practical view of the process of ballasting a ship, and suppose that as master we are told by the naval architect or shipbuilder that a metacentric height of 3 feet in the loaded condition, which gives a displacement of 4000 tons, will put our ship in an excellently seaworthy condition.

In the light condition, however, by placing 500 tons of ballast in the hold, the same metacentric height is secured with a total displacement of 2000 tons.

Heeled to an angle of 10°, the righting lever will be—

$$G M \times \text{sine of angle} = 3 \times 0.1736 = 0.52 \text{ foot.}$$

As regards length of lever at this angle of inclination, the vessels are practically identical at both these draughts. The righting moment, however, is lever multiplied by displacement. Thus, at the load draught the righting moment is 0·52 × 4000 = 2080 foot-tons ; at the light draught the righting moment is 0·52 × 2000 = 1040 foot-tons—only one-half the loaded righting moment. It is clear, then, that with equal sail area and equal wind pressure, the vessel in the light condition would heel to a much greater angle than in the load condition; moreover, the effect of a sudden squall of wind will produce about double the angle of inclination which would otherwise be reached in steady heeling. There is always greatest motion at and near the surface of wave water, so that the lighter the vessel, the more on the surface she will float and be subject to the influence of waves and wind. Possessing great stiffness under such conditions, the more excessive will be the rolling. Righting lever alone, then, does not provide righting moment, but lever multiplied by displacement. To get moment without excessive metacentric height, there is no alternative but to considerably immerse the vessel in order to get displacement.

Understanding this, we proceed to put ballast into our vessel. Supposing the ballast to be sand, it would probably be found that if it were all poured into the bottom of the hold, by the time the ship was sufficiently immersed excessive stiffness would be set up.

This method, therefore, cannot be adopted. We know that to reduce metacentric height, low weights must be raised so as to lessen the distance between the metacentre and the centre of gravity. Part of the ballast, therefore, would require to be carried in the 'tween decks. But here, again, a difficulty arises in many cases, where a vessel with good beam, and a depth to require two tiers of beams instead of having a laid deck on the lower tier, this lower tier is made extra strong and the beams are widely spaced, making it impossible to carry ballast higher than the hold. What is to be done? Very often nothing is done, and only one or two conclusions can be arrived at, either the expense of making provision for efficient ballasting is considered too much for some owners whose vessels are amply covered by insurance, or else out of pure ignorance of the mode of ballasting to ensure safety, this subject receives no consideration. One thing which could be done under such circumstances to produce excellent results, would be to build two tanks at the middle of the length of the vessel, one on each side, between the hold beams and upper deck beams, to contain, say, about 50 tons each, or, altogether, 100 tons. Each tank would, therefore, require to be about 30 feet long, $8\frac{1}{3}$ feet broad, and 7 feet deep. Especially if the ship were fitted with water ballast tanks in the bottom, the size of these upper tanks could be fixed to a nicety; but in any case, the shipbuilder could supply the information as to the exact amount of hold ballast to be used.

This would reduce the stiffness by raising the centre of gravity, but furthermore, having these weights "winged" out to the ship's sides, would still more conduce to steadiness. The application is, therefore, twofold.

This method entails the expense of the plating, additional beams, and pillars for supporting the tanks. Expense is always objectionable, but there is the choice between possible and probable loss of life through ignorance or carelessness in ballasting, not to mention the ship; and the comparatively small additional cost upon the vessel while building. When loading cargo, and these ballast tanks are not required, the space could conveniently be used for cargo also, if a hatch be made on the deck above them.

The same idea could be carried out by constructing these ballast spaces of wood battens, instead of iron plating, and using earth or sand ballast. The former method is, however, preferable, and economical, since the water could be run out by means of a cock on the ship's side.

Means to Prevent Shifting of Ballast.—The other great point in ballasting ships is to see that the ballast is secured so as to render the likelihood of disaster from shifting impossible. If it be water ballast confined in a tank, it is all right if the tanks are full, for it must be remembered, as will be pointed out in the remarks upon "water in the interior of a vessel," free water may create a list if the vessel is inclined to be tender. But supposing the ballast to be sand in the hold, great precaution should be taken to make it immovable as far as possible. The value of shifting boards as applied to cargo applies equally to this also, for, after all, ballasting is just a form of loading.

Another method sometimes adopted is to cover the surface of the ballast with boards and shore them down. This is all well enough if the covering and shoring is thoroughly carried out, rendering no possibility of any ballast shifting, or finding its way between the boards or uncovered spaces, for where such is possible the precaution is useless, as the ballast will all the more readily and easily relieve itself from its confinement.

Water Ballast.—Among a host of important considerations in designing a vessel, draught is one which ever demands careful attention. Limitations upon the depths of dock entrances, harbours, rivers, etc., have in turn fixed the limits of depth and draught for vessels. And thus, in many cases, while owners have vastly increased the size of their vessels to carry a greater deadweight, the length and breadth almost entirely have furnished the additional capacity, little alteration having taken place in the depth.

While earth, sand, or stone possesses certain advantages as agents in ballasting, the cost entailed in loading and discharging hundreds of tons of these materials often causes both serious loss of time and expense. The advantage they possess is that they can be laid wherever desired in order to secure a certain condition of ballasting—on the deck, in the 'tween decks, or in the hold with perhaps, in some cases, precautions to prevent shifting as previously pointed out. But with the adoption of iron for the construction of ships, it was found that the bottom of the vessel could be constructed so that while providing adequate and ample strength, it might form at the same time a most convenient and economical means of carrying water for ballast, for trimming purposes, and for adjusting the stability under certain conditions, not to mention the further important advantage of providing an inner bottom, which has proved the salvation of many a ship, when, by some means or other, the outer bottom has been so damaged as to admit water. No doubt the division of this watertight space into separate watertight compartments gave to them the more correct name of Trimming Tanks. For this purpose they are admirably adapted, and may be fitted into almost any type of cargo or passenger vessel with decided advantage. But when these trimming tanks are used apparently with the expectation that they will also serve as an efficient means of ballasting, one hesitates before calling them a universal success.

In these days of fluctuating freights, many vessels, especially of the tramp class, find themselves without cargoes. To proceed to sea with an empty tramp would be to court disaster. As seamen know from experience, the enormous freeboard, slight immersion, and sometimes deficient stability, would make these vessels both extremely dangerous and utterly unmanageable in bad weather. Ballast, therefore, becomes an absolute necessity. And the tempting convenience of double bottom tanks has caused them to be so widely adopted that their primary use has come to be for ballasting purposes. A variety of forms of double bottom ballast tanks have been built, but the system now generally adopted is that known as the cellular double bottom. See fig. 55.

Inadaptability of Double Bottom Tanks to Provide an Efficient Means of Ballasting.—And yet after we have got this apparently commendable system of ballasting fitted into the modern tramp, to whose lot it oftenest falls to do long, and even Atlantic runs, in ballast, together with possibly a few hundred tons of coal in bunkers, officers on board many of these vessels complain of the miserable existence they endure in bad weather, owing to the heavy rolling, disregard of helm, and general unmanageableness of their ships. Something must be wrong somewhere, and as this double bottom tank does not produce the results required, we had better examine the basis upon which it finds its way into a ship.

Taking the specification of an ordinary cargo vessel which has to be built to Class 100 A1 at Lloyd's, we are pretty sure in most cases to find the phrase "cellular double bottom all fore and aft for water ballast to Lloyd's requirements." But what does "Lloyd's requirements" mean? Requirements for ballasting? No. Ballasting lies entirely outside their province. Neither Lloyd's nor any other of the societies for the registry and classification of vessels, have any requirements for ballasting. Ballasting is a feature in the ship design, and the owner, or whomsoever he appoints to design his vessel, is solely responsible for the ballasting arrangements. The owner may have practically whatever design of vessel he likes, with whatever arrangement for ballasting he chooses, such as, cellular double bottoms extending fore and aft, or, through part of the length; deep tanks; peak tanks; hold, or part of hold spaces; bunker spaces, etc.; providing that, so long as his vessel is built to the standards of strength fixed by these societies, and the spaces intended for the carriage of water as ballast are structurally to their satisfaction, the vessel will be classed with a minimum freeboard.

A vessel may be built without any arrangement whatever for ballasting purposes, and owing to her particular mode of construction, it might be absolutely impossible to carry water as ballast, yet she may still be perfectly eligible for the highest class, and fully satisfy the requirements of the Board of Trade. And when a double bottom system of construction is adopted, it is never asserted that such space is adequate and properly adjusted for the efficient ballasting of a ship, but it is simply offered as a means of carrying water for ballast or for trimming purposes, or for fresh water for the boilers, or for whatever purpose owners may find it useful.

It would have been unnecessary to go into this detail were it not for the fact that it is evidently very often and absurdly assumed that because the double bottom is built to Lloyd's requirements, the arrangement for ballasting is both proper and sufficient. This is proved by the fact that large numbers of vessels are constantly proceeding to sea dependent only upon this ballast, together with a greater or less amount of bunker coal on board. We have only to observe the method followed in, say, Lloyd's rules, for arriving at the dimensions of cellular double-bottom ballast tanks, to see that such could never be intended as a competent mode of ballasting the various types and differently proportioned cargo vessels.

Let us take an ordinary flush deck type of tramp steamer with poop, bridge, and forecastle, built of steel to the three-deck rule, and Classed 100 A1 at Lloyd's. The dimensions are: length, 350 ft.; breadth, mld., 45 ft.; depth, mld., 29 ft.

In order to find the particulars for the construction of the ballast tank, we are instructed in Lloyd's rules to proceed as follows :—

Find the 2nd numeral for scantling, as described on page 63. For this vessel it is 32,506. Turning to Lloyd's rules we find in a table relating to double bottoms a graduated list of similar numerals. Tracing down this table, we arrive at the particulars for tanks for vessels whose numerals are between 28,000 and 33,000. Here we see that the depth of the centre girder is 3' 6", and the minimum breadth of the tank side is 2' 4".

The depth of the centre girder, or centre keelson, being determined—which, by the way, is one of the most important items in the structural strength of the bottom of the ship, since it regulates the depth of the floors, and, to a large extent, the other side girders in the tank also—the plating for the inner bottom is laid over the top of these, and riveted, caulked, and made watertight. And thus we have our ballast tank to Lloyd's requirements.

In fig. 102, curve 1 is for a modern beamy cargo vessel and curve 2 for a vessel of the older and narrower type.

We observe, that at light draughts the metacentre is at the greatest heights above the keel, simply because the extreme fulness of the bottom of especially the beamy class of vessels causes them to float at comparatively very light draughts with very large moment of inertia of the waterplane in comparison with the displacement at the same draught. It is found, however, that by increasing the displacement the metacentre rapidly lowers, the explanation being that the successive waterplanes above the light draught increase very slightly in fulness, and consequently the moment of inertia also is only slightly increased. The displacement, on the other hand, increases much more rapidly in proportion, and thus :—

$$\frac{\text{Slightly increased Moment of Inertia}}{\text{Greatly increased displacement in cubic feet}} = \text{a very much}$$

lower metacentre, notwithstanding the fact that with every increase in draught, the centre of buoyancy has risen, and tended to keep up the metacentre.

The downward tendency of the metacentre continues until, as the vessel approaches her load draught, the ratio between the moment of inertia of the lower waterplanes and the displacement, and the moment of inertia and the displacement at the upper waterplanes, has so altered, that the steady rising of the centre of buoyancy has at last the effect of causing the metacentre to take an upward course again (*see* curves 1 and 2). Here, then, in these natural movements of the metacentre it seems we have the secret of correct ballasting.

BALLASTING.   173

With the deductions and inferences we have made from these simple considerations of the metacentre, we shall now revert to our cellular double-bottom tanks, and examine the complaints lodged against vessels ballasted by means of them only.

First, then, many of them are heavy rollers, and cause extreme

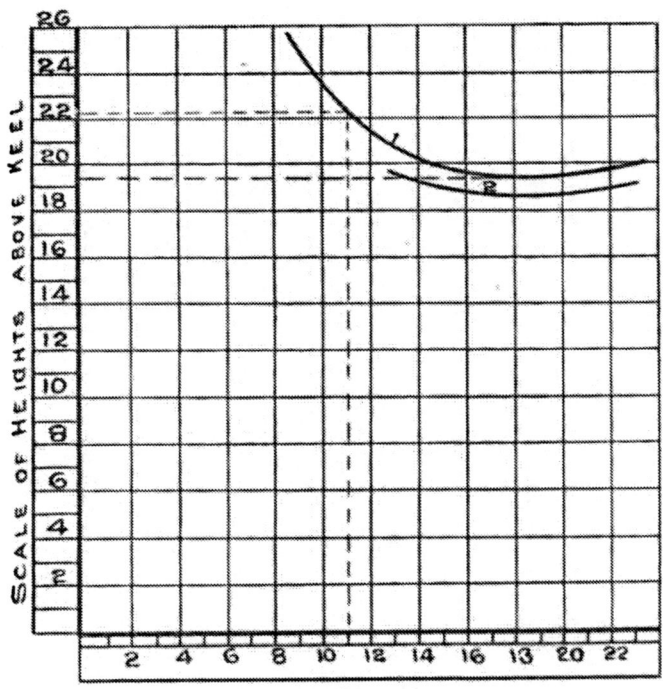

[Graph with vertical axis labeled "SCALE OF HEIGHTS ABOVE KEEL" marked from 2 to 26, and horizontal axis labeled "SCALE OF DRAUGHTS" marked from 2 to 22, showing two curves labeled 1 and 2.]

SCALE OF DRAUGHTS.

No. 1 Steamer, 360′ × 47′ Mld. × 28′ Mld. Light draught, 8′ 6″.
Metacentre, 25′ 8″ above keel.
No. 2 Steamer, 376′ × 43′ Mld. × 29′ 1″ Mld. Light draught, 12′ 11″.
Metacentre, 19′ 7½″ above keel.

FIG. 102.—CURVES OF TRANSVERSE METACENTRES.

discomfort to those on board, in addition to severe straining to the vessels themselves. Some of the narrower types of cargo vessels (notably older ones), have extremely little metacentric height, and in some cases actually a negative metacentric height in the light condition. (Compare heights of metacentre in light condition, Nos. 1 and 2, fig. 102.) By filling the double-bottom ballast

174   KNOW YOUR OWN SHIP.

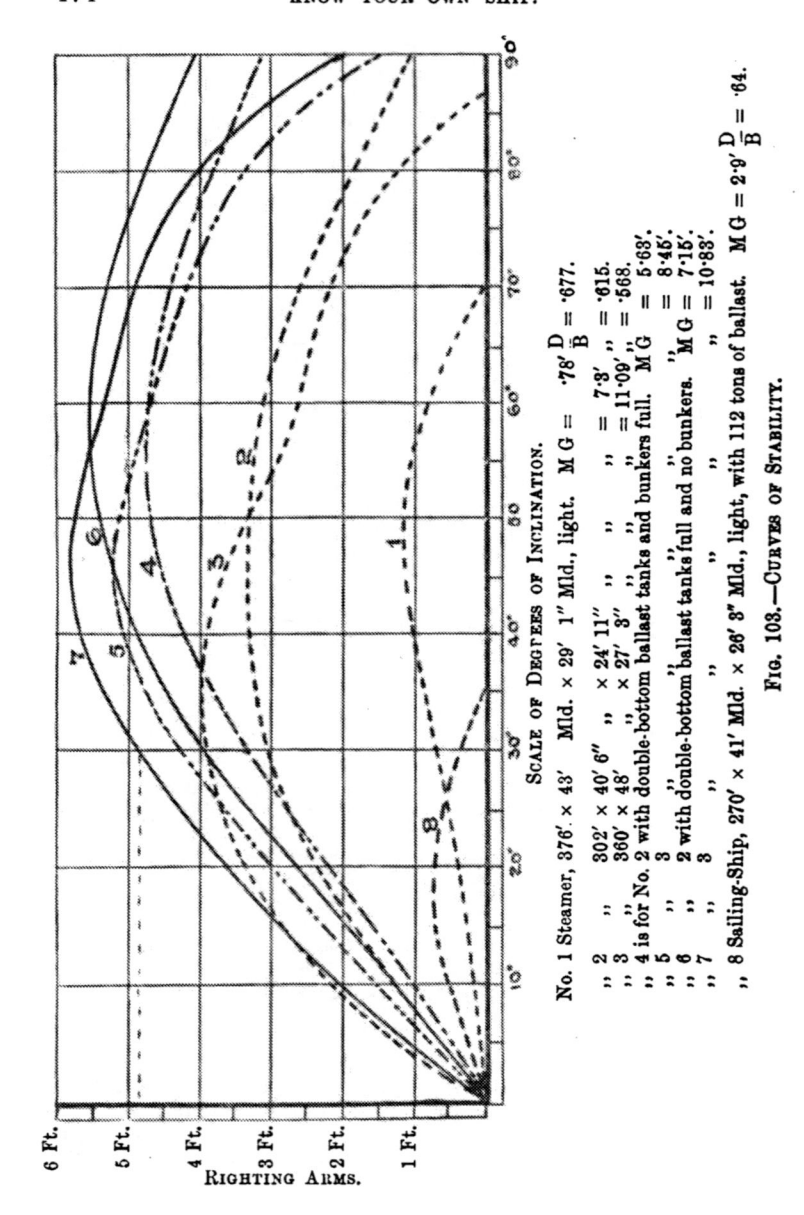

No. 1 Steamer, 376′ × 43′ Mld. × 29′ 1″ Mld., light. M G = ·78′ $\frac{D}{B}$ = ·677.
,, 2   ,,   302′ × 40′ 6″ ,, × 24′ 11″ ,, ,,   = 7·3′, ,, = ·615.
,, 3   ,,   360′ × 48′   ,, × 27′ 3″ ,, ,,   = 11·09′, ,, = ·568.
,, 4 is for No. 2 with double-bottom ballast tanks and bunkers full. M G = 5·63′.
,, 5   ,,   3   ,,   ,,   ,,   ,,   ,,   ,,   = 8·45′.
,, 6   ,,   2 with double-bottom ballast tanks full and no bunkers. ,, M G = 7·15′.
,, 7   ,,   3   ,,   ,,   ,,   ,,   ,,   = 10·83′.
,, 8 Sailing-Ship, 270′ × 41′ Mld. × 26′ 3″ Mld., light, with 112 tons of ballast. M G = 2·9′ $\frac{D}{B}$ = ·64.

FIG. 103.—CURVES OF STABILITY.

tanks, a decidedly positive metacentric height is obtained, which possibly is not excessive, and these vessels are known as steadier and better behaved at sea. But in recent years, considerable additions, amounting to 2, 3, 4 feet, and more, have been made to the beams of vessels with the corresponding increase in the proportion of beam to depth and draught. The result is, that though in both narrow and beamy vessels the centre of gravity may be at the same height above the keel, the metacentre of the broad vessel will be higher, and the metacentric height will be greater, and some of these vessels are actually stiff and stand upright in perfect safety in a light condition. When proceeding to sea light, the ballast tanks are filled; the centre of gravity is thus considerably lowered, and the naturally large metacentric height is slightly, if at all, diminished. It will be remembered that the metacentre lowers with increasing draught more rapidly than in narrower vessels. Fig. 102 illustrates such a comparison. Many of these modern beamy vessels are abnormally stiff, and as a result earn the reputation of being unmistakably lively.

These remarks upon the metacentre may be further illustrated by referring to fig. 103. Here we have a number of curves of stability showing righting arms up to 90° of inclination.

Curve 1 is for a steamer, 376′ × 43′ × 29′ 1″, floating empty with ·78 of a foot metacentric height. This may be taken as a fair representative of the narrower type, the proportion of depth to breadth being ·677. Unfortunately, we have no curve showing the stability of the vessel in ballast. But probably with ballast tanks and bunkers full, she would have approximately 2′ 6″ metacentric height which, judging by the curve for the light condition, would provide ample righting moment at large angles of inclination.

With this curve compare No. 2, which is for a vessel 302′ × 40′ 6″ × 24′ 11″. Here the proportion of depth to breadth is ·615, and the metacentric height 7·3 feet empty, and also compare No. 3, which is for a vessel 360′ × 48′ × 27′ 3″. The proportion of depth to breadth is ·568, and the metacentric height 11·09 feet, empty. The curves for these two vessels are fair representatives of a large number of the more modern beamy cargo vessels. With ballast tanks and bunkers full these two latter vessels have 5·63 and 8·45 feet metacentric height respectively (*see* curves 4 and 5), and with bunker coal consumed and ballast tanks full, 7·15 and 10·83 feet metacentric height respectively. (*See* curves 6 and 7.) It is scarcely necessary to say that such metacentric heights as these indicate enormous stiffness, the effect of which quality is well known to seamen. As already stated, 7·3 feet is the metacentric height for the 302 feet vessel light, and yet after the double-bottom ballast tanks have been filled, and this large weight placed in the lowest possible position, the metacentric height is 7·15 feet, actually less than for the light condition.

It is evident then, that, though the centre of gravity must have lowered considerably, the metacentre has lowered a still greater distance. With both ballast tanks and bunkers full, the metacentric height is only 5·63 feet, owing probably to the fact that, the bunkers being situated considerably higher than the water ballast, the centre of gravity has not lowered very much, if at all, while the metacentre, within these limits of draught, continues to descend very rapidly for beamy vessels. These remarks apply also to the 360 feet vessel, and an examination of the metacentric heights for corresponding conditions will show similar results.

A much more marked difference is found in dealing in like manner with the 376 feet vessel. To begin with, the metacentre is within a foot or so of its lowest position at the light draught (*see* curve 2, fig. 102), and the filling of ballast tanks could not fail to lower the centre of gravity to such an extent as to produce a greater metacentric height. As already approximated, the metacentric height would be about 2·5 feet, and the curve of stability would rise much more steeply and produce longer righting arms.

We may also notice here that though the metacentric heights and freeboards of both the 302 and the 360 feet vessels are reduced by the filling of ballast tanks and bunkers, yet, owing to the lowered centre of gravity, the righting arms are less for small angles of inclination, as indicated by the reduced metacentric height, and greater at larger angles of inclination.

Another complaint, especially against tramps in ordinary ballast, is that of unmanageableness. No one would doubt the validity of such an accusation against a ship sent to sea light without ballast. But when properly ballasted surely a better state of affairs ought to exist. As an example upon which to work, let us take a cargo steamer whose deadweight capacity is 5800 tons. When fully loaded, the freeboard is 5 feet. The cellular double bottom tank has capacity for 900 tons of water, which is only about one-sixth of the total deadweight. To this, say, 300 tons of coal are put on board for bunker use. This gives a total of 1200 tons of deadweight on board in ballast sea-going condition, one-fifth of the maximum deadweight. The draught is now 11' 6" on even keel, against 23' 0" in loaded condition, and the freeboard is 16' 6" against 5' 0". The propeller is slightly more than one-half immersed in still water against more than total immersion when at load draught. Assuming her to be an ordinary full type cargo vessel with a 10-knot speed and considerable stiffness in ballast, she both rolls and pitches. The propeller in the best condition is only partly immersed, and during her lively movements is subject to intervals of immersion and emersion. In her pitching movements, her often massive box-shaped fore end lends little assistance in making headway, but is continually thumping against walls of water, and the vessel actually experiences greater head resistance comparatively than she does in the fully loaded condition. There is, moreover, less expenditure of propeller power in actually driving her. There is the sudden shock produced by the propeller blades striking the water after racing in mid-air, which is no doubt the cause of what has become quite a frequent occurrence, viz., the loss of propellers and the breaking of shafts. The rudder also adds its decreased efficiency through decreased speed and periods of emersion. Moreover, the huge freeboard, together with all erections—poops, bridges, forecastles, deckhouses, etc.,—exposed to the force of a gale, with little or no keel resistance where a flat plate keel is used, introduces more or less leeway into the category of grievances, and where, as in some cases, the speed has become an unknown quantity, actual drifting is a consequent result. To reduce part of these ill features, after peak tanks have been fitted in many vessels. By this means both propeller and rudder are kept at greater immersion, but while the mean draught may have been increased an inch or two, the fore end of the vessel has suffered in emersion, and in attempting to steam with the wind abeam, the effect in causing the vessel's head to fall off will be obvious.

Every seaman knows that the cure for excessive stiffness (large metacentric height) is to raise weight already on board, or add top weight, and thus raise the centre of gravity. Suppose the first of these methods to be adopted. Instead of having a cellular double-bottom tank throughout the length, let it only extend over part of the bottom, and let the difference of the weight be placed, say, in a part of the 'tween decks arranged for water ballast. By such means we can arrive at practically whatever metacentric height we desire. But this is sometimes rather a dangerous experiment to make, and, as already pointed out, more especially so in sailing-ships, for in securing what may appear a desirable metacentric height we may rob our vessel of stability most seriously, when she heels to considerable angles of inclination.

Going back to our curve of metacentres, No. 1, fig. 102, we observed that the metacentre is very high at light draughts. To attempt to approach such heights of metacentre with the centre of gravity by raising ballast in order to make the ship easy, would only increase the possibility of disaster. Every upward movement of the centre of gravity shortens both the lengths of the righting arms and the extent of the range of stability. So that by the time we have so raised the ballast as to produce a moderate and desirable metacentric height, it is possible that the result is an unseaworthy ship, that is, a ship with a good metacentric height under other conditions, but with too little reserve stability at possible angles of inclination. For be it remembered that most carefully designed vessels may at times fall in with such a condition of sea as to produce lurching to considerable angles of inclination. It will easily be seen, then, how dangerous it might be in the case of a beamy ship, if small metacentric height were procured at the cost of decreased range and righting arms (fig. 104). It is therefore evident, particularly in the case of beamy ships floating at light draughts, with the usual amount of water ballast, that there is no choice, in order to ensure safety, but to accept larger metacentric height than required at load draught, which, moreover, cannot be averted by the ordinary double-bottom ballast arrangement, and in a spirit of resignation to put up with the rolling and other accompanying consequences.

To produce safe and desirable results more ballast than is provided by the usual double bottom is required, not necessarily to either raise or lower the centre of gravity to any great extent, if at all, but rather to further immerse the vessel. Why? Simply because, as we have previously seen, by increasing the displacement, the metacentre rapidly lowers, and by this manipulation of the metacentre, we are relieved from indulging in any dangerous experiment of raising the centre of gravity to a height such as might produce disastrous results under the effect of heavy rolling or lurching.

By placing more ballast on board (increasing displacement), a safe compromise is effected between the metacentre and the centre of gravity. Moderate metacentric height with the centre of gravity in the lowest possible position will give the best results, for by this means a slow and easy rolling period may be obtained, combined with ample righting moment at large angles of inclination and sufficient range. This desirable condition could never be attained by the usual method of only building tanks along the bottom, varying from 3 to 4 feet in depth, as per Lloyd's rules. Nor yet could it be accomplished if these tanks were increased to twice their depth, which on no account could be advisable. For though the more than doubled quantity of ballast would have the effect of increasing the draught and the displacement, and thus inevitably bringing down the metacentre, no compromise has been made on the part of the centre of gravity, it having receded into a lower position.

That more ballast is required to make both behaviour and manageableness more satisfactory features many owners are

BALLASTING. 179

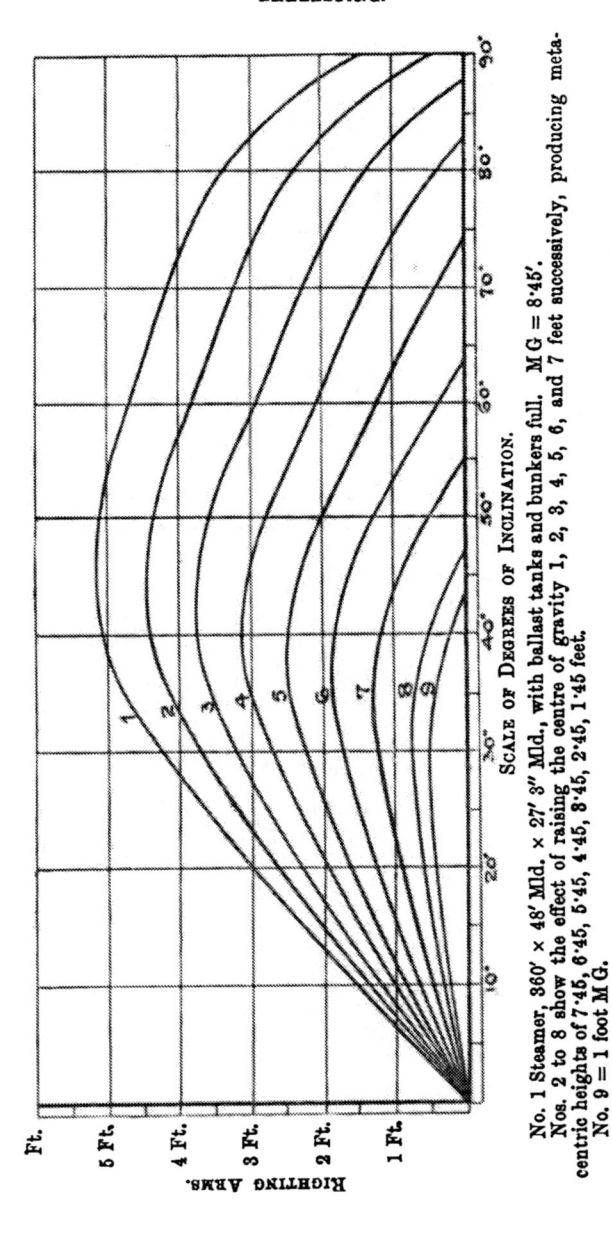

FIG. 104.—CURVES OF STABILITY SHOWING EFFECT OF RAISING CENTRE OF GRAVITY.

No. 1 Steamer, 360' × 48' Mld. × 27' 3" Mld., with ballast tanks and bunkers full. M G = 8'45'.
Nos. 2 to 8 show the effect of raising the centre of gravity 1, 2, 3, 4, 5, 6, and 7 feet successively, producing metacentric heights of 7·45, 6·45, 5·45, 4·45, 3·45, 2·45, 1·45 feet.
No. 9 = 1 foot M G.

perfectly aware, and in their specifications for new vessels they have stipulated that deep tanks of some sort should be fitted in some particular position in their ships. But even these have not in all cases given such results as were expected. It has been stated by some of those who have adopted them that the ships behaved little better, and the tanks were often a source of trouble, on account of leakage.

There is nothing very astounding in the fact that many ships behave little better with such arrangement, simply because it is fitted on much the same principle as the average stevedore loads a ship he knows little about. In some cases these ships are almost as stiff as when depending only upon double-bottom ballast, and their behaviour at sea shows little, if any, improvement.

The trouble of leakage has very often been proved to be the outcome of pure carelessness or ignorance. Instead of the tanks being filled to the uttermost, empty spaces have been left at the top, and, with a rolling ship, it does not require a very large endowment of common sense to foresee what would happen with so large a quantity of free water. An argument against these tanks has been brought forward to the effect that it is difficult to keep them thoroughly watertight, owing to the numerous severe strains borne by a ship in her rolling and pitching movements. A similar argument might apply to fore and after peak tanks, in which localities very severe strains are experienced, and yet these tanks can be kept watertight. Where outrageous proportions of depth and breadth to length are adopted, such working might be experienced as to make watertightness a doubtful quality, but in the ordinary type of cargo vessel, with sound workmanship and proper strengthening, there ought to be no special difficulty.

Minimum Seagoing Draught.—The important questions then are, How much water ballast should be carried, and where should it be placed? As previously stated, it is desirable to get the centre of gravity into the lowest possible position compatible with a moderate metacentric height. Returning again to the curves of metacentres, fig. 102, No. 1, for the beamy ship, shows that the metacentre is over 25 feet above the keel at light draught. With increasing draught it lowers, until, at about 18 feet draught, it reaches its lowest position. Here, then, is probably the ideal draught at which to fix the metacentric height. At no other draught could tenderness—small metacentric height—be obtained with greater safety with its accompanying easy rolling motion than at this draught.

The centre of gravity is in the lowest possible position for such a condition with ample freeboard, 5 feet more than at load draught. The curve of stability would rise gently at first, and at

BALLASTING. 181

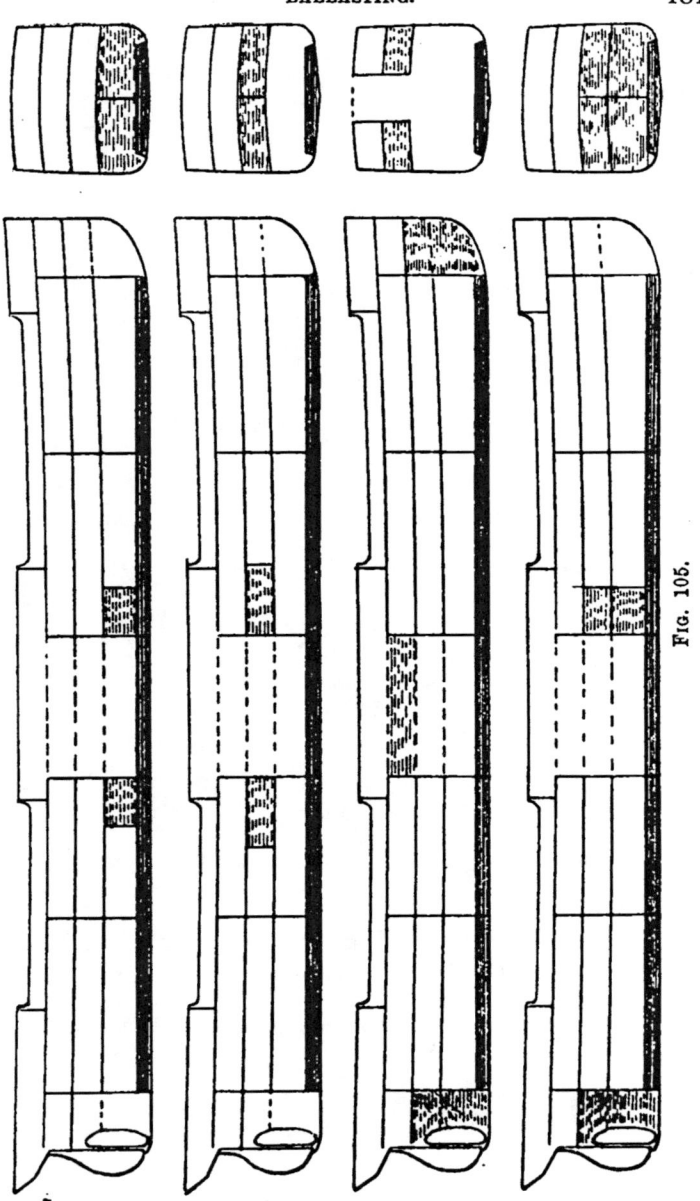

Fig. 105.

182   KNOW YOUR OWN SHIP.

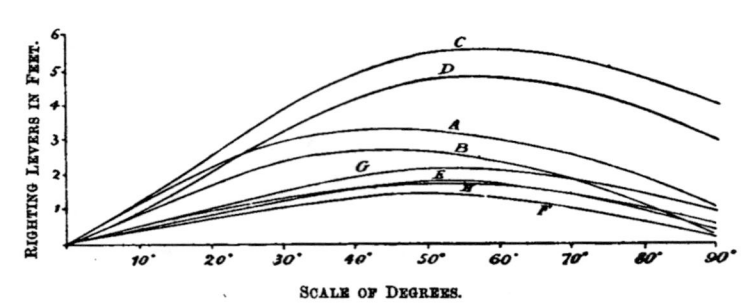

Curves of stability for a steamer. Length, 302 feet; beam, 40 feet 6 inches; depth mld., 24 feet 11 inches. Load displacement, 5183 tons. Freeboard, loaded condition F = 4 feet 10¾ inches. Poop, bridge, and forecastle.

SHIP COMPLETE—STEAM UP.

Conditions.	Bunker Coal, Stores, and Fresh Water.	Water Ballast.	Cargo.	Maximum Stability. Degrees.	Righting Lever.	Height of Metacentre above Centre of Gravity.
					Feet.	Feet.
A	out.	out.	nil.	45	3·31	7·30
B	in.	,,	,,	46	2·68	4·92
C	out.	in.	,,	58	5·58	7·15
D	in.	,,	,,	57	4·77	5·63
E	out.	out.	Homogeneous cargo at 48 cub. ft. per ton.	53	1·85	2·15
F	in.	,,	ditto.	50	1·45	1·99
G	out.	,,	Coal cargo at 45 cub. ft. per ton.	55	2·22	2·62
H	in.	,,	ditto.	52	1·78	2·43

CONDITION F LOADED WITH HOMOGENEOUS CARGO.

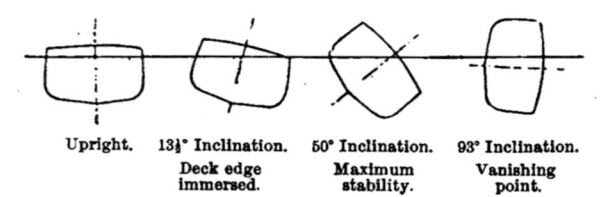

Upright.   13¼° Inclination.   50° Inclination.   93° Inclination.
           Deck edge   Maximum   Vanishing
           immersed.   stability.   point.

FIG. 106.—CURVES OF STABILITY FOR A STEAMER IN VARIOUS CONDITIONS OF BALLASTING AND LOADING.

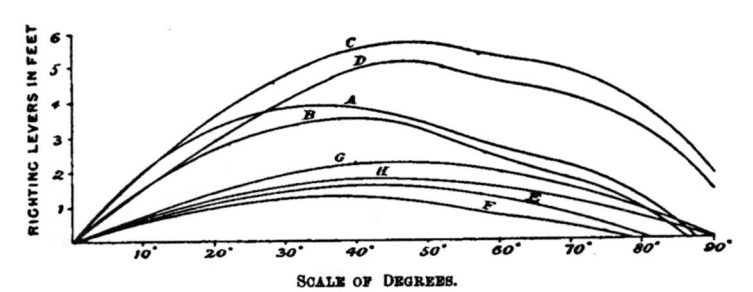

BALLASTING. 183

Curves of stability for a steamer. Length, 360 feet; beam, 48 feet; depth mld., 27 feet 3 inches. Load displacement, 8050 tons. Freeboard, loaded condition F = 5 feet 5 inches. Poop, bridge, and forecastle.

SHIP COMPLETE—STEAM UP.

Conditions.	Bunker Coal, Stores, and Fresh Water.	Water Ballast.	Cargo.	Maximum Stability.		Height of Metacentre above Centre of Gravity.
				Degrees.	Righting Lever.	
					Feet.	Feet.
A	out.	out.	nil.	35	3·96	11·09
B	in.	,,	,,	39½	3·58	8·09
C	out.	in.	,,	45	5·79	10·83
D	in.	,,	,,	46	5·18	8·45
E	out.	out.	Homogeneous cargo at 51 cub. ft. per ton.	39½	1·68	3·16
F	in.	,,	ditto.	37	1·31	3·12
G	out.	,,	Coal cargo at 45 cub. ft. per ton.	43	2·25	4·02
H	in.	,,	ditto.	41	1·82	3·92

CONDITION F LOADED WITH HOMOGENEOUS CARGO.

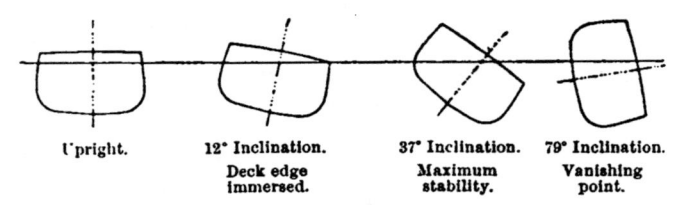

Upright. 12° Inclination. Deck edge immersed. 37° Inclination. Maximum stability. 79° Inclination. Vanishing point.

FIG. 107.—CURVES OF STABILITY FOR A STEAMER IN VARIOUS CONDITIONS OF BALLASTING AND LOADING.

large angles of inclination would show ample righting arms and range, and the large displacement would produce large righting moment. To reach this enviable condition, no less than 3700 tons of weight, including water ballast and bunker coal, would be required in this vessel. Ordinary double bottom and peak tanks provide 1000 tons towards this. The bunkers contain, say, 500 tons of coal, and we find that 2200 tons more are wanted, which would require considerable provision, and entail additional outlay. But as shown by fig. 103, vessels of this class have such enormous reserve stability at large angles of inclination that, in order to obtain a good condition of stability, it is not necessary to get the metacentre into the lowest position, but, as the metacentre lowers much more slowly as the 18 feet draught is approached, by arranging the ballast, the centre of gravity may be so situated as to give moderate metacentric height at a less draught with perfect safety. But we have manageableness to consider as well as stability, and it is probable that in order to get a satisfactory combination of both, the vessel should be immersed to approximately about half the difference between light and load draught, which means about 2750 tons for this vessel. By way of illustrating positions for water ballast, examples are given in fig. 105, though a little consideration will show that many modifications or other combinations might be adopted. Each vessel has a cellular double bottom fore and aft. In all the four examples it has been assumed that the vessels have at least three decks, or are of such depth as to require three decks or an equivalent to three decks. Modification would naturally be necessary in vessels of less depth.

Example 1 shows a deep tank at each end of the engine and boiler space. Where the fore one is shown, a cross bunker in many ships is located. When full of coal, the coal would naturally serve as ballast. By the introduction of additional strength it could be built to carry water as ballast when not occupied by coal. One or both of these tanks might better give the condition required by being made shorter and carried up higher.

Example 2 shows an arrangement where one or both spaces in the lower 'tween decks could be constructed for ballast.

Example 3 is another case showing the ballast higher still. In some vessels this method could be carried out with perfect safety, and a desirable metacentric height could be obtained without sacrificing, to any serious extent, the stability at considerable angles of inclination. A further advantage of such an arrangement of locating the ballast at the sides of the vessel is, that it assists in producing slower and easier transverse movements without making any draw upon the stability, such as always accompanies the raising of the centre of gravity. This space again could be used for bunkers, and would naturally need special support in the form of strong beams, pillars, etc.

Example 4 shows a ballast tank situated at the fore end of the boiler space.

This, again, is in a common position for the cross bunker, and might be used for such purpose when sufficient coal is carried. Fore and after peaks, when properly constructed, may also be used for ballast. The disadvantage of large fore and after peak tanks is that, owing to the fineness of the ends of the vessel, they are almost unsupported, and act as hanging weights along a lever, causing severe straining, which is more excessive in the light condition than any other.

In the foregoing examples the midship ballast spaces should be so constructed, with watertight hatches and doors, as to be thoroughly adapted for cargo or bunker coal, and adjusted to suit trim.

Indeed, in designing such spaces for ballast, a foremost consideration should be to see that they are not rendered useless for the carriage of cargo when ballast is not necessary. Another important consideration is to adopt, when possible, such spaces as by their requirements, apart from ballast, approach nearest to the requirements for carrying water as ballast—such as watertight 'tween decks, bunkers, hold spaces, etc. By this means considerable saving may be made in the course of construction.

It would be absurd to propose the introduction of ballasting arrangements such as have been mentioned for vessels on regular routes, and always sure of cargo, but where there is the possibility and probability of occasional, and sometimes frequent, runs in ballast, it appears that the best sea-going conditions can only be arrived at by some such process. Nor could any of the methods illustrated be recommended before making a thorough investigation of the particular ship under consideration.

It is true that large ballast tanks add somewhat to the cost and weight of a ship, though, as previously shown by the adoption of certain spaces, where the usual construction lends itself for such purpose, both cost and weight are kept down.

However, by the reduction of overdue voyages, and less consumption of bunker coal, together with the greater safety of the vessel, and the often vastly improved conditions of existence for those on board, it would seem that additional ballast to that provided by the ordinary cellular double bottom, carried in tanks properly constructed and adjusted, can only be worthy of commendation.

Note.—Actually fitted water ballast tanks, showing methods of construction and special strengthening, are given in the author's companion volume, *Construction and Maintenance of Steel Ships.*

CHAPTER VI. (Section IV.)

LOADING—HOMOGENEOUS CARGOES.

Contents.—Alteration to Curve of Stability owing to Change in Metacentric Height—Stability of Self-Trimming Vessels—Turret—Trunk.

Loading.—All cargo-carrying vessels are not of the same type, proportions, or form; therefore, they cannot all be loaded alike.

Loading does not mean, as some would imagine, the art of throwing into the smallest possible hold space the greatest amount of cargo in the least time. This method might do all very well, probably, for the loading of railway trucks, when the freight is not of a more damageable nature than sand or rubbish, but to adopt such a method in dealing with so sensitive an object as a ship, simply betrays unwonted ignorance. Certain ships lend themselves more than others to the production of objectionable results, but in the majority of vessels built in these days, the person in charge of the loading or ballasting is often more blameworthy for the bad stability and behaviour of his ship at sea than the ship herself. By the term *bad stability* is not only meant too short righting levers or too short range (this might be called deficient stability), but also too much stability, with too long righting arms for small angles of heel, which produces, as observed in Section II., rapid movements, and probably excessive rolling.

Suppose the shipbuilder supplies a captain with a curve of stability for his ship in her loaded condition, with a certain metacentric height. This, while perfectly safe, he also finds produces easy motion and general steadiness at sea. As far as it is possible, he observes and makes notes of the distribution of weights in the hold, as regards their vertical and horizontal position—that is, if the cargo be of a miscellaneous character, and strives to obtain a similar condition on each succeeding voyage, testing at times, when doubtful, the metacentric height in the load condition, before sailing. Now suppose in going to a strange ship he proceeds to adopt exactly the same methods of loading as in the previous one, it is extremely unlikely that similar results would be obtained, for at sea we should probably find her behaviour widely different, and on testing the metacentric height, that she had either far too much, or else too little, or possibly scarcely any at all; the difference in form, proportions, type, or arrangement of permanent weights being accountable for this result.

Thus we see that efficient loading demands much more knowledge, intelligence, and wise discretion than one would at first imagine. Mere rule-of-thumb methods can only produce uncertainty in the majority of cases. A clear understanding must exist as to what is the best condition of seaworthiness.

For all vessels identical in their proportions, in type, and in internal arrangement, it would be an easy matter to fix upon a freeboard and metacentric height such as would ensure the best possible results at sea; but the immense variety of vessels which are continually being built renders this impossible. However, the officer who superintends the loading is relieved from the former of these responsibilities, as this is fixed either according to the rules of the Board of Trade and Registration Societies, or else by the designers, who in some cases prefer to give more freeboard than the rule minimum. Neither can the officer determine the best metacentric height. Here, again, he is dependent upon the builders or naval architect, who, by calculation, experiment, and varied experience, are in the best position to specify the metacentric heights under various conditions likely to prove most satisfactory. The responsibility which does rest with the ship's officer is the obtaining of the required metacentric height, without which the freeboard determined by the Board of Trade Rules (a freeboard which is calculated on the necessary conditions being fulfilled to provide the ship in a state of seaworthiness) is by no means of itself a guarantee of safety. As some captains would remind us, it is true that probably with the majority of ships built no such information is provided, much less curves of stability. The reason has been given many times before — viz., that in most cases the ship's officer does not know how to use them if he got them. But it would be unjust to blame him for not understanding a subject about which so little attempt has been made to provide the proper means of obtaining a knowledge of their meaning and use. In the past there has been even a worse feature than this. Probably there are no shipbuilders who have more endeavoured to supply information of this nature than Messrs William Denny & Brothers, Dumbarton, and yet their experience is the regrettable one, that only in some cases has their information been used. The happy side, however, is that, when it has been used, most satisfactory results have been obtained. Shipbuilders have not been much encouraged in the past to go to the additional trouble, except in special cases, of providing information upon stability. It is hoped, however, that with the ever-increasing facilities of education for persons in all occupations, the subject will no longer be relegated to the remote position it has hitherto held, especially among officers of the mercantile marine.

The writer's experience in lecturing has proved that the importance of all the points dealt with in this book is already being more fully comprehended, and that seamen generally are desirous of acquainting themselves with them. Undoubtedly, when such information can be used, shipbuilders, as a rule, will be only too ready to supply it, if it were for no other reason than the return which ships' officers would be able to make, in giving reliable information to the shipbuilder, which would greatly assist in the development of vessels thoroughly adapted for their special trades.

At this stage a few observations may possibly be made with profit upon the loading of homogeneous cargoes. Although in some respects less complicated than those of a miscellaneous character, nevertheless their varying densities necessitate that their varying effects upon the vessel's stability be understood.

Homogeneous Cargoes.—By a *homogeneous cargo* is meant one all of the same kind, such as a complete cargo of cotton, or coal, or wool, or grain, or timber. All homogeneous cargoes which exactly fill the holds, and bring a vessel down to her load waterline, have the same effect upon the levers of stability, and produce the same amount and range in every case. This is self-evident, since the centre of gravity of every such cargo must occupy the same position, thereby producing in each case identical levers of stability. As the total weight of each such cargo as brings the vessel to the same load waterline must be equal, it follows that the moment of stability, which equals the displacement multiplied by the righting lever, must also be identical; the other conditions, metacentric height and freeboard, which are necessary in the production of similar stability, remaining constant. But supposing that we have a homogeneous cargo of the nature of timber, which necessitates a part of it being placed on deck in order to sufficiently load the vessel to bring her down to the load waterline, then we alter the previous conditions, and the stability is affected.

Having placed a part of the weight constituting the cargo upon the deck, it follows that the centre of gravity will have risen by the distance of the centre of gravity of the deck cargo from the centre of gravity of the ship (when laden with a homogeneous cargo, which simply fills the holds alone and brings her down to the load waterline), multiplied by the weight of deck cargo, and divided by the total displacement. However, although the metacentre always occupies the same position when the vessel is at her load waterline, and the freeboard is constant, yet the reduction in the metacentric height, due to the deck timber raising the centre of gravity, will result in decreased levers of stability, and, consequently, decreased moment and range. But, again, a vessel laden with a light homogeneous cargo, which, like timber, does not put the vessel down to the load waterline when the holds are filled, and is of such a nature that if exposed to the weather would suffer damage, has, consequently, considerably more than the necessary freeboard. The probable result in most cases would be a reduced metacentric height. This would in all likelihood be caused by a lowering of the metacentre and raising of the centre of gravity. It is true that the metacentre ought, according to the formula, to be at a greater height above the centre of buoyancy, owing to the decreased displacement and very slightly reduced moment of inertia of the new waterplane; but then the centre of buoyancy has lowered also, owing to the lesser draught, which, altogether, may have produced in the region of the load waterline a lowered metacentre (see *Metacentric Curves*, fig. 115). This, however, could readily be ascertained from the curve of metacentres.

Coming to the centre of gravity, we should most likely find that the centre of gravity of a homogeneous cargo filling the holds lies below the centre of gravity of the vessel in her light condition; so that the loading of such cargo must produce a centre of gravity for the loaded ship below the centre of gravity in her light condition. When such is the case, it, moreover, follows that the heavier the homogeneous cargo, the greater its effect in producing the lowest centre of gravity and the greatest metacentric height. The lightest homogeneous cargo would therefore produce the highest centre of gravity and the least metacentric height. Thus the conclusion should not be jumped to that the greater freeboard will compensate for the loss of metacentric height and raised centre of gravity, for it might very possibly be found that the whole range of stability had suffered reduction, which would be further influenced by the smaller displacement.

On the other hand, with a homogeneous cargo which puts the vessel down to her load waterline, and yet does not fill the holds, it follows that the centre of gravity must occupy a lower position than when laden with a homogeneous cargo filling the holds. The result in this case is, that the metacentric height

190   KNOW YOUR OWN SHIP.

being greater and the freeboard unchanged, the lever of stability is lengthened and the moment and range are increased; so that for a vessel engaged in general trade, the value to a shipmaster of a curve indicating her stability when laden with a probable light, homogeneous cargo, which causes her to draw considerably less water than her load draught, and another, when laden with a denser homogeneous cargo, exactly putting her down to the load waterline, must be obvious.

To find Alteration to Curve of Stability owing to a Change in Metacentric Height.—Having given the metacentric height and a curve of stability for a vessel at a certain draught, it is a very simple matter to find the new curve of stability when any change has taken place in the distribution of

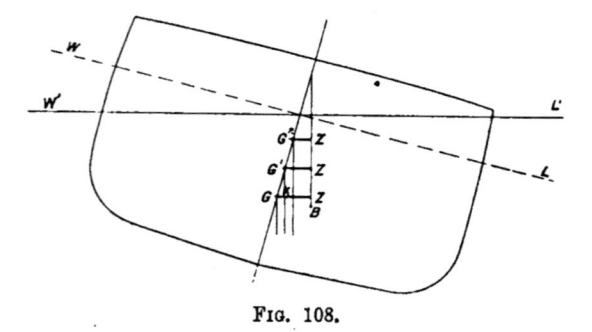

FIG. 108.

the weights of the cargo in loading, and thereby altering the metacentric height, as long as the same draught is maintained.

Fig. 108 shows a vessel floating originally at the waterline W L, but under inclination at the waterline W' L'. B is the centre of buoyancy at the new waterline. G, G^1, G^2 are three positions for the centre of gravity, produced by different arrangements of miscellaneous cargo upon three successive voyages. Let it be assumed that a curve of stability has been provided for the first condition of loading. The distance from G to G^1, and G^1 to G^2, is, say, 1 foot. Under the first condition of loading, the righting lever is G Z, under the second condition it is $G^1 Z$,—that is, G Z reduced by the part G k.

G k = G G^1 (1 ft.) × sine of the angle of inclination.

G Z − (G G^1 × sine of angle of inclination) = G^1 Z.

G Z − (G G^2 × sine of angle of inclination) = G^2 Z.

The new levers for the curve of stability for any condition of loading are thus found by multiplying the distance the centre of

LOADING—HOMOGENEOUS CARGOES. 191

gravity has risen by the sine of the angle of inclination, and deducting the result from the original levers of stability. Should the centre of gravity have lowered, then the correction will require to be added to the levers of stability. (See fig. 104.)

This is both important and exceedingly useful to a ship's officer. For example, his vessel's loaded sea-going draught is fairly constant, and having once been supplied by the designer or shipbuilder with the metacentric height, and a curve of stability for the vessel loaded with a homogeneous cargo, he can always readily ascertain what his curve of stability is for any other condition of loading at the same draught. This is further important, because the metacentric height for one vessel at the load draught may be quite unsuited and even dangerous for another vessel at her load draught. Many Atlantic liners have only a few inches metacentric height, being purposely designed in this way in order to make them easy and comfortable at sea. But there is no danger of their capsizing through lack of stability, for as soon as they begin to heel, their righting levers of stability begin to grow

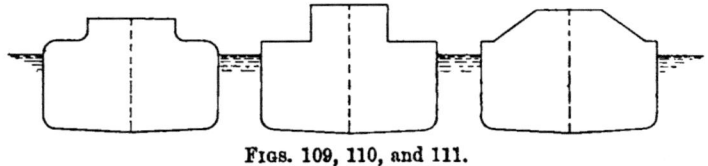

FIGS. 109, 110, and 111.

in length, owing to their great freeboard and true proportions. (See figs. 92, 93, 94, 106, 107.)

In some measure this applies also to cargo vessels with great freeboard, such as awning and spar-decked vessels. The metacentric height in well-proportioned vessels of these classes, when properly loaded, would be satisfactory at about $1\frac{1}{2}$ feet. Where excessive beam is adopted, it becomes almost impossible, and also most dangerous, to attempt to follow the metacentre in its high altitudes by using every possible means to raise the centre of gravity. There is no alternative in such vessels but to put up with their heavy rolling, lurching, straining, and general discomfort.

Vessels with very low freeboard necessarily require greater metacentric height, and more especially so if, in addition, they are very beamy. Naturally, great beam produces large metacentric height, unless it is overruled in the process of loading.

Now, in designing a vessel, unless full particulars are given of the exact nature of the cargo to be carried, the designer works upon the assumption that the holds are exactly filled by a cargo of a homogeneous nature. During recent years, a number of new types of vessels have presented themselves, such as the "Turret," "Trunk," and "Self-trimming" types. (Figs. 109, 110, 111.) There is no doubt that these vessels have certain advantages, especially in the nature of their self-trimming capabilities. The reserve buoyancy afforded by the turret, trunk, or other self-trimming erection is rightly taken into account in determining the freeboard, and, as a result, the deck is brought much nearer to the water level. But with the good beam which it is usual to give to these vessels, ample metacentric height is provided, so that, when loaded with a homogeneous cargo (the only way in which to make a fair comparison of vessels), there is no doubt that their designers and builders have amply provided for all demands upon their stability that wind and weather are likely to make.

CHAPTER VI. (Section V.)

SHIFTING CARGOES.

Contents.—Variations in Stability on a Voyage.

Shifting Cargoes.—Professor Elgar, in a most valuable and instructive paper read before the Institute of Naval Architects, in 1886, upon "Losses at Sea," states that in the three years, 1881, 1882, and 1883, out of 264 British and Colonial vessels registered in the United Kingdom of and above 300 tons gross register, which had been lost at sea under the category of "foundered or missing," one-fourth of these were laden with coal, and one-sixth with grain—very large percentages of the total losses from these causes. It must be remembered, however, that

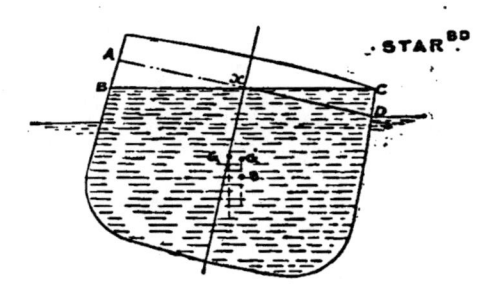

Fig. 112.—Vessel with Cargo Shifted.

these are two of the largest trades in which British vessels are engaged, but at the same time it is obvious that they are two of the trades in which there is most possibility of cargo shifting. Hence the precautions insisted upon by the Board of Trade in the case of vessels carrying grain in bulk, that proper feeding arrangements be adopted in order to keep the hold spaces filled; that proportions of grain to be carried in bags; and also that shifting boards be fitted down the centre of the holds, extending from keelson to deck, and in 'tween decks from deck to deck. Without such precautions, where both grain and coal are carried entirely loose in the hold, the danger is easily perceivable.

For example, let fig. 112 represent a vessel which is so laden. Even if the hold be filled before commencing the voyage, the motion at sea tends to settle the cargo more compactly until there is a space at the top of the hold. Suppose that owing to heavy rolling the cargo shifts as shown, with the result that the vessel takes a heavy list, causing her, in some cases, to be abandoned. The shift of the cargo has affected her stability. Previously, the centre of gravity was at G in the vertical line passing through the centre of the vessel. Now, however, owing to the transference of the wedge of coal, or grain, as the case may be, x A B from the port side, to x C D on the starboard side, the centre of gravity has moved in the same direction. Let the wedge of grain shifted be 40 tons, the distance from the original centre of the wedge to the new centre 20 feet, and the total displacement 3000 tons; then the shift of the centre of gravity in a line parallel to that joining the centres of the wedges will be—

$$\frac{40 \times 20}{3000} = 0\cdot 26 \text{ of a foot to starboard side G}'.$$

The levers of stability will now have become reduced for the whole range of the vessel's stability. If she lies at rest at the angle of heel at which she now is, it is evident that she still possesses some stability, though in the present condition the lever of stability is *nil*, it having all been absorbed by the outward movement of the centre of gravity. But the immersion of a new wedge of buoyancy on the starboard side draws out the centre of buoyancy to that side, and produces some stability at greater angles of inclination. Should the cargo shift still further, the whole range of stability may vanish altogether owing to the shortening of the length of the righting levers; or, in other words, if the total heeling moment of the shifted cargo exceeds the greatest righting moment of the vessel, she will capsize. The most dangerous vessels, with a shifted cargo, as their tendency is to be tender, are the older type of small beam to depth. These vessels, fitted with a double bottom for water ballast, show this objectionable feature in a more marked degree, when the tanks are empty, owing to the raised condition of the cargo. It is, therefore, clear that vessels with good beam have the advantage, in the event of such an accident happening, since their form naturally provides them with greater metacentric height and greater resistance to inclination to large angles.

Possibly, the thought may have struck some reader who has observed grain heaped upon a warehouse floor, as to whether a ship laden with loose grain will heel, until the grain surface in

SHIFTING CARGOES. 195

the hold is at as large an angle as the surface of the heaped grain, before any shift takes place. If the ship could be heeled slowly and steadily, no shift of cargo would occur before the surface of the grain has reached as great an angle as it is possible to heap it upon a floor, as in fig. 113, ab and ab being at the same angle of inclination.

If this answer were correct in all cases, possibly a little less uneasiness might be felt as regards shifting cargo, for the late Professor Jenkins, in a paper read before the Institute of Naval Architects, on this subject, gives the greatest angles to which it is possible to heap a free surface of wheat (or, as it is termed, the *angle of repose*) at $23\frac{1}{4}°$, which is considerably less than that of most grain, and as grain-laden vessels are generally inclined to be tender, and, therefore, usually roll less, the angles of repose

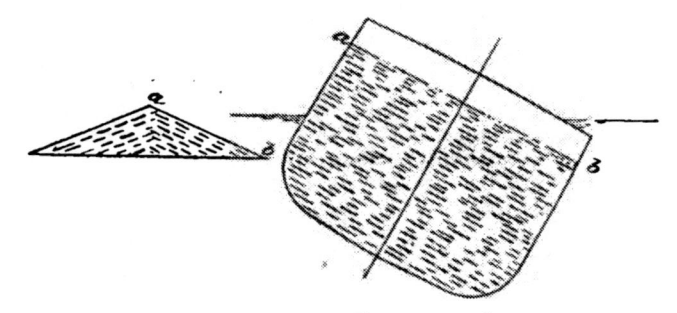

FIG. 113.—ANGLE OF REPOSE FOR GRAIN.

for grain are greater than many such vessels would roll through. But this is not the case, for the effect of rolling, pitching, and blows from the sea is to reduce the angle of repose considerably. Moreover, movement of cargo will take place all the sooner, the greater the distance its surface is situated from the centre about which the ship rolls; hence, cargo in the 'tween decks will shift sooner than cargo in the hold.

When it is known that shifting boards extending down the middle line of a ship reduce the heeling moment of shifted cargo to about one-fourth of what it would be without them, their value will then be better understood.

Variations in Stability on a Voyage. — Steamers are especially liable to considerable change in their stability and behaviour at sea between the time of leaving one port and arriving at another.

This is chiefly due to the consumption of bunker coal.

Take an ordinary cargo vessel which on a six days' voyage from one port to another consumes, say, 100 tons of coal. Naturally the displacement will be reduced by 100 tons, and by referring to the "tons per inch" curve it may be ascertained how many inches the draught has decreased, evidently giving greater freeboard in the first place.

The effect upon the metacentric stability will depend chiefly upon the position of the centre of gravity of the bunker coal. In the case where the bunkers are situated on each side of the boilers, and in a cross bunker extending from the top of the floors to the first deck when there are two or more decks, the centre of gravity of the bunker coal will generally lie below the centre of gravity of the ship, making the ship stiffer. When the coal is consumed the vessel will become more tender, and usually steadier when among waves, due to the reduced metacentric height.

But as is more common in these days among tramp steamers of good beam, where, in order to get the greatest reduction on the tonnage by obtaining the largest possible propelling space (see *Tonnage*), the practice of almost dispensing with lower side and cross bunkers is adopted, and the coal is carried in the 'tween decks at the sides of the engine and boiler casings. Frequently an additional 30 or 40 tons are placed on the bridge deck, and kept there by means of temporary boards secured to the rails, or by closed-iron bulwarks. In such cases the centre of gravity of the coal is much above the centre of gravity of the ship, producing reduced metacentric height and greater steadiness. But on the consumption of this coal, the top weight being removed, the centre of gravity lowers, the ship becomes stiffer, and is found to roll more than previously.

After lecturing on this subject on a certain occasion recently, a captain who was one of the audience related the following uncommon incident in his own experience, an incident which he had never been able to understand :—

Coming up the Irish Sea to Liverpool almost at the end of a homeward voyage, laden both in holds and on deck with esparto grass, the vessel began to indulge in most peculiar movements. There was not much sea, and previous to this time the vessel had been exceedingly steady; but now she commenced to take a slight list, and to move occasionally from side to side with a jerk, and not with a rolling motion. She would then lie still after one of these movements, except for the rising and falling of the waves, until after another interval another wave lifted her up, and she jerked to the other side, there again lying for a time, the motion being repeated at intervals. As this captain now explained, the vessel having a light homogeneous cargo, had been very tender, and all the more so with the deck cargo. The bunkers being situated low down in the vessel, the consumption of coal on the homeward voyage must undoubtedly have still further reduced the metacentric height. Moreover, during the latter part of the voyage the deck cargo became soaked with rain, which made it heavier and more effective in raising the centre of gravity, until at last the centre of gravity actually coincided with or rose above the metacentre. In the latter case an upsetting lever having arisen, the vessel heeled, until by the immersion of a new wedge of buoyancy this heeling force was absorbed. The vessel was now practically in a similar condition to the homogeneous cylinder we considered in fig. 98, except that on being still more inclined, the immersion of another new wedge of buoyancy caused the righting levers of stability to grow again, which resisted further inclination.

Thus the cause of the vessel jerking from side to side was not due to the effect of any righting moment, but simply to the heaving motion of the sea pushing her occasionally to the upright, and then losing her balance, being in a state of neutral equilibrium, she dropped over to the other side.

Any vertical movement of weights, or increase or decrease of weights already on board, will in some degree influence the stability, which, as has been shown, will affect her motion among the waves.

Emptying or filling water ballast tanks may have a similar effect.

CHAPTER VI. (Section VI.)

EFFECT OF ADMISSION OF WATER INTO THE INTERIOR OF A SHIP.

CONTENTS.—Admission through a hole in the Skin into a Large Hold—Curves, showing Variation in Height of Metacentre with Increase of Draught—Buoyancy afforded by Cargo in Damaged Compartment—Longitudinal Bulkheads—Entry of Water into Damaged Compartment beneath a Watertight Flat—Entry of Water into Damaged Compartment above a Watertight Flat—Value of Water Ports—Water on Deck—Entrance of Water through a Deck Opening—Entry of Water into an End Compartment — Height of Bulkheads — Waterlogged Vessels.

Effect of the Admission of Water into the Interior of a Ship.—1. *Through a hole in the Skin below the Load Water-line:*—

Let fig. 114 be a box-shaped vessel 100 feet long, 20 feet broad, 10 feet draught, and 5 feet freeboard. For the sake of example,

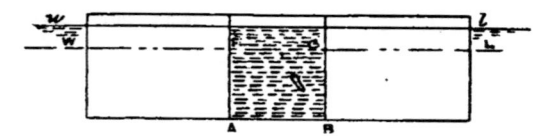

FIG. 114.—EFFECT OF ENTRY OF WATER INTO A CENTRAL WATERTIGHT COMPARTMENT.

let there be a watertight compartment, one-fourth of the length of the vessel—viz., 25 feet—situated at the middle of the length, and bounded at each end by a watertight bulkhead, as shown. We will suppose this compartment to be damaged by collision, and the sea to enter by the hole indicated.

After the damage the draught of water will have increased from W L to *w l*, for, as was pointed out in the Chapter on *Buoyancy*, the buoyancy of the compartment into which the water has entered having been lost, in order to support the weight of the vessel which remains unchanged, the reserve buoyancy in the other intact compartments has had to be drawn upon, hence the increase of draught. But let us see what effect

EFFECT OF ADMISSION OF WATER INTO SHIP'S INTERIOR. 199

this has had upon the metacentric height. The rule for the height of metacentre above centre of buoyancy is—

$$\frac{\text{Moment of inertia of waterplane}}{\text{Displacement in cubic feet}}.$$

Now, in taking the moment of inertia of the new waterplane of this vessel, the compartment into which the sea has entered is entirely ignored, and treated as though it were no part of the vessel, since it affords no buoyancy. The moment of inertia of the new waterplane $w\,l$, less the part F G, divided by the displacement, gives the height of the metacentre above the centre of buoyancy. It will be pretty evident that the moment of inertia will have decreased considerably owing to the loss of a part of the waterplane area, which must result in bringing the metacentre nearer to the centre of buoyancy. The position of the centre of gravity of the ship remaining unchanged, it would appear to follow, at first sight, that in every such case the metacentric height is reduced. But let not the important fact be overlooked that the centre of buoyancy must have risen with the increase of draught. Possibly a simple calculation may more effectively throw further light upon this. Referring to the box vessel before the accident occurred:—

The displacement was $100 \times 20 \times 10 = 20{,}000$ cubic feet.

The moment of inertia of the waterplane $= \dfrac{100 \times 20^3}{12} = 66666$.

The metacentre above centre of buoyancy $= \dfrac{66666}{20000} = 3\cdot3$ feet.

The metacentre above the bottom of the box is $5 + 3\cdot3 = 8\cdot3$ feet.

Let the centre of gravity be, say, $6\cdot3$ feet from the bottom of the box, then the metacentric height in the undamaged condition is $8\cdot3 - 6\cdot3 = 2$ feet.

For curve of stability (see fig. 87, curve No. 3).

After the collision, the draught is found to have increased $3\cdot3$ feet; this is arrived at by dividing the volume of the lost buoyancy F G B A (fig. 114) by the total area of the waterplane, less the part F G.

The centre of buoyancy is therefore—

$$\frac{10 + 3\cdot3}{2} = 6\cdot6 \text{ feet from the bottom of the box.}$$

The displacement remains unaltered at 20,000 cubic feet.

The moment of inertia of the waterplane $= \dfrac{75 \times 20^3}{12} = 50000$.

The metacentre above the centre of buoyancy $= \dfrac{50000}{20000} = 2\cdot5$ feet,

which is lower than formerly.

The metacentre above the bottom of the box $= 6\cdot6 + 2\cdot5 = 9\cdot1$ feet, which is higher than formerly.

Then the metacentric height after the collision is $9\cdot1 - 6\cdot3 = 2\cdot8$ feet, which is greater than formerly.

Thus we see that the vessel has actually a greater metacentric height after the accident with the compartment flooded than she had originally in her intact condition. At first, one is almost tempted to jump to the conclusion that this could scarcely be the result, for if such be the case, why the loss of so many vessels subdivided into numerous watertight compartments? and we may point to the battleship "Victoria," and the more recent disaster to the "Elbe." To understand this, it is necessary to carefully trace the effect of the chief influencing agents, upon which the initial or metacentric stability depends, from the moment the collision happens to its climax. Be it observed, however, that while the metacentric stability is less for this box ship in the undamaged condition, yet the stability for greater angles of inclination is unquestionably superior. This is accounted for by the greater freeboard. Moreover, all ships cannot be classed as boxes, and the difference in form and the fining away of the waterlines (or beams) towards the ends, has a very marked effect upon the height of the metacentre, owing to the great reduction in the moment of inertia of the waterlines.

Variation in Height of Transverse Metacentre owing to Increase of Draught.—It is generally found in merchant vessels that for light draughts the metacentre is highest, owing to the small displacement in comparison with the moment of inertia of the waterlines, but as the draughts increase the immense bulk of the vessel grows more rapidly in proportion than does the moment of inertia. The result often is that the metacentre falls, until, when approaching the load waterline, the lowering effect of the increasing displacement is less than the raising effect of the centre of buoyancy, owing to the increasing draught, and the metacentre rises again.* This leads us to the important fact that dangerous as the shipping of heavy seas may be into large wells upon the decks of many vessels, yet in some cases, owing to the increased draught from such sudden deck weight the metacentre rises, and this greatly tends to avert what might often be a serious condition.

In fig. 115 several examples are given for different types of actual ships, showing the effect of increased draught upon the height of the metacentre, as shown by the curves. The table on the same page gives the particulars for each vessel.

It will also be noticed that a collision happening in way of a moderately large compartment at or near amidships, the loss of this most effective moment of inertia of waterplane (the beam being greatest here), greatly modifies the results obtained from

---

* Flare out on the load waterline at the ends of a vessel increases the moment of inertia and tends to produce the same result.

EFFECT OF ADMISSION OF WATER INTO SHIP'S INTERIOR. 201

vessels of box form, and makes impossible such increase in metacentric height as is obtained from our box vessel with its central compartment flooded (fig. 114). In the box vessel in our illus-

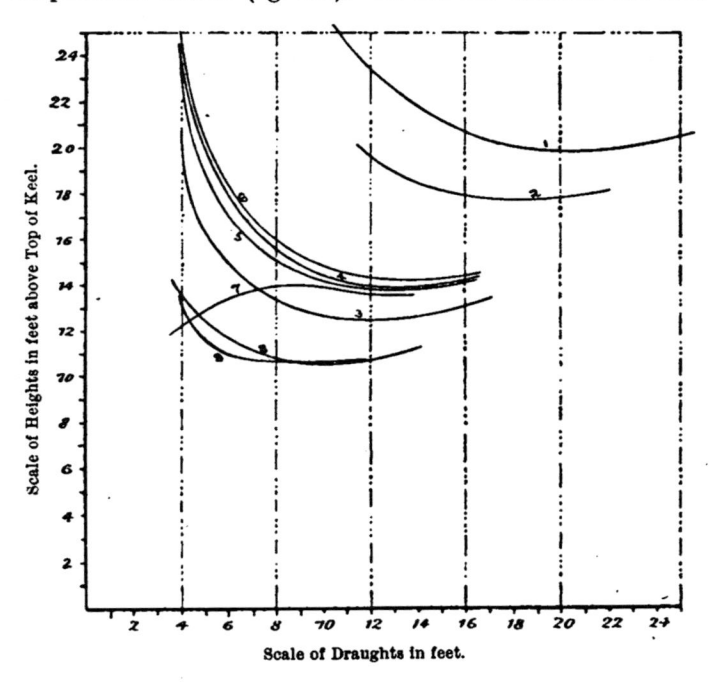

FIG. 115.—CURVES OF METACENTRES, SHOWING UPWARD TENDENCY IN REGION OF LOAD DRAUGHT.

No. of Vessel.	Length.	Breadth.	Depth Moulded.	Load Draught.	Description.
	Feet.	Ft in.	Ft. in.	Ft. in.	
1	320	45 3	28 0	22 0	Full sailing ship.
2	270	41 0	26 3	21 0	Full sailing ship.
3	206	28 6	15 10	14 0	Fine steamer.
4	231	32 0	17 4	15 0	Moderately full steamer.
5	215	31 10	16 3	15 0	Moderately full steamer.
6	245	33 0	17 3	15 9	Moderately full steamer.
7	190	27 6	18 0	12 0	Steam yacht.
8	162	22 0	14 4	11 6	Steam yacht.
9	154	22 0	13 6	9 0	Steam yacht.

202   KNOW YOUR OWN SHIP.

tration the draught was 10 feet, and the freeboard 5 feet, and if these figures be for loaded conditions, the amount of freeboard, and, consequently, reserve buoyancy, is exceptionally great.

Had there been anything less than 3·3 feet freeboard she would have foundered, the weight of the vessel more than swamping the total remaining buoyancy after the collision. But probably neither the "Victoria" nor the "Elbe" sank from this cause, for before the reserve buoyancy had been exhausted they capsized.

Let fig. 116 represent the section of a vessel in way of a

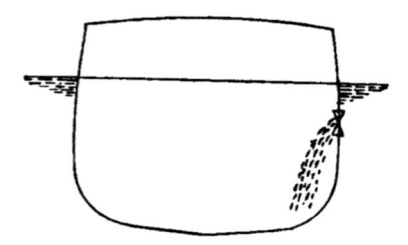

FIG. 116.—INFLUX OF WATER INTO DAMAGED COMPARTMENT.

damaged compartment, the buoyancy of which compartment is less than the remaining reserve buoyancy, thus proving that there is, at any rate, capacity to float after the inflow of water has ceased.

Water flows in through the hole X, and if the hole be large,

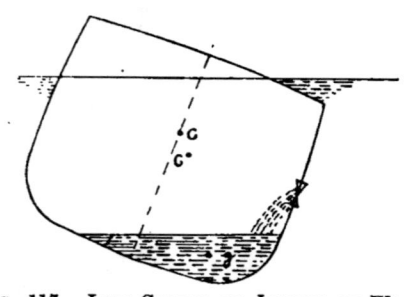

FIG. 117.—LIST CAUSED BY INFLUX OF WATER.

as it usually is under such circumstances, the immense weight of water which rushes in *on the one side*, together with the natural tendency of water to find the lowest possible position, has the effect, for the time being, of drawing out the centre of gravity of the vessel in the same direction—that is, towards the

EFFECT OF ADMISSION OF WATER INTO SHIP'S INTERIOR. 203

side of the ship where the hole is, and the vessel takes a list. The water already in the compartment naturally seeks a horizontal surface, as shown in fig. 117, with its centre of gravity g, well over to the heeling side, adding to the inclination.

The continuous pouring in of water, accompanied by increasing list, and the added effect of the water which has already entered, combine to increase the heeling process, and thus eventually, in many cases, when the heeling moment from these causes exceeds the greatest righting moment indicated by the vessel's curve of stability, she capsizes. If the vessel be very stiff in the upright condition, in some cases she will resist the heeling caused by the water pouring in, and she will do this all the more easily after a moderate quantity has entered, as this has the effect of lowering the centre of gravity and adding to the stiffness and resistance to further inclination, and eventually, when the reserve buoyancy has been drawn upon for the loss of the buoyancy of the damaged compartment, she may remain at rest in a condition of stable equilibrium, unless the change of trim has been so great as to cause the vessel to go down by the head or the stern

Buoyancy afforded by Cargo.—It should be noted that immediately after the collision the weight of water pouring in

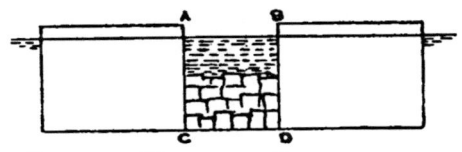

FIG. 118.—EFFECT OF WATER IN A COMPARTMENT WITH CARGO.

acts exactly as deadweight, lowering the centre of gravity and increasing the draught. But after the compartment is filled to the level of the water outside the ship—and the water inside the ship continues to be freely in contact with the water outside—the entered water is no longer considered as weight, any more than the water outside the ship is weight, since this space has ceased to represent floating power (excepting the cubic capacity of the cargo inside, if there be any, for displacement is always actual buoyancy), and the entered water is, therefore, dismissed as being no part of the vessel; in fact, the vessel is now identical with fig. 118.

Here we have two intact end compartments. The middle space, A B C D, has only, say, a bottom upon which rests weights representing cargo. There are no sides to this compartment, but simply the means for holding the structure together. In this example we have an exact illustration of the vessel we have been considering. The whole is being supported by the buoyancy of two intact end compartments, together with the volume of the cargo, which also affords its cubic capacity of buoyancy, though beneath the water level. The centre of gravity is the actual centre of the weight of the ship and her cargo, the water in the space over the cargo neither affecting its position nor adding to the weight.

Thus we see that in an actual ship the entry of water may at first affect the centre of gravity, yet after the inflow has ceased, having reached the water level outside, the centre of gravity is again in its original position. If the vessel is tender in the upright condition, she will heel all the more easily under the effect of water so entering.

Longitudinal Bulkheads.—Should the vessel possess a longitudinal watertight bulkhead down the middle of her length, instead of this providing a means of safety, as is sometimes supposed in such an accident, it simply adds to the effect of the water in heeling the ship, and by robbing the ship of buoyancy upon one side only tends all the more to the production of the often disastrous result. In cases of war and passenger vessels, where a system of exceedingly numerous watertight compartments is adopted, damage to one or more of these compartments may not produce very serious results, though possibly causing a list owing to loss of buoyancy, perhaps, on one side only. Being devoid of this floating support, the centre of buoyancy can no longer lie in the middle line of the vessel, but in the centre of the actual remaining immersed buoyancy, though, if the vessel remain at rest, it follows that the centre of gravity and the centre of buoyancy are in the same vertical line.

It should be observed, however, that the loss of many vessels, thoroughly subdivided into watertight compartments, is often due to the fact that they have so many watertight but unclosed doors. Collisions generally happen at unexpected moments, and often in fine weather, when many of these doors are open, with the result that when the collision occurs, they cannot be reached, or are entirely forgotten, and hence the result.

Entry of Water into a Compartment beneath a Watertight Flat.—Instead of the whole middle compartment being open to the inroad of the sea after damage to the skin, let us suppose that a watertight flat or partial deck situated at one-half the depth of the original draught (5 feet) be fitted as in fig. 119, and the sea to enter through a hole into the lower compartment. Let us now observe the effect upon the metacentric height. The waterline being entirely intact, the moment of inertia

EFFECT OF ADMISSION OF WATER INTO SHIP'S INTERIOR. 205

must be reckoned upon the whole area. The metacentre is 3·3 feet above the centre of buoyancy as for the vessel undamaged.

FIG. 119.—EFFECT OF ENTRY INTO A WATERTIGHT COMPARTMENT BELOW PARTIAL DECK.

The volume of the lost buoyancy of the compartment B, divided by the area of the effective waterplane, gives the increase of draught—

$$\frac{2500}{2000} = 1\cdot25 \text{ feet} = \text{increase of draught.}$$

The centre of buoyancy of the intact end compartments only, from the bottom, is—

$$\frac{10 + 1\cdot25}{2} = 5\cdot62 \text{ feet.}$$

The effect of the buoyancy above the watertight flat is to raise the centre of buoyancy for the whole vessel to 6·01 feet above the bottom.

The metacentre above the centre of buoyancy is 3·33 feet.

The metacentre above the bottom of compartment is 6·01 + 3·33 = 9·34 feet.

The centre of gravity above bottom of compartment is 6·3 feet.

Therefore, the metacentric height is 9·34 − 6·3 = 3 feet, which is greater than for the vessel in the original intact condition, and also, as in the foregoing example, where the whole compartment between the watertight bulkheads is lost buoyancy. Moreover, with intact waterline and upper buoyancy, a little study of the wedges of immersion and emersion will abundantly prove that in every respect its stability is improved.

Entry of Water into an Upper Compartment.—But again, supposing the upper compartment to have been damaged, as in

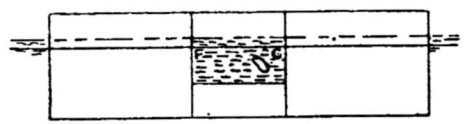

FIG. 120.—EFFECT OF ENTRY OF WATER INTO AN UPPER COMPARTMENT.

fig. 120, we shall see that the effect is greatly different. First of all, the moment of inertia is reduced owing to the loss of the space F G. The moment of inertia of the waterplane, and, consequently, the height of the metacentre above the centre of buoyancy, are now the same as in fig. 114, the latter being 2·5 feet. The loss of buoyancy below the waterline is the same as in the last case, but the increase of draught is more, owing to the effective area of the waterplane being reduced by the space F G.

$$\text{The increase of draught is } \frac{2500}{1500} = 1\text{·}66 \text{ feet.}$$

The centre of buoyancy of the intact end compartments only above bottom of box is—

$$\frac{10 + 1\text{·}66}{2} = 5\text{·}83 \text{ feet.}$$

The effect of the buoyancy below the watertight flat is to lower the centre of buoyancy for the whole vessel to 5·42 feet above the bottom.

The metacentre above centre of buoyancy is 2·5 feet.

The metacentre above the bottom of the box is $5\text{·}42 + 2\text{·}5 = 7\text{·}92$ feet.

The centre of gravity above the bottom of the box is 6·3 feet.

Therefore, the metacentric height is $7\text{·}92 - 6\text{·}3 = 1\text{·}6$ feet.

It will be very clear that this case is vastly different from the previous one, for not only are the metacentric height and freeboard less, but the total loss of the upper buoyancy above the watertight flat will greatly reduce the effect of the immersed wedges in producing the levers of stability.

Value of Water Ports.—*Effect of water on deck, and necessity of ample freeing ports.*

In well deck vessels of the raised quarter deck type, and also those with poop, bridge, and forecastle, special care should be taken that ample means be provided for speedily ridding the space between the forecastle and the bridge, and also between the bridge and the poop, of water accumulated through the shipping of heavy seas. In addition to the discomfort and the strain to the structure from such sudden and heavy deck weights, the vessel's stability may, under certain circumstances, be seriously affected, and if the water be not rapidly cleared, may prove a source of real danger. This is easily perceived when we consider the enormous weight which from time to time is poured on the deck during heavy weather, and this weight being considerably above the centre of gravity of the ship, the effect is to raise the centre of gravity *possibly* as high as the metacentre.

Suppose a vessel of 2000 tons load displacement has a well between the bridge and the forecastle 60 feet long, 34 feet broad, and with bulwarks 4 feet high. This space would hold $\frac{60 \times 34 \times 4}{35} = 233$ tons. That shipped seas would often entirely fill this space would be unlikely; but suppose, for example, that it be three-fourths filled, the water shipped would weigh 175 tons. The centre of gravity of the weight being, say, 10 feet above the centre of gravity of the vessel, the effect is to raise the position of the latter $\frac{175 \times 10}{2000 + 175} = 0.8$ of a foot.

Supposing the vessel's metacentric height to be ·6 of a foot—not at all uncommon in many cases of loading—then the centre of gravity is now approximately 0·2 of a foot above the metacentre, supposing no rise to have occurred in the height of the latter. In this condition it follows that the vessel is unable to remain upright, apart from the influence of the waves, and commences to heel, the water all rushing towards the inclining side. The original metacentric height is only regained by the water pouring over the bulwarks and through the freeing ports, which latter should therefore be amply sufficient to speedily clear the deck.

The danger will be all the more obvious in vessels possessing deep wells, the bulwarks of which extend to the height of the bridge and forecastle, usually about 7 feet. It will be evident that such a vessel, unless capable of speedily relieving herself of the water through the water ports, will heel to a greater angle before getting rid of much of it over the bulwarks. The concentration of this water to one side, when inclined, and the possible synchronism of a wave, may lead to disastrous results. Certainly, as some reader may suggest, this is an extreme case, but the fact that so many vessels are, in ignorance, sent to sea unseaworthy, and so many are recorded annually as "foundered," and "unheard of," demands attention, and furthermore proves that extreme cases are not so uncommon as is often imagined.

Entrance of Water through a Deck Opening.—*Effect upon Stability owing to the Admission of Water into the Interior, either through a Damaged Deck or Deck Opening.*—This case differs from the previous ones in the fact, that the skin being perfectly intact, there is no free communication between the water inside and outside of the vessel. There is, therefore, no loss of buoyancy, the entered water acting directly as deadweight, and thereby increasing the draught. If this operation were to continue until the total weight of displacement exceeded the total buoyancy, the vessel would sink. Should the entered water be only moderate in quantity, through, perhaps, leakage in the deck during heavy weather, let us briefly observe the possible effect upon the stability.

208   KNOW YOUR OWN SHIP.

If the vessel be very flat on the bottom, the effect, as Professor Elgar very aptly points out in the paper previously referred to, will be similar to that of water poured slowly into a box. Water naturally seeks the lowest position, and this will only be accomplished by the vessel inclining, and lowering one of the bottom corners. At first, only a slight list is perceived, but as the amount of water in the interior increases, the list continues to increase until the centre of gravity of the water finds a position in the vertical line through the centre of buoyancy, after which, the water, which was previously the inclining element, now resists further inclination.

Let fig. 121 be a vessel which has gone through these stages of inclination. If she then remains at rest it proves at once that the vertical lines through the centre of gravity and the centre of buoyancy coincide, the centre of gravity having moved from

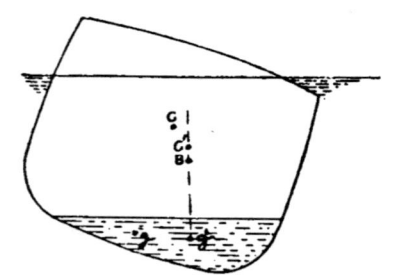

FIG. 121.—EFFECT OF WATER COLLECTED IN THE BOTTOM OF A VESSEL.

G to G′, owing to the effect of the shifted water. If a vessel has more righting moment of stability (righting lever × displacement) at any angle of heel than the inclining moment at corresponding angles (weight of water × perpendicular distance between $g\ g'$),* then the vessel will return to the upright. The vessel in our example (fig. 121) was more tender, for as the water entered slowly she inclined gradually, thus showing that the heeling moment of the water was exactly absorbing the righting moment of the vessel, and thus leaving her, at each successive angle of inclination, in a state of neutral equilibrium. She could not heel all at once to her greatest angle of inclination, simply because the heeling moment was not sufficient to absorb the righting moment of the vessel until more water had entered.

In fig. 121 let the weight of water in the hold be 50 tons, and

* g is centre of gravity of water in upright condition, and g' in inclined condition.

the displacement of the vessel in this condition 1000 tons. In the upright condition the centre of gravity of the water was at g, but after the inclination it is found to have travelled 6 feet, to g'. The effect is to draw out the centre of gravity of the vessel in the direction of the moved water, which distance is $\dfrac{50 \times 6}{1000} = 0.3$ of a foot.

Owing to the shift of the centre of gravity a distance of 0·3 of a foot, to, say, starboard side, the effect is the same as deducting 0·3 of a foot from the lever of stability with no water in; and thus where the curve of stability shows 0·3 of a foot of lever, there would in reality be no lever at all, and at this angle the vessel would lie at rest if undisturbed. If the vessel were forcibly inclined further, she would probably return to this position again immediately on the inclining force being removed, proving that beyond this angle her righting moment exceeded the heeling moment of the water.

If the vessel were now forcibly heeled to the port side and again left undisturbed, she would lie at the same angle of inclination as on the starboard side. Should it happen that the greatest righting lever of the vessel at any angle of inclination did not exceed 0·3 foot, she would undoubtedly capsize.

The more rise there is on the bottom of the vessel, the less danger or likelihood is there of water thus taking a permanent shift, since this form of bottom makes the upright the lowest position that entered water can find. In such a case, the entry of a moderate quantity of water is to lower the centre of gravity and give more metacentric height. But at the same time it will be remembered that great metacentric height conduces to heavy rolling, which motion will be transmitted to the water inside, tending, it is possible, to incline the vessel further than she otherwise would. These latter remarks apply also to free water in partially filled water ballast tanks, and show why vessels which are very tender when light—which tenderness is sometimes further augmented in the operation of coaling—so easily incline and sometimes take a sudden lurch. This generally happens at the unexpected moment, so that, as in the case of the "Orotava," which capsized in Tilbury Dock in December 1896, coaling ports, or side scuttles, or other openings are suddenly brought below the water surface, capsizing ultimately ensuing.

Entry of Water into an End Compartment.—Thus far it will be observed that only the effect upon transverse motion has been dealt with, and that in the foregoing examples the water is supposed to have entered into a compartment in the middle of the length, and thus to have increased the draught uniformly all fore and aft. However, where a compartment is damaged at or near the ends of the vessel, not only may there be tendency to transverse inclination, but longitudinal also.

Thus in fig. 122 let G be the centre of gravity and B the centre of buoyancy in the intact condition. Should the vessel be now damaged through collision in the fore compartment, C D, and the sea find free entrance, then the buoyancy of this compartment being lost, the draught must increase, and the centre of buoyancy will endeavour to shift into the centre of

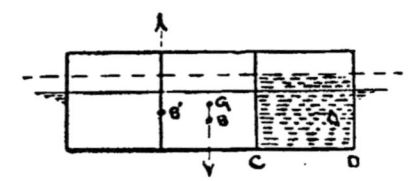

FIG. 122.—EFFECT OF ENTRY OF WATER IN FORE COMPARTMENT.

the remaining buoyancy, B'. But the centre of gravity and the centre of buoyancy being no longer in the same vertical line, a condition of unrest is the result, there being a lever set up between the vertical lines through these two forces. The vessel will, therefore, heel longitudinally and go down by the head, and will not come to rest till the centre of buoyancy is again in the same vertical line with the centre of gravity. The larger the compartment damaged, and the greater its distance from the original centre of buoyancy, the greater will be the change of draught and trim, with, possibly, disastrous results, for it will be seen that immediately the stem has gone under water, when such calamities happen there are possibly several deck openings or companions uncovered, and the sea finding inroad through these, adds to the effect by making its way to that end of the vessel which is inclined.

Height of Bulkheads.—It will now be better understood why a watertight bulkhead should be placed at a short distance from each end of all vessels, for these localities are most likely to suffer in case of collision, and, moreover, being at so great a distance from the centre of buoyancy, their loss of floating moment is great. Hence the necessity of carrying all bulkheads, and especially watertight ones, as high as possible, in order to prevent water finding its way over the tops of them into the next compartments.

Waterlogged Vessels.—It is not an unfrequent sight during

EFFECT OF ADMISSION OF WATER INTO SHIP'S INTERIOR. 211

bad weather in the Baltic, where such a large percentage of the export trade is in timber, to see an old wooden ship, laden with timber in both holds and on deck, in a waterlogged condition. There she lies with a heavy list, rising and falling with each successive wave, with scarcely ever a movement of her own (fig. 123).

She has sprung a leak, and water has found its way into her hold and risen to the level of the sea outside. There being now free communication between the sea and the water inside, the whole of the interior space *unoccupied* by timber—every crevice

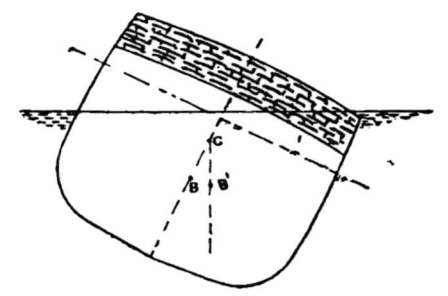

FIG. 123.—WATERLOGGED TIMBER-LADEN VESSEL.

and corner—is lost buoyancy. The result is the draught increases, but not the displacement, until a volume of water has been displaced by the timber and the framework of the vessel equivalent to the original volume of displacement. Undoubtedly the centre of buoyancy has risen. It will now be situated in the new centre of displacement, which no longer constitutes the portion of the ship which is immersed and within the skin, but the centre of the immersed portion of ship and timber unoccupied by water. The centre of gravity has remained stationary. If the metacentre had remained in the same position relatively to the centre of buoyancy, the ship, since the centre of buoyancy has risen, would actually be stiffer than before the leak took place, when she was probably very tender. The permanent list proves that such is not the case, and we can rightly come to the conclusion that a considerable reduction has occurred in the height of the metacentre.

It has not been caused by the displacement, for that is unchanged. It can only lie in the moment of inertia of the new waterline, which is no longer for the whole area of the waterplane at which the vessel floats, but simply for the sectional area of *actual* timber (both cargo and ship) at that waterline. It will now be found that the moment of inertia is considerably less, and when divided by the displacement, gives a position for the metacentre below the centre of gravity. The vessel is, therefore, unstable, and heels over, and possibly in some cases, were it not for the volume of wood on deck, she might capsize altogether. But she is prevented from this by the effect of the immersed wedge of timber, which acts as freeboard, giving buoyancy exactly where it is most needed to keep the vessel afloat.

The effect has been to draw out the centre of buoyancy from the original position B to B', fig. 123, until the vertical line through the new centre of buoyancy coincides with the vertical line through the centre of gravity. The vessel now rests in a state of neutral equilibrium, having no righting lever of stability at this angle of inclination.

CHAPTER VI. (Section VII.)

SAILING, SAIL AREA, etc.

Sail Area.—In briefly considering the subject of the capability of a vessel for carrying sail, our study of "Moments," Chapter II., again proves of great assistance; moreover, we shall also discover that sail area is inseparably connected with stability.

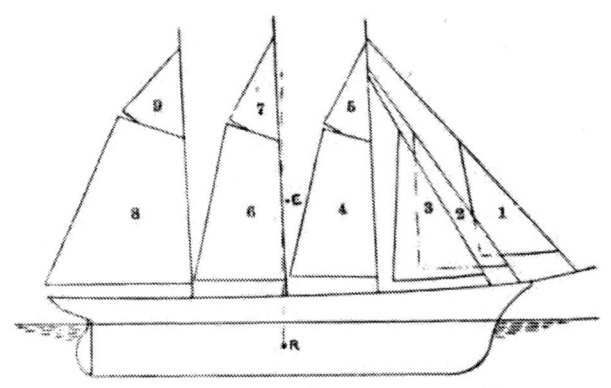

FIG. 124.—THREE-MASTED SCHOONER-RIGGED VESSEL.

Fig. 124 is an outline sketch of a three-masted schooner-rigged vessel, which will serve as an example for reference.

Sails might aptly be compared to a number of weights ranged miscellaneously along a lever; the lever would, therefore, be an imaginary line passing from the region of the hull to beyond the topmast sails. Before we could possibly calculate the moment of pressures about a point at the end of a lever, we should have to determine the position of the end of the lever from which the moments are calculated. Exactly the same thing takes place in dealing with sails; we must determine the end of the imaginary lever. This point is called the *centre of lateral resistance*, and is the centre of the resistance of the water to lateral or broadside motion. Such movement would, therefore, be square, or at right angles to the forward motion of the vessel, and is usually termed *leeway*. The centre of lateral resistance varies in position; but it is approximately and sufficiently correct for all practical purposes at the centre of the immersed longitudinal section, passing through the middle line of the ship. In comparing pressures on sails to weights on a lever, it is not meant that the actual sail acts as a weight, but rather the moment of the wind pressure on each sail, varying according to its distance from the centre of lateral resistance, has the same effect as a weight on a horizontal line in giving a bending or heeling moment. The next operation is to find the centre of this wind pressure for all the sail area, or, as it is usually termed, *the centre of effort*. This is done by multiplying the area of each sail by the height of its centre from the centre of lateral resistance. Then the sum of all these moments, divided by the sum of the areas of all the sails, gives the vertical height of the centre of effort above the centre of lateral resistance. But it is also necessary to have the fore and aft position of the centre of effort, and this is obtained by multiplying the area of each sail forward of the vertical line through the centre of lateral resistance by the distance of its centre from the vertical line. Having found the sum of all these forward moments, the same operation gives the sum of all the moments aft of the centre of lateral resistance. The difference between the forward and after moments, divided by the total sail area, gives the distance the centre of effort is forward or aft of the centre of lateral resistance. It will, therefore, be on that side on which the moments preponderate. (See position of centre of lateral resistance R, and centre of effort E, in fig. 124.)

We might just notice here that when the centre of effort is before the centre of lateral resistance, the tendency of the vessel is to fall off from the wind. This is termed *slackness*. On the other hand, when the centre of effort is abaft the centre of lateral resistance, the tendency of the vessel is to fly up to the wind. This is termed *ardency*.

In calculating the position of the centre of effort, only such sail as could safely be carried in a fresh breeze is calculated upon, and the sails are all *supposed* to be braced right fore and aft. A fresh breeze* is reckoned to blow with a force of 4 lbs. to a square foot of canvas. Thus the total sail area, multiplied by the distance of its centre of effort above the centre of lateral resistance, multiplied by 4 lbs., gives the moment of wind pressure in foot-lbs. For example, in fig. 124, if the sail area is 10,000 square feet, and the centre of effort above the centre of

* The pressure, equal to a fresh breeze, is taken from the British Meteorological Office Tables.

lateral resistance is 50 feet, and the wind pressure on the sails is 4 lbs. per square foot, the heeling moment of the wind pressure is 4 lbs. × 10,000 square feet × 50 feet = 2,000,000 foot-lbs., or $\frac{2,000,000}{2240}$ = 892 foot-tons.

Now, supposing the displacement of our vessel to be 2000 tons, and the metacentre above the centre of gravity, G M, to be 2 feet, what will be the angle of inclination with this force of 892 foot-tons wind pressure? We know that the moment to hold a vessel inclined at any angle is the righting lever of stability, G Z, multiplied by the displacement, D, = foot-tons. We have already got the foot-tons of heeling moment—viz., 892, so that

$$\frac{892}{D} = GZ = \frac{892}{2000} = 0.44 \, GZ.$$

We also know that G Z = G M × sine of the angle of heel for small angles, so that $\frac{GZ}{GM}$ = sine of angle,

$$\frac{.44}{2} = 0.22,$$ sine of the angle of inclination, and reference to the table of natural sines at the end of this book shows the angle of inclination to be about 13°.

The effect of setting top sails and top-gallant sails, and all such light sails, will now be evident.

The higher the sail above the centre of lateral resistance, the more effect it has in producing large heeling moments, though it gives no increased propelling power. Thus, when signs of a squall appear, and all topsails are set, these are taken in first, for two reasons—

1. If the vessel has *a great metacentric height*, and is, therefore, very stiff, when the squall strikes her she offers great resistance to heeling, with the result that, in such a case, the topmasts are often carried away, or the sails torn to shreds.

2. If the ship is tender, and has, therefore, *a small metacentric height*, the effect of a squall of wind is to incline her to about twice the angle to which she would have gone if she had steadily and slowly inclined. Indeed, this would happen in any case. Many a vessel has been capsized through no other cause than this. A vessel with heavy yards, etc., is made stiffer by lowering these spars and stowing them on deck. The result is simply to lower the centre of gravity for the whole ship, and give greater metacentric height.

A point deserving of notice here is, that as the vessel inclines, the effect of a horizontal wind pressure will not be so great as in the upright condition, since the surface upon which it blows is at an angle less than a right angle. The same applies to sail which is not braced right fore and aft. However, our reasoning is sufficiently correct for all practical purposes, and the slight error, moreover, is on the safe side. Under such conditions, the angle of inclination is slightly less than it would otherwise be.

Thus far we have simply considered the effect of a certain amount of sail in heeling a particular vessel with a given metacentric height. But let us view the subject from a different aspect—viz., in determining some features in the design to carry a certain amount of sail at a given angle of inclination. From our previous consideration of the causes affecting metacentric height, we find that this may be done in two ways. It is known that great beam produces a high metacentre, and also, that in a deep vessel with comparatively small beam and low metacentre, the centre of gravity can be brought down by arranging the ballast to obtain the position of the centre of gravity required. The opinion is still in some degree prevalent that a vessel with a narrow midship section is better adapted for speed than one with considerable breadth. These erroneous ideas have been entirely upset by the valuable experiments conducted by the late Mr. Froude, who clearly showed that of two vessels of similar length and equal displacements, the one with the greater beam, and consequently of finer ends, was the better design for speed. At the same time, this design of vessel, with good beam and fine underwater form and ends, is the type which can be made stiffest and most capable of carrying great sail area, for it not only sends the metacentre high, but in yachts, by placing the ballast low, the centre of gravity can be brought down.

Frictional Resistance.—From a point of speed, however, it must be remembered that short vessels encounter greater resistance from the friction of the water upon the immersed skin than long ones, *the first 50 feet of length being the part which suffers most from frictional resistance.* This is more especially so when high speeds are attained. After this length the particles of water, having partaken of the onward motion of the vessel, naturally offer a diminished resistance. Up to about 8 knots, from 80 per cent. to 90 per cent. of the total resistance is due to skin friction.

The necessity, therefore, of keeping the bottoms of vessels clean is most important, as a rough surface adds immensely to the amount of resistance, a surface of the smoothness of calico offering approximately twice the resistance of a varnished surface; and, of course, sand about two-and-a-half times the resistance of varnish. Beyond these speeds, other resistances become very important, frictional resistance diminishing to 50 per cent. or 60 per cent. of the total.

Wavemaking Resistance.—This is found most in vessels of bluff form at the ends, simply because, especially at the fore end, the head resistance offered by the bluffness, corresponding to a flat board moved as shown in fig. 125, causes the water to rise up in front of the vessel, and the making of this wave means expenditure and loss of power. It is not uncommon in vessels of the bluff cargo type, of not more than 9 knots per hour, to create a considerable wave 6 or 8 feet in front of them.

In vessels bluff at the after end, waves are also created at the stern, though not to such a marked extent; but something else occurs which gives rise to another resistance—viz. :—

Eddy-making Resistance.—This is similar to the effect of the flat board moved as shown in fig. 125. After the water has poured round the edges of the board it eddies behind it, and travels in the direction in which the board is moved. All the energy needed to produce this is loss of power. Just so is it with our actual ship. The eddying water being dead, the rudder is rendered less efficient, and if the vessel be a screw steamer the propeller is also less effective; so that, while it is most essential for speed to have the entrance of the fore end fine, it is of paramount importance to have the run of the aft end fine, as there is the danger of a double resistance—wave and eddy-making—which for high speeds assumes enormous proportions.

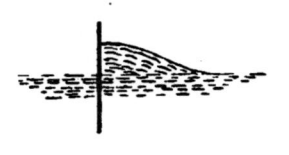

Fig. 125. — Wave and Eddying made by Moving a Flat Board through Water.

Coming back to the subject of design, let us take a yacht 100 feet long, with a displacement of 200 tons, and a required sail area of 5000 square feet. This would represent, at the rate of 4 lbs. per square foot, a sail pressure from wind of

$$\frac{4 \text{ lbs.} \times 5000}{2240} = 8\cdot 9 \text{ tons.}$$

Let the estimated height of the centre of effort above the centre of lateral resistance be 40 feet, then $40 \times 8\cdot 9 = 356$ foot-tons heeling moment. This must be balanced by the stability at, say, a required angle of inclination of 15°.

First, we must find the righting lever of stability at this angle—

$$\text{Displacement} \times \text{lever, or } GZ = \text{moment},$$

then, $\dfrac{\text{Moment}}{\text{Displacement}} = \text{lever, or } GZ$;

therefore, $\dfrac{356}{200} = 1\cdot78 = GZ.$

The next operation is to find the height of the metacentre, GM, above the centre of gravity, which will give $1\cdot78 = GZ$.

The rule is, $GM \times$ sine of the angle of heel $= GZ$,

then $\dfrac{GZ}{\text{sine of angle}} = GM,$

the sine of $15° = 0\cdot258$;

therefore, $\dfrac{1\cdot78}{0\cdot258} = 6\cdot9$ feet metacentric height.

In yachts there is the alternative, where the displacements are equal, as to whether it is better to get metacentric height by beam, with more or less draught and less ballast, or by increased depth and draught, and less beam and more ballast. On this point, as the designs of recent fast racing yachts prove, naval architect experts differ in opinion in some degree.

However, in sailing-ships, where it is not possible to regulate and fix the centre of gravity as on a yacht, whose conditions are fairly constant, but which, on succeeding voyages, carry cargoes of varying densities, it becomes necessary to give sufficient metacentric height and stiffness by means of beam.

In some cases double bottoms are fitted for water ballast, but this is not the common practice, since, when the holds are full and the tanks empty, the effect of the double bottom is to raise the centre of gravity of the cargo, and tend to make the vessel tender. The combined effect of lofty masts and great sail area unite to necessitate greater stiffness.

STABILITY INFORMATION. 219

CHAPTER VI. (Section VIII.)

STABILITY INFORMATION.

What stability information should be supplied to a commanding officer?

It would be rather a difficult matter to answer this question to the satisfaction of every naval expert, and non-expert, since the purposes for which vessels are built vary so much in their

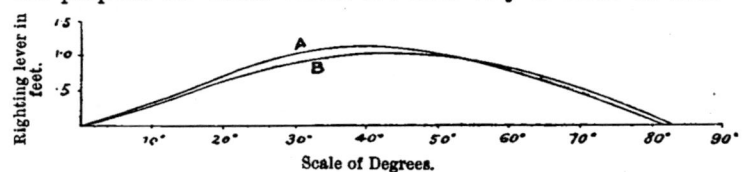

A, Curve for fully equipped condition.
B, Curve for coal and stores consumed.

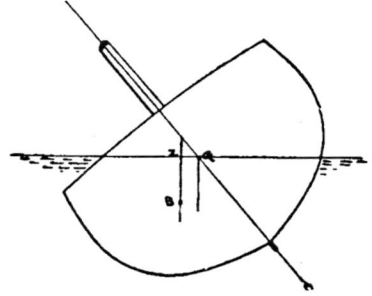

Angle of Maximum Stability. A, condition.

Fig. 126.—Curves of Stability of a Steam Yacht.

nature. But it appears that the commanding officer, after making himself thoroughly acquainted with the principles of the subject, and fully aware of the possible exigencies which his vessel may meet, is the fittest and most capable person to decide what information is required. It is scarcely necessary at this stage to state that officers of all vessels do not need the same information. In the case of a yacht whose condition is practically constant, the only change arising from the consumption of bunker coal and stores, scarcely any information is required, as such a vessel can be designed and built to fulfil certain conditions and to possess a definite amount of stability. At most, therefore, two curves, showing her righting levers and range, in the light and equipped condition, can be of any value in satisfying her captain of her reserve safety. Fig. 126 represents the curves of a steam yacht 152 feet long, 22 feet beam, and $13\frac{1}{2}$ feet deep.

Then, again, other vessels are built for special trades, and the exact natures and densities of their cargoes being thoroughly understood, the naval architect is able to produce a design such as will ensure certain conditions of seaworthiness in the loaded condition.

But by far the majority of vessels are built to carry miscellaneous cargoes, those known as *tramps* carrying anything and everything, anywhere. So that while it is impossible to design and build a ship specially adapted for every trade, it lies in the power of the naval architect or shipbuilder to provide such information as will greatly help the commanding officer in understanding what condition of loading and ballasting to avoid, and what to adopt, to best make his particular vessel seaworthy. The method of supplying curves of stability, adopted by Messrs J. L. Thompson & Sons, Ltd., of Sunderland, is certainly a most commendable one. Figs. 106 and 107 show sets of curves provided by them for two of their steamers.

In addition to the information on these curves in the case of general carrying vessels, there might be provided with great advantage and profit to the ships' officers the following, viz. :—

(1) *A curve of stability when floating at the load draught* with the least metacentric height compatible with safety, as a guard against making the vessel too tender.

(2) *A curve showing the stability when loaded to the hatches* with the lightest homogeneous cargo likely to be carried of specified density or cubic capacity per ton, with a statement of the amount of ballast to be carried in order to ensure safety and produce the best behaviour. (But it does not follow that the vessel should be ballasted down to her load waterline to obtain such a condition.)

(3) *A curve of stability with the heaviest cargo* likely to be carried of specified density or cubic capacity per ton weight, together with a statement of how such cargo should be stowed in the holds and 'tween decks, so as to ensure sufficient stability, and the best behaviour at sea.

(4) *An intermediate curve for a cargo of medium density*, and a statement of the quantity of ballast, if such is required, to produce the best results.

(5) A most important *curve* is that *showing the best condition of a vessel light and in ballast,* with an exact statement of the amount and position of ballast to be carried to ensure such a condition.

In these days, when water ballast is so extensively used, a great amount of ignorance is evinced as to *how* and *where* it should be stowed to secure the desired results. In some cases water ballast is carried only in fore and after peaks, with as little temporary ballast as possible in the holds. The effect of this in straining the vessel has already been pointed out. The commonest method is to fit water ballast tanks throughout the whole or part of the length of the bottom, and to fill them all up when light, regardless of the difference in type, proportions, etc., and the result is that officers often have to complain of their vessels almost rolling their masts overboard. A glance at figs. 106 and 107 very clearly indicates what would be the probable behaviour of these vessels at sea, light or with their ballast tanks full. Curves A and B show the stability levers of the vessels light, and C and D with the ballast tanks full. In all these cases, and especially in fig. 107, the levers are very long for small angles of inclination; in fact, the vessels have too much stability, with the natural and consequent result that heavy rolling would most probably be experienced. We are thus driven to the truth that the bottom of the ship is not always the proper place for all the ballast. As previously explained, a better principle sometimes adopted is to fit very deep tanks in the region of amidships, which, while increasing the displacement, also mitigates the evil of excessive stability.

CHAPTER VI. (Section IX.)

CLOSING REMARKS ON STABILITY.

General Results.—Before concluding our considerations of the various aspects of the subject of stability, it may be advisable to gather up, as briefly as possible, a few of the most important facts which have been revealed.

The *centre of buoyancy* is the centre of gravity of the displaced water, at whatever angle the vessel may be inclined, and through which the upward vertical pressures of buoyancy act.

The *transverse metacentre* is the point where the vertical line through the centre of buoyancy for indefinitely small angles of inclination intersects the middle line of the vessel (such middle line being the line through keel and masts), and is also the point above which the centre of gravity must not rise, in order that a condition of stable equilibrium may be maintained. It is influenced chiefly by beam.

The *metacentre* and the *centre of buoyancy* vary in position with every variation of draught, but always occupy the same respective positions at any particular draught. Their exact positions, when floating upright at any draught, are found from curves obtainable from the shipbuilder or naval architect.

The *centre of gravity* is the centre of the total weight (not bulk) comprising a ship and her entire equipment and cargo. It may, therefore, occupy widely different positions for the same draught, according to the high or low positions of the weights carried. The centre of gravity is found by experiment, as previously explained in this chapter.

The distance between the centre of gravity and the metacentre is termed the *metacentric height*.

Metacentric height in feet, multiplied by the sine of any angle not exceeding 10° or 15° for ship-shaped vessels, gives approximately the righting lever of stability, in feet, for that angle. Righting lever in feet, multiplied by displacement in tons, equals righting moment in foot-tons.

Metacentric height, or metacentric stability, is held to be unreliable beyond the small angles previously stated, as it may continue to grow both in the length of the levers and in the

CLOSING REMARKS ON STABILITY.

extent of the range, or it may rapidly decrease and vanish altogether.

The *lever of stability* can be ascertained at any angle of inclination by finding the position of the centre of buoyancy at that angle. The horizontal distance between the vertical lines through the centre of buoyancy and the centre of gravity indicates the righting lever. If the vertical line through the centre of buoyancy intersects the centre line of the vessel (that is, the line passing through the keel and masts) above the centre of gravity, the lever is a *righting* one; if it intersects below the centre of gravity, the lever is an *upsetting* one.

A diagram, showing the levers of stability at all angles of inclination when in any particular condition, indicating both the angles of maximum levers and vanishing point of stability, is termed a *curve of statical stability*, and is also obtainable from the naval architect or the shipbuilder.

Large metacentric height indicates great *stiffness*, and small metacentric height *tenderness*.

Stiffness can be obtained in designing a ship by increasing the beam, and thereby raising the metacentre; and in an actual vessel already built, by lowering the weights on board and bringing down the centre of gravity.

If the range of a vessel's stability is known at the load draught with a certain metacentric height, any reduction in that height, which can only be effected through the centre of gravity, means reduced range and reduced levers of stability, and any increase produces both increased levers and range.

Similar metacentric heights at different draughts produce widely different conditions of stability.

Freeboard is also a powerful agent in influencing a vessel's stability, but by itself is no guarantee for either the range or the length of righting levers. The greater the freeboard, combined with a fixed position for the centre of gravity and a fixed draught, the longer become the righting levers at considerable angles of inclination, and also the greater the range.

At light draughts, even good metacentric height (such as might be admirably adapted for load draught) and great freeboard may only produce small range and levers of stability.

Great stiffness creates rapid movements among waves, and tends to heavy rolling, while tenderness conduces to slow, easy movements, and general steadiness.

All vessels endeavour to float among waves in a position perpendicular to the wave surface. How far they succeed, depends chiefly upon the ratio of their period of oscillation in still water to the period of the waves among which they are moving.

By a vessel's *period of oscillation* is meant the time taken to perform one complete roll, that is, say from port to starboard, when forcibly inclined in still water. The period for large and small angles of roll is approximately the same.

By the *wave period* is meant the time occupied in the passage of two successive wave crests past a stationary object.

When the period of oscillation is less than half the wave period, the vessel will roll with the waves, approximately maintaining herself perpendicular to their surface, being upright at their summit and trough, and at her greatest inclination on the steepest part of the wave slope. This could only be accomplished by vessels of extraordinary beam to depth, and also when the beam is comparatively small in relation to the length of the waves, but is exactly the manner in which a small raft would float among waves.

Although it is most uncommon to find vessels so rapid in their movements as to have their period of oscillation less than the half wave period, it is not so uncommon to find vessels which have been made so stiff as to have a roll period of probably half the period of the waves they are likely to fall in with, producing *synchronism* with the waves, or, in other words, keeping time, and thus when at their greatest angles of inclination, they receive the impulse of the wave, they answer by taking a tremendous lurch.

Synchronism may also occur periodically on a series of waves.

When it is remembered that waves in different localities, and at different times, vary both in length and period, it will be seen that it is scarcely possible to always escape the effect of synchronism, however a vessel may be loaded. It should be noted that in vessels made excessively stiff, the probability of synchronism is greatly increased, and on the other hand, with the slow rolling ship, though it is not unlikely that occasionally she may fall in with waves producing the same effect, nevertheless, taking her all round, she is certainly the steadier and more comfortable ship.

Slow rolling motion, which is conducive to steadiness, may be obtained chiefly in three ways:—

1. By small metacentric height.
2. By means of resistance agents on the immersed skin in the form of bilge keels.
3. By winging out the weights from the centre to the sides of the vessel.

The two latter of these are very safe methods, since they make no deduction from the vessel's stability. The first, however, necessitates a correct understanding of the vessels dealt with, for, as previously stated, any reduction of metacentric height produces both shorter levers and range. Moreover, what is safe and good for one ship may be exceedingly dangerous for another, hence the necessity of correct information being provided by the designer, and received by the commanding officer, in order to ensure safety.

Synchronism among waves upon a vessel of very small metacentric height and slow rolling period may cause her to reach large angles of inclination, and if the levers and range are very short, will greatly increase the possibility of her capsizing.

Bad loading may thus produce both the dangerous extremes—excessive stiffness with heavy rolling, and too much tenderness with great lack of necessary stability.

Synchronism may be destroyed by altering the speed or course.

CHAPTER VII.

TRIM.

CONTENTS.—Definition—Moment to Alter Trim—Change of Trim—Centre of Buoyancy of Successive Layers of Buoyancy at Successive Draughts—Longitudinal Metacentre—Longitudinal Metacentric Height—Moment to Alter Trim One Inch—Practical Examples showing how the Change of Trim is Ascertained.

UNDER certain circumstances—cases of emergency, accidents to propeller and shafting, entering docks, etc., it often becomes necessary to decrease the draught of a vessel with or without an alteration to trim, or to simply alter the trim by putting the ship down by the head or the stern as may be required. (Trim is the difference between the draughts as indicated at the stem and stern.)

Trim is a subject of every-day consideration to the scientific designer, and is so inseparably related to stability that, in order to obtain the best results, they must be dealt with simultaneously. However, in dealing with the subject it is assumed that due regard has been paid to stability in the vertical distribution of the weights of the cargo, and our attention is directed to the longitudinal distribution of weights, whether they be in the form of cargo in the holds, or water in the trimming tanks.

Sometimes, in a specification for a new vessel, a shipowner inserts a clause to the effect that his vessel when loaded with a homogeneous cargo, must trim in a certain condition, say, 1 foot by the stern.

At first sight this might appear an awkward condition for a shipbuilder to guarantee. But with a thorough grasp of the principles of buoyancy and gravity it is very simple, though it may entail considerable labour and time in order to get an accurate result.

Where a high speed is required, the under-water form of the vessel is designed to the stipulated draught regardless of internal arrangement of holds, position of engines and boilers, etc. When this is completed, having secured the displacement required and the best form of lines for high speed, the longitudinal position of the centre of buoyancy is calculated, and the designer knows that the weights constituting the hull and equipment of the ship, including engines and boilers, and the holds loaded with homogeneous cargo, must be so arranged that the centre of gravity of the whole must be vertically over the centre of buoyancy. When such a condition is fulfilled, no doubt rests as to the trim of the ship when she is built and loaded.

Everyone knows that to shift a weight forward or aft along a ship's deck will produce some alteration to the trim. If the weight is moved forward, as in fig. 127, where 10 tons are shifted 60 feet,

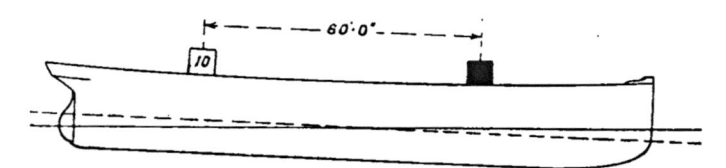

FIG. 127.—CHANGE IN TRIM CAUSED BY MOVING A WEIGHT FORWARD ALONG A SHIP'S DECK.

the vessel must have increased in draught at the stem, or, as we say, "gone down by the head," and, consequently, as the mean draught remains practically the same, she must have risen by the stern. As the effect upon the trim for a particular ship at a certain draught depends almost entirely upon the amount of the weight and the distance it is moved, the moment obtained by multiplying the one by the other (10 × 60 = 600) is termed the "*moment to alter trim.*" However, this quantity gives little idea as to what the exact change in trim will be, especially when it is remembered that the same moment will produce different results at different draughts, and by itself it is therefore practically useless. Could we discover a moment that would give 1 inch change in trim at the required draught, then the problem is solved, for it would be simply a matter of finding how often the moment to change trim 1 inch could be got out of the total moment to alter trim, and the result would be change of trim in inches.

It should be clearly understood that "change of trim" means the change in draught at stem added to change at stern and *vice versâ*. A ship floating at, say, 10 feet draught forward and 11 feet aft, changes trim owing to the shifting of weights on board and the new draught is 10 feet 1 inch forward and 10 feet 11 inches aft. She has gone down 1 inch by the head and risen 1 inch by the stern. The change in trim is 2 inches.

How to obtain the moment to change trim 1 inch we shall investigate shortly.

Let fig. 128 be a ship floating at the waterline *a b*. The centre of buoyancy is at B. One condition for a vessel to float in a state of equilibrium is that the centre of buoyancy and the centre of gravity be in the same vertical line, considered both transversely and longitudinally. If this be a ship's condition as regards the transverse centre of buoyancy, and with the metacentre above the centre of gravity, she floats upright in a condition of stable equilibrium, and the vessel will incline to neither one side nor the other, except by the application of exterior force. Sometimes,

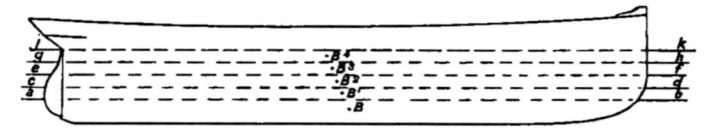

FIG. 128.—ILLUSTRATING THE ALTERATION IN THE FORE AND AFT POSITION OF THE CENTRE OF BUOYANCY OF THE SUCCESSIVE LAYERS OF BUOYANCY CAUSED BY INCREASED IMMERSION.

however, probably owing to the shifting of weights on board, the ship will only find a condition of rest after she has inclined to an angle of greater or less degree; in short, till the centre of gravity and the centre of buoyancy are again in the same vertical line.

When a vessel is launched, she at once seeks a position of rest such as we have described. Transversely, she most likely floats upright, but, looking at her fore and aft, we often find that she heels, perhaps several degrees either towards the stem or the stern, or, as we say, "trims" either by the stem or the stern.

Let fig. 128 represent the condition of a ship just launched, trimming, say, 2 feet by the stern at waterline $a\,b$, and B is the fore and aft centre of buoyancy. It is clear, then, since the vessel floats in a condition of equilibrium, that the centre of gravity is in the same vertical straight line as the centre of buoyancy, and thus we at once have the longitudinal position of the centre of gravity.

Suppose we wish to load such a vessel without altering the trim. Naturally the draught increases as the loading proceeds. The question is:—Where should the additional weight of the cargo be placed? and it is exactly at this point where many blunder. The assumption often worked upon is, that the centre of gravity of the added weight should be placed in the same vertical line as the centre of gravity and the centre of buoyancy when floating at the original draught. Were a ship as regular in shape as a box, such an assumption would be quite right; but for ships it is utterly wrong, for were such a method carried out, it is not at all unlikely, in many cases, that the trim would have altered—the reason is not difficult to ascertain. Let us suppose that the cargo has to be loaded in four instalments of equal weight, putting the vessel down successively to the waterlines cd, ef, gh, jk. Though B is the fore and aft position of the centre of buoyancy at the waterline ab, it does not follow that it will occupy the same fore and aft position at the waterline cd. If the ship trims considerably by the stern, it may be found that the centre of buoyancy of the layer of displacement between the waterlines ab and cd may be some distance aft of B, so that to immerse this buoyancy, the centre of the added weight must be placed over the centre of its own particular buoyancy B' or equally distributed on each side of B'. Similarly, the centre of buoyancy of the next layer of displacement may not be over either B or B^1, but at B^2, and thus the next instalment of cargo must be placed over B^2 in order to put the ship down bodily and not alter the trim, and so on up to the load waterline.

In order to make further comprehensive these counterbalancing forces of weight and buoyancy, let us take another illustration. Suppose our ship to be floating at rest in a light condition at the waterline ab, fig. 128. The centre of buoyancy is at B, and the centre of gravity is at a point vertically above it. Here we have the whole weight of the empty ship supported by all the buoyancy below ab. One hundred tons of cargo in quantities of 10, 15, 30, 20, and 25 tons are to be placed on board in such positions as to cause no alteration to trim, the vessel having to float at the waterline cd, which is parallel to ab. It is evident, therefore, that the volume between ab and cd must measure 100 tons of floating power or buoyancy. So that what we have to deal with is not the original buoyancy below ab, nor the light weight of the ship. These drop out of our consideration altogether. Our attention is directed to the 100 tons of weight, and the 100 tons of buoyancy which has to support it. Let B^1 be the centre of this 100 tons of buoyancy; the centre of the 100 tons of cargo must be placed in the vertical line which passes through B^1, it matters not whether the weights be on deck or in the bottom of the hold, the centre only must be, as stated, in the same vertical line with B^1.

Let us proceed to load.

The 10 tons are placed 5 feet abaft of B^1.
,, 15 ,, 30 ,,
,, 30 ,, 24 ,,
,, 20 ,, 40 ,, forward ,,

The question arises, "Where should the 25 tons be placed in order to get the vessel to the trim required?"

Moments aft.	Moments forward.
10 × 5 = 50	20 × 40 = 800 foot-tons.
15 × 30 = 450	
30 × 24 = 720	
1220 foot-tons.	

1220 − 800 = 420.

We still require a moment forward of 420 foot-tons in order to equalise the moments of each side of B^1, and this has to be obtained by placing the remaining weight of the cargo, 25 tons, in such a position as to produce this.

Weight × distance from B^1 = moment in foot-tons, therefore, $\frac{\text{moment}}{\text{weight}}$ = distance from B^1 that the 25 tons must be placed.

$\frac{420}{25}$ = 16·8 feet, distance required.

To the average ship's officer considerable difficulty presents itself in carrying out a method such as that just described. He asks, how is he to find the centre of buoyancy of the layer of buoyancy which is immersed, owing to the addition of a certain weight. It is true that there are ships' officers who, far ahead of their profession in the knowledge of ship theory, have had themselves provided with the lines plans of their ships, from which they can ascertain the form of the whole or any section of their vessels. Without such a plan the centre of buoyancy of any particular layer of buoyancy could not be found. However, it is not intended to propose any such process as an every-day, practical, handy method for seamen. But, understanding the fundamental principle, a simple, practicable, and reliable method may be deduced. Where the increase in draught is not very great, instead of finding the centre of buoyancy of such a layer as that between $a\,b$ and $c\,d$ in fig. 128, it is sufficiently correct for all practical purposes to find the centre of gravity of the waterplane $a\,b$, and take this as the approximate position of the centre of buoyancy. From this point calculate the moments of the weights of cargo placed on board. Here, again, for any seaman to have to calculate the centre of a waterplane,* whenever he desired to estimate the trim, would be out of the question. But the second difficulty can readily be overcome, for just as easily as a curve of displacement, or a curve of "tons per inch" is made, can a curve of centres of gravity of waterplanes be constructed, and, by this means, the centre of any waterplane between the light and load draughts can be ascertained. This could, with little trouble and at the expense of little time, be supplied by the shipbuilder or the

---

* Rules for exact calculations and worked examples are found in Chapter X. on "Ship Calculations."

designer. Curve 1, fig. 129, is such a curve for a steamer 352 feet by 44 feet by 28 feet. The vertical scale represents draughts, the horizontal scale, distances of centre of gravity of waterplanes from after side of after stern-post.

It must be understood that all centres of gravity so found will be for waterplanes parallel to the keel only, and are only strictly correct when taken for a vessel floating upon even keel. When a ship trims slightly by the stem or the stern, the results obtained by using the distances from this curve would be practically correct.

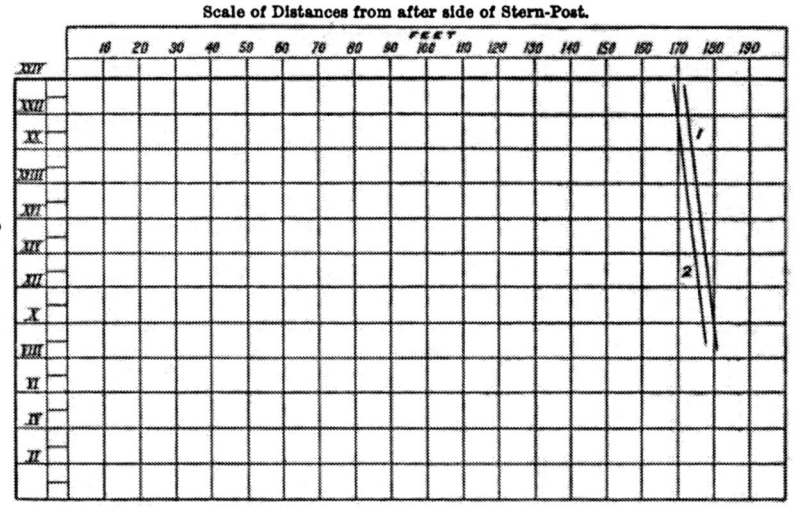

Fig. 129.—CURVES OF CENTRES OF GRAVITY OF WATERPLANES.
No. 1. ON EVEN KEEL.
No. 2. TRIMMING 6 FEET BY THE STERN.

Where, however, a vessel trims very much by the stern, these distances may be considerably in error. In order to come nearer the truth, it might be advisable to have another curve, No. 2, on our diagram, calculated for the same mean draughts, but with the ship trimming 6 feet by the stern. This curve would show the centres of gravity further aft. With these two curves, and the exercise of a little judgment, it is possible to fix the position of the centre of gravity of any waterplane in any condition of trim with an exactitude amply sufficient for all practical and working purposes.

Let it be understood that when only a weight already on board is shifted, the moment altering the trim is simply the product of the weight and the distance it is shifted. But when a weight or weights are placed on board in certain positions, or are taken out of a ship, then these weights are multiplied by the distance between their centres and the centre of the waterplane. If the weights are placed forward of this point, then the moment obtained tips the ship forward; if the weights are aft, the moment tips the ship aft. If some of the weights are placed aft and some forward, then the difference between the fore and aft moments gives the moment altering the trim. When weights are discharged from a ship forward, the result is a moment putting her down by the stern, and *vice versâ* when weights are discharged aft. Naturally, discharging weights decreases the mean draught, while alteration to trim may also be caused.

The centres of gravity of the waterplanes, from which are taken the distances in order to get the moment to alter trim, have been very aptly called the "tipping centres," and in future we shall adopt this handy and expressive term.

Having, it is hoped, made clear how to ascertain the moment to alter trim, our next step will be to find what the exact alteration to the trim is.

When a ship is moderately inclined transversely, the stiffness is measured by the metacentric height, and thus, the greater the metacentric height, the less inclination will be caused by the transverse shifting of a weight on board. A vessel, considered either transversely or longitudinally, will have the centre of gravity at the same height from the keel, in any particular condition of loading or ballasting. The centre of buoyancy also is at the same height in both cases.

But, in turning to the metacentres, there is a great difference in their positions. Transversely, the metacentre varies from a few inches to a few feet above the centre of gravity for passenger and cargo steamers. But longitudinally, for the same vessel, it may be as high, or higher, than the vessel is long.

Metacentre above centre of buoyancy

$$= \frac{\text{Moment of Inertia of Waterplane.}}{\text{Displacement in cubic feet.}}$$

We have only one displacement, as we have only one ship with which to deal, so that the difference must lie in the moment of inertia of the waterplane. And yet we have only one waterplane, and one area for both transverse and longitudinal considerations. Then it is evident that the moment of inertia for longitudinal inclination cannot be similar to the moment of inertia for transverse inclination, and this is so.

TRIM. 233

For transverse inclination, the moment of inertia is calculated about the fore and aft middle line of the particular waterplane. For a rectangular waterplane the moment of inertia is $\frac{\text{Length} \times \text{Breadth}^3}{12}$.

But for longitudinal inclination, the moment of inertia is calculated about the transverse axis passing through the centre of gravity of the same waterplane, and for a rectangular figure the moment of inertia would be $\frac{\text{Length}^3 \times \text{Breadth}}{12}$.

Taking the length of a rectangular waterplane as 100 feet and the breadth 20 feet the moment of inertia for transverse inclination would be 6 6 6 6 6 (draught 10 feet), and metacentre above centre

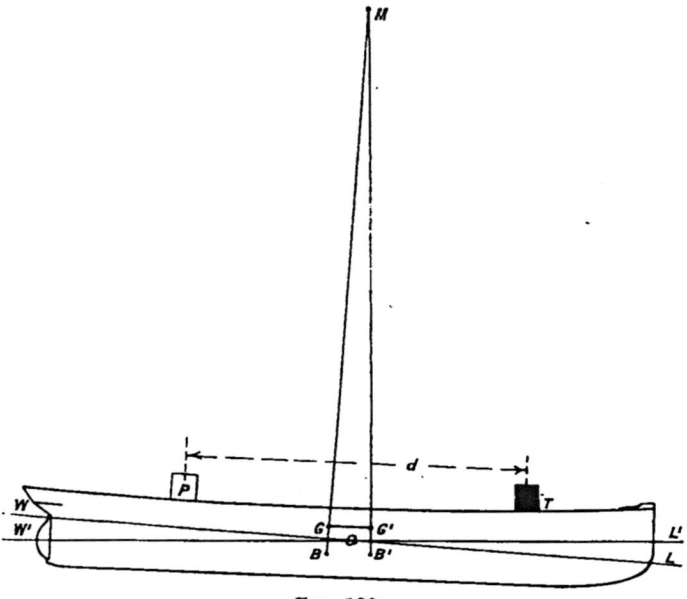

FIG. 130.

of buoyancy, 3·3 feet; and for longitudinal inclination, 1 6 6 6 6 6 6, and metacentre above centre of buoyancy, 83·3 feet. This enormous longitudinal stiffness possessed by ordinary cargo and passenger vessels explains why they are so much more easily inclined transversely than longitudinally, either by the application of exterior force, or the shifting of weights on board. Let fig. 130 represent a vessel floating at the waterline W L, in which condition G is the centre of gravity, B the centre of buoyancy, and M the metacentre. A weight at P is shifted horizontally a distance of d to T. The vessel alters trim so that $W^1 L^1$ is the new waterline, B^1 the new centre of buoyancy, and G^1 the new centre of gravity which has travelled in a direction parallel to a line joining the centre of the weight in the original and new position. The vertical line through B^1 and G^1 intersects the originally vertical line through B and G at M, the metacentre. Let the waterplanes intersect at the point O, which for moderate inclinations is the tipping centre. Here we have now three triangles $G M G^1$, $W O W^1$, $L O L^1$, whose angles are similar.

$$G G^1 = \frac{P \times d}{\text{Displacement in tons}}.$$

$G G^1$ also $= G M \times \tan$ angle M,

then tan of angle M $= \dfrac{G G^1}{G M}$ or $\dfrac{P \times d}{D \times G M}$

D = displacement in tons.

Similarly in triangle $L O L^1$ we have:—
$L L^1 = O L \times \tan$ of angle O, but angle O = angle M,
therefore $L L^1 = O L \times \dfrac{G G^1}{G M}$ or $\dfrac{P \times d}{D \times G M}$,
and similarly in triangle $W O W^1$ we have
$W W^1 = W O \times \tan$ of angle O,
or $W W^1 = W O \times \dfrac{P \times d}{D \times G M}$.

And for a total change of trim we have
$W W^1 + L L^1 = W L \tan$ of angle O
$= W L \times \dfrac{P \times d}{D \times G M}$.

Assuming that a certain weight P moved a certain distance gives 1 foot change of trim, then $\dfrac{1}{12}$ of $P \times d$ will equal the moment to alter the trim 1 inch, and this is the moment it is usual to find.

The formula will now be:

$$\frac{1}{12}\left(W L \times \frac{P \times d}{D \times G M}\right) \text{ or } P \times d = \frac{D \times G M}{W L \times 12} = \text{moment to}$$

alter trim 1 inch.

The only factor in this formula for moment to alter trim 1 inch which presents any difficulty to a ship's officer is G M—the longitudinal metacentric height.* But this can be overcome by

---

* See Chapter X. for rules for the calculation of the height of the longitudinal metacentre.

TRIM. 235

obtaining from the builders or designers a curve of longitudinal metacentres. This is interpreted in exactly the same manner as the curve for transverse metacentres. Fig. 131 is such a curve.

Knowing the position of the centre of gravity in a light, and perhaps one or two other conditions, it will be an easy matter to roughly approximate the position of the centre of gravity in any other condition. With so great a distance between the longi-

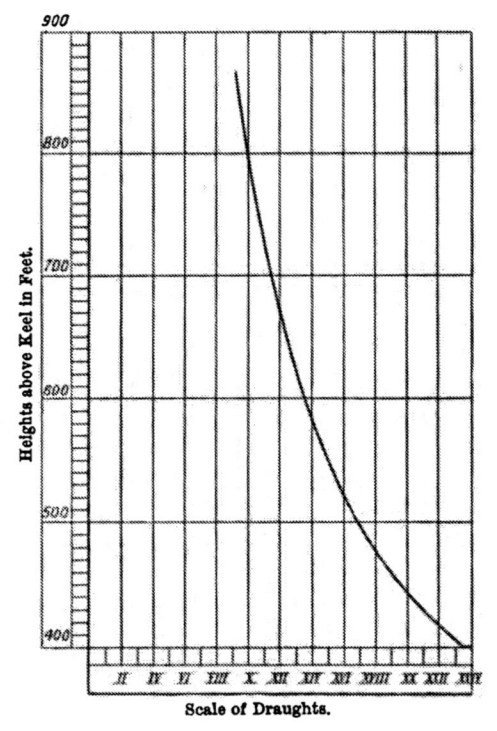

FIG. 131.—CURVE OF LONGITUDINAL METACENTRES.

tudinal metacentre and the centre of gravity, an error of a foot or so in approximating the position of the centre of gravity would make practically little difference in the final result. A few examples are now given illustrating the application of these methods.

To calculate the effect upon the trim of filling or emptying trimming tanks, the centre of gravity of each separate tank ought to be known; this could easily be supplied by builders, together with the exact capacity of each. The same remark applies to the filling and emptying of bunkers.

Example I.

A ship is 352 feet long. The displacement is 6200 tons on a draught of 20 feet aft and 18 feet forward. The mean draught is therefore 19 feet. At this draught, it is found by referring to the metacentric diagram (fig. 131) that the metacentre is 460 feet above the keel. By approximation, the centre of gravity is 16 feet above the keel. $460 - 16 = 444$ feet metacentric height. The length of the waterline is about 352 feet. Then $\dfrac{6200 \times 444}{352 \times 12} = 651$ moment to alter trim 1 inch.

One hundred tons of cargo are taken out of an after-hold and placed in a fore-hold, the distance moved being 100 feet.

Required—alteration to trim.

$100 \times 100 = 10,000 =$ Moment to alter trim. Then $\dfrac{10,000}{651} = 15\frac{1}{4}$ ins. = Total change of trim.

Should the tipping centre be at or near the middle of the length, then the amount of immersion at the fore end and the amount of emersion at the after end will be equal, the draught forward being $18' \ 0'' + 7\frac{5}{8} = 18' \ 7\frac{5}{8}''$, and aft $20' \ 0'' - 7\frac{5}{8}'' = 19' \ 4\frac{3}{8}''$. However, as previously shown, this could be ascertained with considerable accuracy by referring to the curves of centres of gravity of waterplanes.

It is found, fig. 129, that the tipping centre is 175 feet from the after side of the stern-post, and it will be 177 feet from the fore side of the stem. Under such circumstances the immersion at the fore end and the emersion at the after end will be practically equal. Where the difference is considerable the immersion and emersion at the ends is proportional to the two lengths making up the whole length of the waterline.

Supposing an error of 2 feet had been made in estimating the position of the centre of gravity, and instead of the metacentric height being 444 it should have been 442, the effect upon the total alteration of trim would have been less than $\frac{1}{2}$ inch, which in practice would not be a serious error.

Example II.

We will take the same vessel under similar conditions, floating at the same original draught, 20 feet aft, 18 feet forward.

Suppose we were asked how far to shift the 100 tons forward to bring the ship upon even keel. That is, the trim has to be altered 24 inches.

We found that the moment to alter trim 1 inch was 651, therefore the moment for 24 inches will be 24 × 651 = 15624, $\frac{15624}{100}$ = 156 feet, distance 100 tons has to be moved forward.

EXAMPLE III.

Again, taking the same vessel floating at the same waterline under the same conditions, let the main double bottom tank now be filled. The tank contains 210 tons of water. Required—the effect upon the trim. It is clear that the draught must have increased. In making the estimation for alteration to trim caused by placing a weight on board a ship, it is always assumed that the weight is first placed over the centre of gravity of the waterplane at which the vessel is floating. This, as we have seen, increases the draught uniformly, when the weight is small compared with the displacement of the ship. The weight is then shifted to its proper position, and the alteration to trim found, as shown in the foregoing examples. Referring to the scale or curve of "tons per inch," which, along with capacities and deadweight scales, is often, or should be, supplied to ships, the tons to increase draught 1 inch is found. Let the "tons per inch" be 30, then $\frac{210}{30}$ = 7 inches increase in draught by filling the tank.

The new draught is 18 feet 7 inches forward, and 20 feet 7 inches aft. At this draught, we find, from the curve of longitudinal metacentres, the height of the metacentre, and again we approximate the position of the centre of gravity, taking account of the 210 tons in the bottom tank.

The metacentric height is now 434.

Then $\frac{6410 \times 434}{352 \times 12}$ = 658 moment to alter trim 1 inch.

The distance from the centre of gravity of the tank to the tipping centre is 60 feet. Then, 210 × 60 = 12600 = moment to alter trim.

$\frac{12600}{658}$ = 19 inches total alteration to trim.

Supposing the ratio of the length of the waterline before the tipping centre to the length abaft be as 8 to 7, then the alteration to draught at each end of the vessel will be 10 inches forward, and 9 inches aft, and the ultimate draught will be 18' 7" + 10" = 19' 5" forward, and 20'-7" − 9" = 19'·10" aft. The effect upon trim of loading a moderate quantity of cargo or filling bunkers would be calculated in a similar manner.

Example IV.

Again, taking the same vessel, it is desired to find the effect upon trim caused by consuming 200 tons of coal out of a cross bunker. In this case we first assume in calculation that the 200 tons of coal are carried from the cross bunker and placed vertically over the tipping centre, and the alteration to trim is then found. Let the 200 tons be situated in a bunker 30 feet forward of the tipping centre. The moment to alter trim 1 inch we found in Example I. to be 651, and the moment to alter trim is $200 \times 30 = 6000$.

$$\frac{6000}{651} = 9 \text{ inches total change of trim.}$$

Let it again be divided in the proportion of 8 to $7 = 5''$ and $4''$. The draught will be $18' 0'' - 5'' = 17' 7''$ forward, and $20' 0'' + 4'' = 20' 4''$ aft.

It is now assumed that the 200 tons weight is lifted from its position in the vertical line passing through the centre of gravity of the waterplane, and got rid of (consumed). The draught will be decreased uniformly all fore and aft. The "tons per inch" is 30.

$\frac{200}{30} = 6\frac{3}{4}''$ decrease in draught, and the ultimate draught is $17' 7'' - 6\frac{3}{4}'' = 17' 0\frac{1}{4}''$ forward, and $20' 4'' - 6\frac{3}{4}'' = 19' 9\frac{1}{4}''$ aft.

Where a moderate quantity of cargo, or water in trimming tanks, is discharged, the effect upon trim is found similarly to the method given for consumption of bunker coal.*

For further illustrations and actual trim calculations see Chapter X.

---

* With the aid of such diagrams as proposed, the effect upon trim of loading large quantities of cargo, or other weight, producing considerable increase in draught would be calculated in a very similar way.

1st. Find increase in draught parallel to present draught from displacement scale.

2nd. Find position of centre of buoyancy of new layer of displacement from curves of tipping centres.

3rd. Distribute cargo into intended positions in holds.

4th. Ascertain moment to alter the trim.

5th. Ascertain moment to alter trim 1 inch at this new load-line, then $\frac{4\text{th}}{5\text{th}} = $ Alteration of trim.

CHAPTER VIII.

TONNAGE.

CONTENTS.—Importance to Shipowners from an Economical Point of View—Under Deck Tonnage—Gross Tonnage—Register Tonnage—Deductions for Register Tonnage—Importance of Propelling Deduction in Steamers—Deep Water Ballast Tanks—Deck Cargoes—Examples of Actual Ship Tonnages—Sailing Vessels—Suez Canal Tonnage—Yacht Tonnage.

MERCHANT VESSELS.

Importance to Shipowners.—Ships' dues, such as pilotage, dock, river, etc., are paid upon the register tonnage. Tonnage is, therefore, a subject of great importance to the shipowner, from an economical point of view. Nevertheless, considerable misunderstanding is prevalent as to what tonnage really is.

Register tonnage does not, as some would imagine, give any idea of the size of a vessel, for, an ordinarily proportioned vessel, 250 feet long, may have a register tonnage of 700, and another of identical proportions may have a register tonnage of only 300, and yet both vessels may have equal displacements. Thus we gather that register tonnage gives no criterion of comparison between one vessel and another as to their dimensions, displacements, or deadweights.

A glance at a Board of Trade Certificate of Survey for any vessel, or a Register of Shipping, shows three distinct tonnages, viz. :—*Under-deck, gross,* and *register.* One ton of under-deck, or gross tonnage, is equal to 100 cubic feet of capacity, so that these tonnages may convey some idea of the entire internal capacity of a vessel. Register tonnage, however, is a number having no dependence upon the internal capacity as a whole, as already stated, but is modified by the arrangement of the vessel, as affected by the space occupied by the propelling machinery and the crew, as well as other deductions allowed under the Merchant Shipping Act of 1894.

Under-deck tonnage is the total tonnage up to the tonnage deck, and is the first part of the vessel measured for tonnage.

Note.—The tonnage deck is the upper deck in all ships which have less than three decks, and the second deck from below in all other ships.

Under-deck tonnage is measured as follows :—

If the vessel is constructed with ordinary floors, the *depth* at any part of the length of the vessel to the tonnage deck is measured from the top of the floors, afterwards deducting the average thickness of the ceiling, which is generally about 2½ inches, to one-third of the camber of the beam, down at the centre of the beam (see fig. 132).

Should the vessel be built with a double bottom, the depth is taken from the height previously described (fig. 133) down to the plating forming the top of the double bottom where no ceiling is laid ; or, where there is ceiling, the average thickness (usually

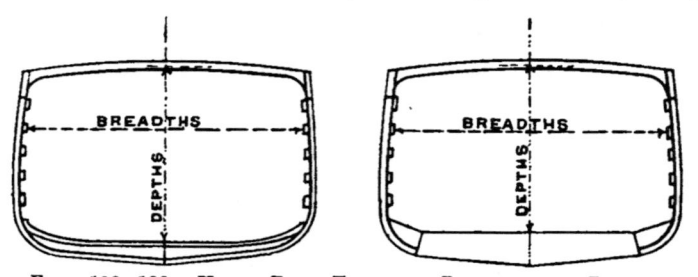

FIGS. 132, 133.—UNDER-DECK TONNAGE ; BREADTH AND DEPTH.

about 2½ inches) is deducted from this depth—no allowance whatever being made for grounds fitted between the tank top plating and the ceiling.

Where a vessel is constructed with ordinary floors at one part of her length, and a double bottom at another, the tonnage in the range of each part is computed separately, and the depths in each part are measured as just described for vessels with ordinary floors and double bottoms.

The tonnage measurements for *breadths* are taken from the inside of the sparring in the hold, 2½ inches being about the usual thickness. Sometimes, instead of wood sparring, half round iron is riveted to the reverse frames, especially in colliers. In this case the breadths are taken from the half round iron, side to side. Where neither sparring nor half round iron is adopted, the breadths are taken to the reversed frames, or to the inside of the framing.

The length for under-deck tonnage is measured from the points at the extreme ends of the vessel where the inside lines of the sparring unite, or in the case where no sparring of any sort is fitted, to the points, where lines forming the inside of the framing unite (fig. 134).

The Board of Trade Surveyor then measures the inside of the

TONNAGE. 241

vessel to the positions indicated, and by means of Simpson's Rules the cubic capacity in feet is found. If anything, this method of finding the cubic contents by these rules, gives the capacity rather under the actual, so that the difference is slightly in favour of the shipowner. In the case of vessels of the raised quarter-deck type, the tonnage deck is the main deck, and where

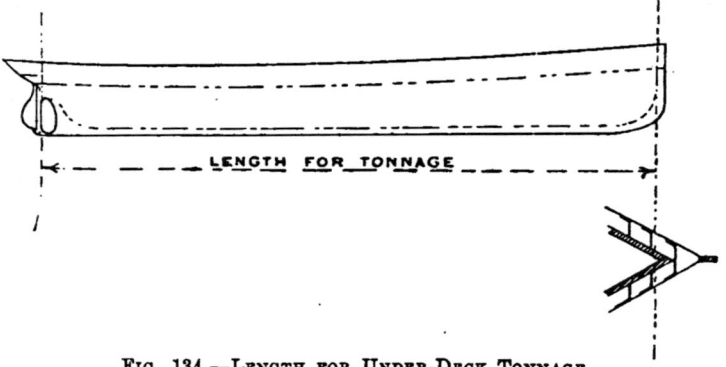

FIG. 134.—LENGTH FOR UNDER-DECK TONNAGE.

the main deck stops, and the raised quarter deck begins, the line of the main deck is taken as the tonnage deck, as shown in fig. 135, and the capacity of the raised quarter deck is computed separately.

Gross Tonnage comprises the under-deck tonnage, together with all enclosed erections in the form of poops, bridges, fore-

FIG. 135.—TONNAGE DECK.

castles, spar decks, awning decks, raised quarter decks, deck houses, engine and boiler casings,* etc. The chief exception to this rule is that the crew's galley, crew's w.c.'s, and companions are usually omitted in the calculations for tonnage.

By enclosed erections is meant spaces closed in on all sides; for example, poops with closed fronts of wood or iron, bridges with closed ends, and forecastles with closed ends. Open-ended

* Engine and boiler casings form part of the gross tonnage, only, when, under paragraph 78 of the Merchant Shipping Act of 1894, the owner desires these spaces included in the calculation of the actual engine room.

Q

poops, bridges, forecastles, or deck shelter, used only for the protection of passengers from the sea and weather, are not included in the gross tonnage. Should, however, any houses or storerooms of any kind be constructed beneath an open-ended erection, they are reckoned in the gross tonnage.

Register Tonnage is obtained from the gross tonnage after certain allowed deductions have been made, and as the various dues and charges are levied upon this tonnage, the nature of the deductions will form the subject of our next consideration. First of all, it must be remembered that no deduction for any space whatever is made, unless it be first included in the gross tonnage. The deductions allowed, and the conditions required to ensure the same, are as follows :—

1. *Crew Space.*—This must be available for the proper accommodation of the men who are to occupy it, protected from sea and weather, properly ventilated and lighted; must contain 72 cubic feet and 12 square feet of floor room per man; must be occupied exclusively by the crew and their personal property in use during the voyage, and reasonable w.c.'s provided.

2. *Master's Accommodation.*—This space must be used exclusively by him, and be reasonable in extent.

3. *Engineers' and Officers' Accommodation.* — This includes berths, mess rooms, and reasonable w.c.'s used by them alone.

Note.—A mess room, used by both the captain and the officers, is not deducted for register tonnage.

4. *W.C.'s.*—In passenger steamers one w.c. for fifty passengers, and not more than twelve w.c.'s altogether are allowed, but only when situated above the tonnage deck.

5. *Sail Room.* — This refers to sailing ships only, and on condition that the room does not exceed $2\frac{1}{2}$ per cent. of gross tonnage. If it does exceed this percentage, $2\frac{1}{2}$ per cent. only of the gross tonnage is allowed.

6. *Boatswain's Store.*—10 to 16 tons is about the usual allowance for the average cargo steamer, varying according to the size of the vessel.

7. *Wheel Room, etc.* — Space occupied exclusively for the working of the helm, the capstan, the anchor gear, or for keeping charts, signals, and other instruments of navigation.

8. *Donkey boiler*, when not connected with the engine room.

9. *Deck shelter* for passengers, when used only for this purpose, and closed at the ends. A sketch of this space must be submitted to the Board of Trade, in order to obtain the approval necessary for its exemption from the tonnage.

10. *Propelling Deduction.* — The propelling space includes

engine and boiler rooms, tunnel, donkey boiler space, if connected with, and forming part of the main engine space, light, and air space. This last comprises all space over or about engines and boilers of reasonable extent, and used exclusively for the admission of light and air;* it must also be safe and seaworthy. Any such space admitting light and not air, or air and not light, will not be included in the deduction. No stores of any sort must be carried in the propelling space. Store rooms or bunkers at the sides of the engines or the boilers will be excluded in making up this deduction. These restrictions being fully complied with, paddle steamers with a propelling space of from 20 to 30 per cent. of the gross tonnage will have an allowance for propelling space of 37 per cent. of the gross tonnage. Screw steamers with a propelling space of from 13 to 20 per cent. of the gross tonnage will have an allowance for propelling space of 32 per cent. of the gross tonnage. If the propelling space is less than 20 per cent. in paddle vessels, and 13 per cent. in screw vessels of the gross tonnage of the ship, the Board of Trade have the option of either estimating the deduction at 37 per cent. in the case of paddle vessels, and 32 per cent. in screw steamers of the gross tonnage, or, if they think fit, allowing in the case of paddle vessels, once and a-half the propelling space tonnage, and once and three-quarters the propelling space in screw vessels. The latter method is usually adopted.

When, however, the propelling space amounts to 30 per cent. or more of the gross tonnage in the case of paddle vessels, and 20 per cent. or more of the gross tonnage in the case of screw steamers, the owner has the option of having the deduction estimated according to the 37 or 32 per cent. respectively, or, if he desires, the deduction may be once and a-half the propelling space in paddle vessels, and once and three-fourths the propelling space in screw vessels. The question arises as to which of these methods is more advantageous to the shipowner. Let us see. Suppose a paddle vessel has a given gross tonnage of 100, and a propelling space of $30\frac{1}{2}$ tons. If the 37 per cent. method is adopted, the deduction will be 37 tons. If the once and a-half method is chosen, the deduction will be $30\frac{1}{2} + 15\frac{1}{4} = 45\frac{3}{4}$, which is certainly preferable from an economical point of view to the 37 tons deduction. Again, let the gross tonnage of a screw steamer be, say, 100, and the propelling space $20\frac{1}{2}$, which proportion is not at all uncommon in vessels of average speed. Choosing the deduction of one and three-fourths of the propelling space, since the propelling space is over the 20 per cent., the

* Only light and air spaces above the upper deck are referred to here. (See also footnote on page 241.)

actual deduction would be $1\frac{3}{4}$ of $20\frac{1}{2} = 35\frac{7}{8}$, which again is considerably more preferable than 32 as a deduction, and thus when we include the other deductions already enumerated, crew space, etc., we can easily understand how, in some vessels of comparatively large dimensions, with large propelling space, the register tonnage is abnormally small. Take a fine high speed passenger paddle steamer with a gross tonnage of, say, 100, and a large propelling space (situated amidships, and occupying the bulkiest part of the vessel) of, say, 40 tons of actual cubic measurement. The deduction would be once and a-half of $40 = 60$ for propelling space, which without the other deductions is more than one-half the gross tonnage. A similar screw steamer with a gross tonnage of, say, 100, and propelling space by actual measurement of 30 tons would have a deduction of once and three-fourths of $30 = 52\frac{1}{2}$ for propelling space alone, also more than one-half of the gross tonnage. These are by no means exaggerated examples. It should be noted, that the higher the actual propelling space tonnage is above the 30 per cent. of gross in paddle vessels, and 20 per cent. of gross in screw vessels, the greater is the proportion of deduction.

In the ordinary tramp type of cargo steamers of comparatively low speed, and where all available space is required for cargo and bunkers, in many cases it might be found unwise to endeavour to get a propelling space of 20 per cent. of the gross tonnage, as this would require a sacrifice of too much space, which might be better utilised for bunkers or cargo. In vessels of this type, to obtain the 20 per cent. generally implies that it is only done by fitting no side bunkers in engine or boiler space, or at most, small pockets. However, this is a point needing the careful consideration of the shipowner or designer of the vessel.

From the study of these deductions, it is evident that the Board of Trade have given every encouragement to the providing of suitable accommodation, or at least, reasonable berthing for officers, and crew especially, for perhaps of all the comfortless and forbidding human habitations, the forecastles of some vessels would claim a foremost place. We have also seen the advantage gained by providing, when practicable, spacious and well-ventilated engine and boiler rooms.

Before giving a few examples of the tonnage of actual vessels, it may be well to point out one or two items from the Board of Trade Tonnage Rules, which might be misunderstood.

In vessels constructed with double bottoms for water ballast, the measurement for tonnage is only taken from the inner bottom, plating (or ceiling), when the space between the inner and the outer bottom, *of whatever depth,* is certified by a Board of Trade

Surveyor to be not available for the carriage of cargo, stores, or fuel.

Deep Water Ballast Tanks.—Should, however, a vessel be built with a raised platform in the bottom, or in other words, a deep tank, thereby making it *possible* for cargo, stores, or fuel being carried in this space, the depths for tonnage are taken down through the platform, or the deep tank, to the height of the ordinary floors, deducting the average thickness of ceiling (if any).

Fore and after peak tanks, though constructed and intended only for water ballast, are *included* in the measurement for tonnage, unless, after submitting sketches to the Board of Trade showing the construction, means of entering, and position of the peak top plating in relation to the load waterline, exemption of these spaces from the gross tonnage be granted.

Deck Cargo.—Ships engaged in the foreign trade, carrying deck cargoes in the form of timber, etc., in spaces not included in the measurement of the tonnage of the vessel, have the space occupied by such cargo measured when the vessel arrives in port, and the cubic capacity being computed, and divided by 100, gives the tonnage which is deemed register tonnage, and charged upon accordingly. Between the 31st of October and the 16th of April, vessels from any port out of the United Kingdom are not allowed to carry deck cargoes exceeding 3 feet in height above the deck, under penalty.

Examples.—

Example I.

Steam screw collier, 228 feet long. Raised quarter-deck type.

Gross Tonnage—
Under-deck tonnage,	680
Erections—	
Raised quarter deck,	60
Poop,	47
Bridge,	120
Lamp room under open forecastle,	5
Casings and deck houses on bridge,*	36
Excess of hatches over $\frac{1}{2}$ per cent. of gross tonnage,	16
Total,	964

Propelling space, 210 = 21·7 per cent. of gross tonnage, therefore the $1\frac{3}{4}$ of actual propelling space is chosen.

$$1\frac{3}{4} \text{ of } 210 = 367·5.$$

---

* It should be understood that where the actual propelling space is over 13 per cent. and considerably less than 20 per cent. of the gross tonnage, it

Had the propelling space been less than 20 per cent. of the gross tonnage the deduction would only have been 32 per cent. of the gross tonnage = 308·4, which makes a difference in the ultimate register tonnage of 367·5 − 308·4 = 59·1. In tramp vessels of this type it is not very usual for the propelling space to exceed the 20 per cent. of gross tonnage. However, it will be seen how spacious light and air space in the casings above the tonnage deck will greatly assist towards this end. Sometimes boiler casings are made wide enough for unshipping the boilers without interfering with the deck.

The deductions for the register tonnage in this vessel were as follows :—

Propelling space,	367·5
Chart room,	5
Bridge accommodation for officers,	54
Crew space,	23
Boatswain's store,	10
Total,	459·5

Gross tonnage,	964
Total deductions,	459·5
Register tonnage,	504·5

Example II.

Passenger steamer, 200 feet long, with a combined poop and bridge, and a topgallant forecastle.

In vessels of this type, where a high speed is required, the propelling space usually exceeds the 20 per cent. of gross tonnage considerably, as seen by the following figures :—

Under-deck tonnage,	530
Poop, bridge, and forecastle,	285
Deck houses and casings,	40
Gross tonnage,	855

Propelling space measured 235 tons = 27 per cent. of gross tonnage.

The allowance, therefore, is 1¾ of 235 = 411 (32 per cent. of gross = 273·6).

is unnecessary to include the light and air casings above the upper deck in the tonnage. Sometimes where the actual propelling space is slightly less than 13 per cent. of the gross the addition of the light and air casings or part of these spaces may secure this percentage.

Deductions for register tonnage are as follows :—

Propelling space,	411
Crew,	40
Passengers' w.c.'s,	4
Master's accommodation and chart room,	7
Boatswain's store,	10
Total,	472

Gross tonnage,	855
Total deduction,	472
Register tonnage,	383

Example III.

Sailing-ship, 320 feet long, with a poop and a topgallant forecastle.

Under-deck tonnage,	2900
Poop,	110·5
Forecastle,	20·4
Houses on deck,	8·0
Gross tonnage,	3038·9

Deductions for register tonnage are as follows :—

Crew space,	115·9
Boatswain's store,	15
Chart house,	4
Sail room,	20
Total,	154·9

Gross tonnage,	3038·9
Total deduction,	154·9
Register tonnage,	2884

In sailing-vessels we always find the gross tonnage large as compared with steamers. This is accounted for by the fact, that there being no engines and boilers, and, therefore, no propelling space deduction, the entire hold space is at the disposal of cargo.

Suez Canal Tonnage.—For ships intended to navigate the Suez Canal, a special tonnage certificate is required, since the method of computing the nett register tonnage differs in a few of its details from the ordinary system.

The extract given on the next page is taken from the *Regulations for the Navigation of the Canal*, and may be of interest to those unacquainted with the canal requirements :—

"When a ship intending to proceed through the canal shall have dropped anchor either at Port Said or Port Thewfik, the captain must enter his ship at the Transit Office and pay all dues for passage, and when there is occasion, for pilotage, towing, and berthing ; a receipt for the same shall be delivered to him, which will serve as a voucher whenever required.

"The following information must be handed in by the captain :—

"Name and nationality of the ship, to be identified by exhibiting the ship's papers respective thereto.

"Name of the captain.

"Names of the owners and charterers.

"Port of sailing.

"Port of destination.

"Draft of water.

"Number of passengers as shown by the passage list. Statement of the crew as shown by the muster roll and its schedules. (Sailors occasionally taken on board of vessels passing through the Suez Canal are not considered as forming part of the crew, and are taxed in conformity with the present regulations.)

"Capacity of the ship according to the legal measurement ascertained by producing the special canal certificate, or the ship's official papers established in conformity with the Rules of the International Tonnage Commission, assembled at Constantinople, in 1873."

Upon arrival at the canal the captain of every vessel receives a copy of the *Regulations*.

The under-deck tonnage is measured, as shown in figs. 132, 133, 134, in the case of vessels constructed throughout with ordinary floors or a cellular double bottom with horizontal top. In vessels with a break or breaks or other irregularities in the construction of the bottom, a slight difference arises owing to the method of computation. Vessels with cellular double bottoms, with the tank top rising from the fore and aft middle line to the bilges, average about 3·5 per cent. less than the British for under-deck tonnage, owing also to a modification in the mode of computation. In vessels built with Macintyre tanks, the depths for tonnage are taken to the top of the inner bottom plating, as in British. Under no circumstances are peak or any other tanks exempted from the under-deck tonnage.

The gross tonnage includes, in addition to the under-deck tonnage, every permanently-covered and closed-in space on or above the tonnage deck without any exception. Such a space as a shelter under a shade deck, open at the sides and supported by means of stanchions, would, therefore, be excluded from the gross tonnage.

The deductions for nett register tonnage are as follows :—

1. *Propelling Space.*

The owner has the option of either of the following methods :—

(*a*) The deduction may be one and three-fourths of the actual engine room as measured for screw steamers, and one and a-half for paddle vessels ; or, (*b*) The actual measurement of the engine room, together with the actual measurement of the permanent bunkers.

Note 1.—Bunkers which are portable, or from which coal cannot be directly trimmed into the engine room or stokehole, or into which any access can be obtained otherwise than through the ordinary coal shoots on deck, and from doors opening into the engine room or stokehole, are not included in the measurement in paragraph (*b*). In no case, except in that of tugs, is the actual engine room allowance to exceed 50 per cent. of the gross tonnage of the ship.

Note 2.—Light and air spaces over the engines and boilers and above the uppermost deck do not form a part of the actual engine room, except when situated in a permanently-enclosed bridge space, poop, or other erection.

2. *Crew spaces*, exclusively and entirely occupied by the crew and ship's officers, with the exception of the master, stewards, cooks, passengers, servants, purser, clerk, etc., in short, only such spaces are deducted as are occupied and used by those persons engaged in the navigation and propulsion of the vessel, with the exception of the doctor's cabin, when he is actually on board and occupying such space. Also the covered and closed-in spaces above the uppermost deck employed for working the ship.

Not more than 4 tons are allowed for an officers' and engineers' mess room.

For a second mess room for boatswain, carpenter, etc., not more than 2½ tons are allowed.

Should passengers be carried and no eating room be provided for them, no deduction whatever is allowed for officers' and engineers' mess room.

When no passengers are carried, a bathroom, used entirely for the officers and engineers, is reckoned as a deduction, and even when passengers are on board, if there be more than one permanent bathroom, one of such spaces is subject to deduction, being considered as specially for the use of officers and engineers.

Not more than 2 tons are allowed for a bathroom.

3. *W.C.'s*, exclusively for the use of the crew.

4. *Wheel house, chart house, winch house, look-out house.*

Should the captain be lodged in the chart room, an allowance of 3 tons is made for the space occupied by charts.

5. *Cooking houses*, used only for the crew. Passengers' galleys are therefore not deducted from the gross tonnage.

6. A *donkey boiler house* in a closed space on the upper deck. Should, however, the donkey boiler be used for hoisting cargo no deduction is allowed.

In no case is the sum total of these deductions, with the exception of propelling space, to exceed 5 per cent. of the gross tonnage of the ship.

No deduction is made for spaces used, or which may possibly be used, for passengers' accommodation; captain or passengers' w.c.'s or lavatories, luggage storerooms, boatswain's store, or sailroom.

In no case is any space to be deducted from the tonnage, which is not first included in the gross tonnage.

YACHTS.

Yachts are measured for tonnage in exactly the same manner as ordinary merchant vessels, the same deductions being allowed for register tonnage.

For example, take a steam yacht 180 feet long. The gross tonnage would be comprised as follows:

Under-deck tonnage,	520·5
Deck houses,	10
Monkey forecastle,	6·1
Total,	536·6

Deductions for register tonnage:

Propelling space,	220·6
Crew space,	49·5
Master's room,	3·5
Chart room,	3·0
Boatswain's store,	3·2
Total,	279·8

Gross tonnage,	536·6
Total deduction,	279·8
Register tonnage,	256·8

Measurement of Yachts for "*Royal Thames*" *Yacht Club.*

Measure the length of the yacht in a straight line at the deck from the fore part of the stem to the after part of the stern post. From this length deduct the extreme breadth. If the vessel be iron or steel, this breadth is taken over the plating, and if composite or wood, over the planking. The remainder is the length for tonnage. If there be any projection of the stem or stern posts beyond this length, such projection must be added to the length already mentioned for tonnage purposes. Multiply the tonnage length by the extreme breadth, then that product by half the extreme breadth, and divide the result by 94. The quotient will be the tonnage.

Example.—Steel yacht 200 feet extreme length over stem and stern posts 28 feet extreme breadth.

$200 - 28 = 172$ feet, length for tonnage.

$\dfrac{172 \times 28 \times 14}{94} = 717\frac{11}{47}$ tonnage Thames measurement.

CHAPTER IX.

FREEBOARD.

Contents.—Definition—Method of Computing Freeboard—Type of Vessel—Nature of Deductions, and Additions to Freeboard—Examples of Estimating Freeboard for Different Types of Vessels.

Definition.—By the term *Freeboard* is meant the height of the side of a ship above the waterline at the middle of her length, measured from the top of the deck at the side. Should a wooden deck be laid, it is taken from the top of the wooden deck. In fig. 3, x shows the amount of freeboard, which, as will be observed, is 2 feet.

Buoyancy, Structural Strength, Stability, and Freeboard are subjects closely related to each other; indeed, the latter depends almost entirely upon the other three. Hence the necessity, in order to deal intelligently with freeboard, that the reader should make himself acquainted with the contents of Chapters III. to VI. inclusive.

Freeboard is given to a vessel as a margin for safety. For instance, it would be possible for a ship to float with her deck edge level with the water; but in such a condition, having no reserve bouyancy she possesses no rising force, and would, therefore, be submerged beneath every wave, to say nothing, moreover, of the possible effect of such a condition upon her stability.

The amount of freeboard for a particular vessel is modified by the type, structural strength, erections on deck, sheer, camber, etc. It must be borne in mind, however, as previously shown, that such freeboard can only fully perform its chief function, and be a real resource of safety, after a proper adjustment of cargo has been carried out in the operation of loading. The effect of freeboard on stability has already been dealt with in Chapter VI.

Flush-decked steamers, other than spar- and awning-decked, and equal in structural strength to 100 A1 at Lloyd's, or equivalent strength at Bureau Veritas, or the British Corporation or other classifying association, require from about 20 per cent. to 35 per cent. reserve buoyancy, according to their dimensions.

As an example, the tonnage under the upper deck of a certain vessel is 78·4 ;

78·4 × 100 = 7840 cubic feet.

The length is 100 feet,
The extreme breadth is 16 feet,
The depth of hold is 7 feet,
100 × 16 × 7 = 11,200.

$$\frac{7840}{11,200} = 0\cdot 7 \text{ Coefficient of fineness,}$$

and this coefficient is that used, in conjunction with the moulded depth, to find the freeboard for this particular vessel from the Tables.

The greater the depth of the vessel, the greater the freeboard. The depth from which the freeboard, as ascertained by the Tables, is measured, is the moulded depth, which is taken from the top of the keel to the top of the upper deck beam at the side, at the middle of the length. When a wooden deck of extra thickness is fitted, the excess of thickness is added to the moulded depth, and the freeboard taken upon this new depth. Now, as we have already pointed out, the structural strength is a most important consideration in determining freeboard, the strongest ship with suitable proportions having the least freeboard. The strongest vessel is the one-, two-, and three-deck type, classed 100 A1 at Lloyd's, or any other vessel, classed or unclassed, but of equivalent strength. Next, we have the spar-decked, and, lastly, the awning-decked vessel.

In assigning freeboard, the term "spar-decked vessel" applies to all vessels equal to, or in excess of, the strength of Lloyd's Spar-deck Rule, but do not reach the structural standard of the three-deck requirements. The freeboard, therefore, depends upon their strength. The standard height for a spar-deck is 7 feet; therefore, since the freeboard is measured from the spar-deck, it will be increased if the 'tween deck height is more, and decreased if it is less than 7 feet.

In like manner, an awning-decked vessel is one equal to, or exceeding Lloyd's structural requirements for awning-decked vessels, but which does not reach the standard for the spar-deck type. In this case also, the freeboard varies with the strength. When the strength of the superstructure above the main deck is less than is required by Lloyd's 100 A1 Awning-deck Rule, the freeboard is increased. The awning deck may, therefore, be classed as simply an erection above the main deck, and since the freeboard is measured from the main deck, no modification is necessary in respect to the height of the awning deck above the main deck. The freeboard of awning-decked vessels classed 100 A1 at Lloyd's is approximately one-twelfth the moulded per cent. reserve buoyancy. To find this accurately, it would be necessary to construct a curve of capacity, identical in its construction with a displacement curve, except that it shows the whole external volume up to the deck, and instead of a scale of tons at the top, the scale would represent cubic feet of capacity.

Fig. 136 represents such a curve for a vessel 16 feet depth moulded. It, therefore, comprises the total buoyancy of the vessel up to the deck. This is found to be, say, 20,000 cubic feet total buoyancy. 20 per cent. of 20,000 = 4000 cubic feet to be left above the load waterline = reserve buoyancy.

20,000 − 4000 = 16,000 cubic feet from the keel to the load waterline. This equals the vessel's load displacement, which gives by the scale a draught of 14 feet, and a freeboard of 2 feet.

But, as is evident, this method entails considerable labour, and would necessitate a copy of the vessel's lines being supplied to the Board of Trade or Registration Society fixing the freeboard, and, therefore, in order to obviate this, another method is adopted, which, though not strictly accurate, is sufficiently correct for all practical and working purposes.

In the last chapter we saw how the Board of Trade Surveyor measured the vessel for tonnage, and the results obtained provide one of the principal factors in estimating the requisite freeboard.

In Chapter I. it was shown how coefficients of fineness of displacement were obtained, and in a similar manner a coefficient of fineness is obtained from the tonnage under the upper deck, and this, combined with the dimensions of the ship, as will be explained, serves the same practical purpose, in referring to the Freeboard Tables, as would the actual volume of the ship.

The coefficient of fineness is ascertained as follows :—For one-, two-, and three-deck, and spar-decked vessels, divide 100 times the total tonnage (1 ton measurement being 100 cubic feet) below the upper deck (exclusive of any deductions) by the product of the extreme length over the stem and the stern posts on the load waterline, the extreme breadth over the plating, and the depth of the hold.* In awning-decked vessels, divide the tonnage to the main deck, by the product of the length, the breadth, and the depth of the hold to the main deck. In the case of vessels built with cellular double bottoms, a modification or correction has to be made in estimating the coefficient. It should, however, be remembered that the coefficient obtained is that of a vessel built with ordinary floors.

* The depth of hold used in ascertaining the coefficient of fineness is taken to the top of the ceiling in iron and steel sailing-vessels, and to the top of the floors in steamers.

As an example, the tonnage under the upper deck of a certain vessel is 78·4 ;

78·4 × 100 = 7840 cubic feet.

The length is 100 feet,
The extreme breadth is 16 feet,
The depth of hold is 7 feet,
100 × 16 × 7 = 11,200.

$$\frac{7840}{11,200} = 0\cdot 7 \text{ Coefficient of fineness,}$$

and this coefficient is that used, in conjunction with the moulded depth, to find the freeboard for this particular vessel from the Tables.

The greater the depth of the vessel, the greater the freeboard. The depth from which the freeboard, as ascertained by the Tables, is measured, is the moulded depth, which is taken from the top of the keel to the top of the upper deck beam at the side, at the middle of the length. When a wooden deck of extra thickness is fitted, the excess of thickness is added to the moulded depth, and the freeboard taken upon this new depth. Now, as we have already pointed out, the structural strength is a most important consideration in determining freeboard, the strongest ship with suitable proportions having the least freeboard. The strongest vessel is the one-, two-, and three-deck type, classed 100 A1 at Lloyd's, or any other vessel, classed or unclassed, but of equivalent strength. Next, we have the spar-decked, and, lastly, the awning-decked vessel.

In assigning freeboard, the term "spar-decked vessel" applies to all vessels equal to, or in excess of, the strength of Lloyd's Spar-deck Rule, but do not reach the structural standard of the three-deck requirements. The freeboard, therefore, depends upon their strength. The standard height for a spar-deck is 7 feet; therefore, since the freeboard is measured from the spar-deck, it will be increased if the 'tween deck height is more, and decreased if it is less than 7 feet.

In like manner, an awning-decked vessel is one equal to, or exceeding Lloyd's structural requirements for awning-decked vessels, but which does not reach the standard for the spar-deck type. In this case also, the freeboard varies with the strength. When the strength of the superstructure above the main deck is less than is required by Lloyd's 100 A1 Awning-deck Rule, the freeboard is increased. The awning deck may, therefore, be classed as simply an erection above the main deck, and since the freeboard is measured from the main deck, no modification is necessary in respect to the height of the awning deck above the main deck. The freeboard of awning-decked vessels classed 100 A1 at Lloyd's is approximately one-twelfth the moulded depth of the vessel less than would have been required had it been built to the three-deck Rule, and flush decked.

No account is taken of erections above the awning deck of vessels of this type. This also applies to spar-deck vessels with but one exception. Since the longitudinal strains of vessels are greatest in the region of amidships, it follows that any efficient erection over this part of the length must add greatly to the strength. Such an erection receives full credit in assigning freeboard in the three-decked heavy deadweight carrier. A similar erection in the form of an efficient bridge on a spar-decked vessel, extending over and protecting the engine and boiler openings for at least two-fifths of the vessel's length, also receives recognition on account of its addition to strength and protection, though not in so great proportion as in the three-decked type. Thus such an erection on a spar-decked vessel of 20 feet moulded depth to main deck, merits a reduction of 3 inches on the freeboard.

For the sake of comparison, let us see the differences of freeboard in a vessel of each of these types—three-decked, spar-decked, and awning-decked, classed 100 A1 at Lloyd's. Let the length in each case be, say, 300 feet, the coefficient of fineness ·7, and the depth to main, spar, and awning deck 25 feet, the spar and awning deck each being 7 feet above the main deck.

The vessel of the first type would have a freeboard of 5 feet 0½ inches.
The spar-decked would have a freeboard of 6 feet 2 inches.
The awning-decked would have a freeboard of 8 feet 7¼ inches.
} For summer voyages.

But even after the coefficient of fineness is found, there are certain modifications to be carried out before the exact freeboard is arrived at, and these modifications will next be dealt with.

Let it be understood that only vessels equal in strength to Lloyd's 100 A1 will be considered. Vessels below this standard have increased freeboard:—

1. In spar-decked vessels, having iron-spar decks, and in awning-decked vessels having iron main decks, the freeboard required by the tables should be measured as if those decks were wood-covered. Also, in vessels where $\frac{7}{10}$ths, or more, of the main deck is covered by substantial enclosed erections, the freeboard found from the tables should be measured amidships from a wood deck, whether the deck be of wood or of iron. In applying this principle to vessels having shorter lengths of substantial enclosed erections, the reduction in freeboard, in consideration of its being measured from the iron deck, is to be regulated in proportion to the length of the deck covered by such erections.

Thus, in a vessel having erections covering $\frac{6}{10}$ths of the length, the reduction is $\frac{6}{10}$ths of $3\frac{1}{2}$ inches (the thickness of the wood deck), or 2 inches.

2. In flush-deck vessels of the one-, two-, and three-deck type, and those of the same type with erections extending over less than $\frac{4}{10}$ths of the length, having iron upper decks not sheathed with wood, the usual thickness of a wood deck is deducted from the moulded depth of the vessel, and the freeboard taken from the column in the tables corresponding with the diminished depth. Thus, a vessel of this type with 19 feet 10 inches moulded depth, with no laid wood deck, would be reckoned as 19 feet 6 inches depth in the tables. Taking the coefficient of fineness of this vessel at ·7, referring to the freeboard tables, the freeboard at 19 feet 10 inches depth would have been 3 feet $7\frac{1}{2}$ inches, where, owing to the reduction of the depth to 19 feet 6 inches, the freeboard is 3 feet $6\frac{1}{2}$ inches. When the erections in vessels of this type cover more than $\frac{4}{10}$ths, and less than $\frac{7}{10}$ths of the length, the correction for the wood deck is made as explained towards the end of the previous paragraph.

3. *Correction for Length.*—The freeboard tables show that in addition to the coefficient given with every depth, a fixed length is assigned. For example, a vessel with a coefficient of 0·7, and a depth of 16 feet, a length of 192 feet is assigned, but if the length of our vessel with the same coefficient and depth be, say, 212 feet, a correction must be made for the additional 20 feet of length. Wherever the standard length is exceeded, the freeboard is increased, and wherever it is less, the freeboard is diminished. The greater the proportion of length to depth, the greater the freeboard. The correction varies from about ·7 to 1·7 inches for a change of 10 feet in length. In the case before us, the correction is 1 inch per 10 feet, and thus for 20 feet it is 2 inches, which has to be added to the specified freeboard in the table. Had the vessel been 10 feet less than 192, the freeboard would have been reduced by 1 inch. The reason for this is easily understood when we remember that the vessel with greatest depth to length is the one most capable of resisting longitudinal bending, and therefore best adapted to carry the most deadweight. Where, however, steam-vessels with top-gallant forecastles, having long poops or raised quarter decks connected with bridge houses, the whole extending over $\frac{8}{10}$ths, or more, of the length of the vessel, the correction for excess of length should be half that specified in the tables; so that if the vessel we have taken as an example complied with these conditions, the freeboard would only have been increased by 1 inch, simply because with so great a length of substantial

erections, the vessel has practically been increased in depth, and consequently the proportion of depth to length is decreased.

The correction for length in spar-decked vessels varies from about ·9 to 1·5 inches per 10 feet, and for awning-decked vessels from about ·5 to ·8 of an inch.

4. *Sheer.*—The Board of Trade Tables specify a mean sheer for all types of vessels. Mean sheer is the sum of the sheers at the ends of the vessel, or at whatever part of the length it is specified, divided by 2 (see fig. 18).

Any increase in the mean sheer means an increase in the reserve buoyancy, and exactly where it is much needed—viz., at the ends, giving additional rising power when the ship dips into the trough of a sea, not to mention the increase of freeboard. This excess of buoyancy is recognised by the Board of Trade, and a deduction allowed in the freeboard according to the amount of the excess of sheer.

For all flush-decked vessels, the mean sheer is found by dividing the length by 10, and adding 10 to the result. Thus a vessel 300 feet long will have a mean sheer of $\frac{300}{10} + 10 = 40$ inches.

Flush-decked vessels of the one-, two-, and three-decked type, with or without a short poop, a topgallant forecastle, and a bridge house completely closed in at the ends, or a long poop, or a raised quarter deck connected with an efficiently closed-in bridge house, where the sheer is greater than in the Table, and is of a gradual character, the reduction in freeboard is found by dividing the difference between the actual sheer and the mean sheer provided for in the Table, by 4. For example, a vessel 300 feet long, with a mean sheer of 46 inches, has a reduction in the freeboard of $\frac{46 - 40}{4} = \frac{6}{4} = 1\frac{1}{2}$ inches. No allowance is given to spar- and awning-decked vessels for excess of sheer; but in any of these types, where the actual mean sheer is less than the mean sheer by the rule, the difference divided by 4 gives the increase of freeboard required for reduced reserved buoyancy.

Line 1 in the Table (p. 258) is for vessels having short poops or forecastles only, or when, in addition, there is a bridge house, with alleyways open at one or both ends. In these vessels the important point is not the amount of sheer at the ends of the vessel, for great additional buoyancy is given there already by the erections, but over the length uncovered by substantial erections. Therefore, in this particular case, the sheer is measured at one-eighth the vessel's length from stem and stern. One-fourth the difference between the sheer and the actual mean is approximately the amount of increase or decrease of freeboard, as the case may be.

LENGTH OVER WHICH SHEER IS MEASURED.

	100	150	200	250	300	350	400
	Mean Sheer in Inches over the Length specified.						
No. 1,	14	18	22	26	30	34	38
No. 2,	14½	18½	23	27	31	35½	40

Line 2 in the Table is for vessels having short forecastles only, and in this case the sheer is measured at points (1) one-eighth from the stem, and (2), at the stern post, and correction for additional sheer is made as in the previous case.

Note.—In flush-decked vessels, and in vessels having short poops and forecastles, the excess of sheer for which an allowance is made is not more than one-half the total standard mean sheer for the size of the ship.

5. *Round of Beam.*—The stipulated round of midship beam is a quarter of an inch for every foot of the length of the midship beam. Here, again, any excess in the round of beam means an increase of reserve buoyancy, and an allowance is made in the freeboard. When the round of beam in flush-decked vessels is greater or less than that given by the Rule, divide the difference in inches by 2, and diminish or increase the freeboard accordingly by this amount.

EXAMPLE.—The beam of a vessel is 40 feet, and the round of beam as measured is 12 inches. This is 2 inches more than is required by the Rule, therefore $\frac{2}{2} = 1$ = amount of decrease of freeboard. Where the deck is partially covered with erections, the amount of the allowance for round of beam depends upon the extent of the upper deck uncovered.

This rule for round of beam does not apply to spar- and awning-decked vessels.

6. *Corrections for Erections on Deck.*—By erections on deck is meant all closed-in or partially closed-in structures erected above the upper deck of vessels built to the one-, two-, or three-deck rule,—for example, bridges, poops, forecastle, raised quarter decks, partial awning decks, etc., and also strong bridges in spar-decked vessels covering the engine and boiler openings if such erections extend over at least two-fifths of the vessel's length. Now, erections such as those enumerated add greatly, according to their proportions, to the reserve buoyancy, and, as in the case of sheer and camber, the merits of the values of these erections

and additions to the reserve buoyancy and structural strength, and the protection afforded to vulnerable localities such as deck openings, are fully considered by the Board of Trade, and deductions in the freeboard allowed accordingly. In assigning allowances for erections, the complete awning deck is the standard worked from.

As has already been pointed out, in taking out the freeboard for a vessel with an awning deck, the coefficient of fineness and the depth are both taken to the main deck, and not to the awning deck, and the freeboard is assigned and measured down from the main deck. The awning deck thus comes to be what might be termed a complete erection, extending all fore and aft, and covering the main deck. All other erections partially covering the vessel have their deductions on the freeboard made as a fractional part of a percentage of the allowance for a complete awning deck. The reader will easily see that a complete erection such as an awning deck is always better than a partial one, comparing length for length; for instance, an erection covering three-fourths of the vessel's length cannot be taken on equal merits with three-fourths of the awning deck, since it is evident that a well has been created on the weather deck in some part of the vessel's length, forming a break in the longitudinal strength of greater or less importance, and a means thus provided, to some extent, for lodging seas on board, or the possibility created of bulkheads being damaged, etc., therefore, three-fourths' erections do not receive the same credit as three-fourths of the awning deck, but a less fractional part. Again, the allowance varies with the nature of the erections. This we shall endeavour to show by arranging the erections according to their respective values.

Taking a vessel 204 feet long, and, therefore, with an awning-deck erection of the same length, we shall see what credit is given, in assigning the freeboard according to the Board of Trade Tables of Freeboard, for the various kinds of erections in comparison with the allowance for the complete awning-deck erection. This, it is hoped, will better enable the reader to grasp the comparative values of the erections.

Let the moulded depth of the vessel be 17 feet, and the coefficient of fineness ·7.

The freeboard, had the vessel been flush-decked, to the two-deck rule would have been 2 feet 10½ inches.

The freeboard, had the vessel been awning-decked, would have been 1 foot 4 inches.

2 feet 10½ inches − 1 foot 4 inches = 1 foot 6½ inches = 18½ inches, this difference being the allowance for the complete awning-deck erection.

Note.—For a raised quarter deck, 4 feet high, and connected with a bridge house, covering the engine and boiler openings with an efficient bulkhead at the fore end, the allowance is made as though it were a part of the actual bridge, and equal in height. A decrease is made in the deduction, if the raised quarter deck extends over the engine and boiler openings, or if it is less than 4 feet high.

The deductions for erections are as follows :—

1. When the combined length of poop, or raised quarter deck, connected with a bridge house covering in the engine and boiler openings, and with an efficient bulkhead at the fore end, and topgallant forecastle is—

(a) $\frac{9}{10}$ or ·9 of the length of the vessel,
the deduction is ·85 of the allowance for a complete awning deck. Freeboard is $\frac{85}{100}$ of 18½ inches = 15¾ inches.

(b) $\frac{8}{10}$ or ·8 of the length of the vessel,
the deduction is ·75 of the allowance for awning deck. Freeboard is $\frac{75}{100}$ of 18½ inches = 13¾ inches.

(c) $\frac{7}{10}$ or ·7 of the length of the vessel,
the deduction is ·63 of the allowance for awning deck. Freeboard is $\frac{63}{100}$ of 18½ inches = 11¾ inches.

(d) $\frac{6}{10}$ or ·6 of the length of the vessel,
the deduction is ·5 of the allowance for awning deck. Freeboard is $\frac{50}{100}$ of 18½ inches = 9¼ inches.

These comprising the maximum reductions, it is essential that the erections be of a most substantial character, the deck openings effectually protected, the crew berthed in the bridge house, or with satisfactory arrangements to enable them to get backward and forward to their quarters, and sufficient clearing ports in the bulwarks to speedily clear the deck of water. Vessels of this type having no topgallant forecastle are allowed a less deduction in the amount of freeboard than would otherwise be given for the same length of erections. And if the bridge be a short one, in front of, and only partially covering the engine and boiler openings, again a less deduction is made. No allowance is granted for a monkey forecastle which is less in height than the main or topgallant rail.

Note 1.—A special allowance is made on the freeboard of vessels of the foregoing or *well-decked* type when their erections extend over $\frac{7}{10}$ of their length, when their bridge bulkheads are specially strengthened, and when the area of their water clearing ports in the bulwarks is at least 25 per cent. in excess of the rule requirements. Such additional allowance must not exceed 2 inches.

Note 2.—Special reductions in the freeboard may be obtained in strong well-decked vessels of the modern type, having erections covering at least $\frac{85}{100}$ths of the length of the ship, the bridge house alone covering at least $\frac{4}{10}$ths of the length when extra strength is introduced, as given in section 44, Lloyd's *Rules for* 1889, for iron and steel vessels. But, in no case must the freeboards assigned to these vessels be less than would be assigned for a complete awning deck.

2. In vessels with topgallant forecastles, short poops, and bridge houses covering engines and boilers in steamers, with efficient iron bulkheads at their ends, when the combined length of erections is—

(a) $\frac{6}{10}$ or ·5 of the length of the vessel,
the deduction is ·4 of the allowance for awning deck. Freeboard is $\frac{2}{5}$ of $18\frac{1}{4}$ inches = $7\frac{1}{4}$ inches.

(b) $\frac{4}{10}$ or ·4 of the length of the vessel,
the deduction is ·33 of the allowance for awning deck. Freeboard is $\frac{1}{3}$ of $18\frac{1}{4}$ inches = $6\frac{1}{4}$ inches.

3. In vessels with topgallant forecastles and bridge houses only, covering engines and boilers in steamers, with efficient iron bulkheads at the ends, when the combined length of erections is—

(a) $\frac{4}{10}$ or ·4 of the length of the vessel,
the deduction is ·3 of the allowance for awning deck. Freeboard is $\frac{3}{10}$ of $18\frac{1}{4}$ inches = $5\frac{1}{4}$ inches.

(b) $\frac{3}{10}$ or ·3 of the length of the vessel,
the deduction is ·25 of the allowance for awning deck. Freeboard is $\frac{1}{4}$ of $18\frac{1}{4}$ inches = $4\frac{5}{8}$ inches.

4. In vessels with only topgallant forecastles and poops, the latter with an efficient bulkhead at the fore end, when the combined length of erections is—

(a) $\frac{2}{5}$ of the length of the vessel,
the deduction is $\frac{1}{10}$ of the freeboard for the vessel flush decked. Freeboard is $\frac{1}{10}$ of $34\frac{1}{4}$ inches = $3\frac{1}{4}$ inches.

(b) $\frac{1}{4}$ of the length of the vessel,
the deduction is $\frac{8}{100}$ of the freeboard for the vessel flush decked. Freeboard is $\frac{8}{100}$ of $34\frac{1}{4}$ inches = $2\frac{3}{4}$ inches.

5. In vessels with topgallant forecastles only, the deduction in freeboard is only one-half that prescribed in the previous paragraph. Thus, were the erection $\frac{1}{5}$ of the length of the vessel, the deduction would be $\frac{4}{100}$ of freeboard for the vessel flush decked = $1\frac{3}{8}$ inches.

6. In vessels with poops only, the allowance is one-half that for the previous paragraph (No. 5) for forecastles only of the same length. Thus, did the length of the poop equal $\frac{1}{5}$ erection, the deduction would be $\frac{2}{100}$ of freeboard for the vessel flush decked = $\frac{3}{4}$ inch.

7. In vessels with raised quarter decks only, not less than 4 feet high, the deduction is at the same rate as in the preceding paragraph (No. 6).

8. In all vessels when the topgallant forecastle is not closed at the after end by an efficient bulkhead, the length is never to be estimated at a greater full value than $\frac{1}{8}$ of the length of the ship, but any extension beyond this may be estimated at one-half the value. For example, a vessel 200 feet long has an

open forecastle ¼ of the length of the ship, or 50 feet; its value for deduction is $25 + 12\tfrac{1}{2} = 37\tfrac{1}{2}$ feet. When the topgallant forecastle has an efficient bulkhead with an elongation abaft that bulkhead, the full allowance is given on the entire length of the closed-in portion, and afterwards according to the previous example.

9. When the poop has no bulkhead, one-half its length is allowed for at the rate of a closed poop.

10. For bridge houses extending from side to side of the vessel, when closed at the fore and open at the after end, with all deck openings, doors, etc., properly protected, ¾ of the length is estimated as the value for deduction. When both ends of the alleyways are open, one-half of the length is estimated as the value for deduction.

Note.—Although it is possible to have occasional gales in the summer season as severe as any in winter, yet it is quite unnecessary to remind anyone acquainted with the sea that it is in the latter of these seasons that boisterous weather is looked for. And thus in summer, when the danger from the weather is decreased, there is no reason why a vessel should not be allowed, to some extent, to carry more cargo than in winter. The Board of Trade, therefore, allow a deduction from the winter freeboard for summer voyages, amounting from 1 to 9 or more inches. Summer voyages from European and Mediterranean ports are to be made from April to September inclusive. In other parts of the world, the reduced freeboard should be used during the corresponding or recognised summer months. Double the above reduction to be allowed for voyages in the fine season in the Indian seas, between the limits of Suez and Singapore. Vessels, up to and including 330 feet in length, engaged in North Atlantic trades, are required to have an addition of 2 inches to their freeboard, from October to March inclusive.

A few examples illustrating how the freeboard is ascertained for different types of vessels may be helpful in showing the practical application of the freeboard rules, etc.

Let it be understood that by Tables A, B, C, and D is meant the freeboard as specified by the Board of Trade for vessels of various types.

Table A.—Flush-decked vessels of the one-, two-, and three-deck type.
Table B.—Spar-decked vessels.
Table C.—Awning-decked vessels.
Table D.—Sailing-ships.

Example I.

Flush-decked screw steamer, 300 feet long, 38 feet broad, and 21 feet depth moulded (no account is taken for any erection required for this vessel).

Sheer forward, 8 feet.
Sheer aft, 3 ,,
Coefficient of fineness, 0·8.
Wood deck on upper deck, 4 inches thick.

FREEBOARD.

	Ft.	In.
By Table A, for 21 feet depth moulded, the freeboard is,	4	3

The mean sheer, by rule, for a vessel 300 feet long is,

$$\frac{300}{10} + 10 = 40 \text{ inches.}$$

In the example the vessel has 8 feet 0 inches sheer forward,
and 3 ,, 0 ,, sheer aft,

2) 11 ,, 0 ,,

5 ,, 6 ,, = 66 inches mean sheer,

which is an excess of 66 − 40 = 26 inches (only half of 40* = 20 inches is allowed), therefore,

$$\frac{20}{4} = 5 = \text{reduction in freeboard,} \qquad \qquad 5 \;.$$

                                                                                3 10

The freeboard 4 feet 3 inches from Table A was for a vessel 252 feet long. A correction of 1·2 feet per 10 feet of additional length must now be made and added to the freeboard.

300 − 252 = 48

$$\frac{48}{10} \times 1 \cdot 2 = 5 \cdot 76, \text{ about } 5\tfrac{3}{4} \text{ inches,} \qquad 5\tfrac{3}{4}$$

Winter freeboard from top of wood deck,   4   3¾

Deduct the thickness of the wood deck less the thickness of the stringer plate (½ inch), ...   3½

                                                                                 4   0¼

Less for summer, ...   2¾

Summer freeboard measured down from the top of the stringer plate at the side of the vessel at the middle of the length, }   3   9¾

EXAMPLE II.

This vessel is identical in every respect with Example I., except that she has a raised quarter deck 4 feet high and 100 feet long, connected with a bridge house 80 feet long, with closed ends, and a topgallant forecastle 30 feet long closed at the after end. The upper deck is iron uncovered with wood.

These erections altogether measure 100 + 80 + 30 = 210 feet.

$$\frac{210}{308} = \frac{7}{10} \text{ erections.}$$

If we turn back to the remarks on *Erections* in this chapter we find that $\frac{7}{10}$ erections are allowed a reduction in the freeboard equal to $\frac{63}{100}$ of the allowance for a complete awning deck.

* See *Sheer*.

	Ft.	In.
By Table A, the freeboard of the vessel flush decked is,	4	3
The correction for additional sheer is,		6½
	3	8½
By Table C, for a depth of 21 feet, the freeboard is,	2	2
(No account is taken of additional sheer in awning-decked vessels.)		
The allowance for a complete awning deck is,	1	6½

$\frac{63}{100}$ of 1 foot 6½ inches = about 11½ inches = allowance for $\frac{7}{10}$ erections.

Now let us total up all the deductions.

1. Deduction for excess of sheer = ... 6½
2. Deduction for $\frac{7}{10}$ erections = ... 11½
3. Deduction for the thickness of a wood deck (less thickness of stringer plate) when the upper deck is of iron, and covered by $\frac{7}{10}$ or more of substantial erections, ... 3½

Deductions, ... 1 9½

We have still to make the correction for excess of length.
$\frac{48}{10} \times 1\cdot 2 = 5\cdot 76$. But only one-half of this should be taken, since the deck is covered by over $\frac{6}{10}$ erections.

$5\cdot 76 \div 2 = 2\cdot 88 = 2\frac{3}{4}$ inches to be added to the freeboard, therefore the nett deduction will be 1 foot 9½ inches less 2¾ inches = 1 foot 6¾ inches.

	Ft.	In.
The freeboard by Table A is,	4	3
The deductions amount to,	1	6¾
Winter freeboard,	2	8¼
Less for summer,		2½
Summer freeboard measured down from the top of the stringer plate at the side of the vessel at the middle of the length,	2	5¾

EXAMPLE III.

This vessel has the same dimensions as Example I., with a poop 50 feet long, a bridge house 60 feet long, and a topgallant forecastle 40 feet long.

Iron upper deck uncovered with wood.
Coefficient of fineness = 0·8.

The bridge house is closed at the fore end with an efficient bulkhead, but open at the after end. Three-quarters of its length will, therefore, be reckoned in the erections. ¾ of 60 = 45 feet.

FREEBOARD.

Total erections, 50 + 45 + 40 = 135.

$$\frac{135}{300} = \frac{4}{10} \text{ erections.}$$

The mean sheer at one-eighth the length from each end of the vessel is 40 inches.

	Ft.	In.
By Table A, the freeboard is,	4	3
The mean sheer by the Rule is 30 inches.		
40 − 30 = 10 excess of sheer, $\frac{10}{4}$ = 2½ reduction in freeboard,		2½
	4	0½
Correction for length (added),		5¾
	4	6¼
By Table C, the freeboard is,	2	2
Correction for length, at the rate of 0·6 foot per 10 feet, is $\frac{48}{10} \times 0\cdot6 = 2¾$ to be added,		2¾
	2	4¾
By Table A, after corrections for sheer and length have been made, the freeboard is,	4	6¼
By Table C, after correction for length only has been made, the freeboard is,	2	4¾
Allowance for complete awning deck =	2	1½

The reduction in freeboard for $\frac{4}{10}$ erections is ⅓ of the allowance for a complete awning deck = ⅓ of 2 feet 1½ inches = 8½ inches.

By Table A, after the corrections for sheer and length have been made, the freeboard is,	4	6¼
Reduction for $\frac{4}{10}$ erections,		8½
	3	9¾
Reduction for wood deck $\frac{4}{10}$ of 3½ =		1½
Winter freeboard,	3	8¼
Less for summer,		2½
Summer freeboard measured down from the top of the stringer plate at the side of the vessel at the middle of the length,	3	5¾

Example IV.

Spar-decked steamer 260 feet long, 36 feet broad, and 24 feet depth moulded.

Height of spar deck above main deck, 7 feet.

Spar deck laid with wood deck 4 inches thick.
Coefficient of fineness = 0·8.
Depth moulded to main deck = 17 feet.

	Ft.	In.
By Table B, for 17 feet depth moulded, the freeboard is,	6	4¼

This is for a length of 288 feet. A correction must be made for decrease in length at 1 inch per 10 feet, and subtracted from the freeboard.
$$288 - 260 = 28.$$
$$\frac{28}{10} \times 1 = 2\cdot8 = \text{about } 2\tfrac{3}{4} \text{ inches,} \qquad 2\tfrac{3}{4}$$

| Winter freeboard from top of wood deck, | 6 | 1¾ |
| Reduction for wood deck less thickness of stringer plate, | | 3¼ |

| Winter freeboard, | 5 | 10½ |
| Reduction for summer, | | 3 |

| Summer freeboard measured down from the top of the stringer plate at the side of the vessel at the middle of the length, | 5 | 7½ |

Example V.

Awning-decked vessel 230 feet long, 32 feet broad, and 17 feet depth, moulded to main deck.

Height from main deck to awning deck, 7 feet.
Coefficient of fineness, 0·8.
Mean sheer, 35 inches.
Iron awning deck.

	Ft.	In.
By Table C, for a depth of 17 feet, the freeboard is,	5	1½

The mean sheer by the Rule is $\frac{230}{10} + 10 = 33$, and this vessel having sheer in excess of the Rule, no correction is made.

This freeboard is for a vessel 204 feet long. A correction of 0·5 per 10 feet of additional length must be made, and added to the freeboard.
$$230 - 204 = 26.$$
$$\frac{26}{10} \times 0\cdot5 = 1\tfrac{1}{4} \text{ inches,} \qquad 1\tfrac{1}{4}$$

| | 1 | 6¾ |
| Deduction for thickness of wood deck less the thickness of stringer plate when awning deck is of iron, | | 3½ |

| Winter freeboard, | 1 | 3¼ |
| Less for summer, | | 2¼ |

| Summer freeboard measured down from the top of the stringer plate at the side of the vessel at the middle of the length, | 1 | 0¾ |

Example VI.

Sailing-vessel 200 feet long, 34 feet broad, 19 feet deep.
Mean sheer, 40 inches.
Wood deck, 3½ inches thick.
Coefficient of fineness, 0·7.

	Ft.	In.
By Table D, for 19 feet depth the freeboard is, ...	3	10

The mean sheer by the Rule is $\frac{200}{10} + 10 = 30$ inches.

40 − 30 = 10 inches excess of mean sheer.

$\frac{10}{4} = 2\frac{1}{2}$ inches reduction in freeboard, . .   2½

                                                           3   7½

The length by the Rule for 19 feet depth is 190. A correction must be made at the rate of 1·2 inches per 10 feet excess of length, and added to the freeboard.
200 − 190 = 10 feet excess of length.

$\frac{10}{10} \times 1\cdot 2$ = about 1¼ inches to be added to the freeboard,   1¼

Freeboard from top of wood deck, ...   3   8¾
Deduction for thickness of wood deck less the thickness of stringer plate, ...   3

Summer freeboard measured down from the top of the stringer plate at the side of the vessel at the middle of the length, ...   3   5¾

Note.—Wherever definite rules have been quoted in the chapters on Tonnage and Freeboard, it will be clearly understood that they have been "extracted" from the Board of Trade instructions to surveyors.

TABLE OF NATURAL SINES AND COTANGENTS.

Degree.	Sine.	Cotangent.	Degree.	Degree.	Sine.	Cotangent.	Degree.	Degree.	Sine.	Cotangent.	Degree.
0	·0000	Infinite.	90	12	·2079	4·7046	78	24	·4067	2·2460	66
¼	·0043	229·1817	89¾	12¼	·2121	4·6057	77¾	24¼	·4107	2·2199	65¾
½	·0087	114·5887	89½	12½	·2164	4·5107	77½	24½	·4146	2·1943	65½
¾	·0130	76·3900	89¼	12¾	·2206	4·4193	77¼	24¾	·4186	2·1691	65¼
1	·0174	57·2899	89	13	·2249	4·3314	77	25	·4226	2·1445	65
1¼	·0218	45·8293	88¾	13¼	·2292	4·2468	76¾	25¼	·4265	2·1203	64¾
1½	·0261	38·1884	88½	13½	·2334	4·1653	76½	25½	·4305	2·0965	64½
1¾	·0305	32·7302	88¼	13¾	·2376	4·0866	76¼	25¾	·4344	2·0732	64¼
2	·0349	28·6362	88	14	·2419	4·0107	76	26	·4383	2·0503	64
2¼	·0392	25·4517	87¾	14¼	·2461	3·9375	75¾	26¼	·4422	2·0277	63¾
2½	·0436	22·9037	87½	14½	·2503	3·8667	75½	26½	·4461	2·0056	63½
2¾	·0479	20·8188	87¼	14¾	·2546	3·7982	75¼	26¾	·4500	1·9839	63¼
3	·0523	19·0811	87	15	·2588	3·7320	75	27	·4539	1·9626	63
3¼	·0566	17·6105	86¾	15¼	·2630	3·6679	74¾	27¼	·4578	1·9416	62¾
3½	·0610	16·3498	86½	15½	·2672	3·6058	74½	27½	·4617	1·9209	62½
3¾	·0654	15·2570	86¼	15¾	·2714	3·5457	74¼	27¾	·4656	1·9006	62¼
4	·0697	14·3006	86	16	·2756	3·4874	74	28	·4694	1·8807	62
4¼	·0741	13·4566	85¾	16¼	·2798	3·4308	73¾	28¼	·4733	1·8610	61¾
4½	·0784	12·7062	85½	16½	·2840	3·3759	73½	28½	·4771	1·8417	61½
4¾	·0828	12·0346	85¼	16¾	·2881	3·3226	73¼	28¾	·4809	1·8227	61¼
5	·0871	11·4300	85	17	·2923	3·2708	73	29	·4848	1·8040	61
5¼	·0915	10·8829	84¾	17¼	·2965	3·2205	72¾	29¼	·4886	1·7856	60¾
5½	·0958	10·3854	84½	17½	·3007	3·1715	72½	29½	·4924	1·7674	60½
5¾	·1001	9·9310	84¼	17¾	·3048	3·1239	72¼	29¾	·4962	1·7496	60¼
6	·1045	9·5143	84	18	·3090	3·0776	72	30	·5000	1·7320	60
6¼	·1088	9·1309	83¾	18¼	·3131	3·0325	71¾	30¼	·5037	1·7147	59¾
6½	·1132	8·7768	83½	18½	·3173	2·9886	71½	30½	·5075	1·6976	59½
6¾	·1175	8·4489	83¼	18¾	·3214	2·9459	71¼	30¾	·5112	1·6808	59¼
7	·1218	8·1443	83	19	·3255	2·9042	71	31	·5150	1·6642	59
7¼	·1261	7·8606	82¾	19¼	·3296	2·8635	70¾	31¼	·5187	1·6479	58¾
7½	·1305	7·5957	82½	19½	·3338	2·8229	70½	31½	·5224	1·6318	58½
7¾	·1348	7·3478	82¼	19¾	·3379	2·7852	70¼	31¾	·5262	1·6159	58¼
8	·1391	7·1153	82	20	·3420	2·7474	70	32	·5299	1·6003	58
8¼	·1434	6·8968	81¾	20¼	·3461	2·7106	69¾	32¼	·5336	1·5849	57¾
8½	·1478	6·6911	81½	20½	·3502	2·6746	69½	32½	·5373	1·5696	57½
8¾	·1521	6·4971	81¼	20¾	·3542	2·6394	69¼	32¾	·5409	1·5546	57¼
9	·1564	6·3137	81	21	·3583	2·6050	69	33	·5446	1·5398	57
9¼	·1607	6·1402	80¾	21¼	·3624	2·5714	68¾	33¼	·5482	1·5252	56¾
9½	·1650	5·9757	80½	21½	·3665	2·5386	68½	33½	·5519	1·5108	56½
9¾	·1693	5·8196	80¼	21¾	·3705	2·5065	68¼	33¾	·5555	1·4966	56¼
10	·1736	5·6712	80	22	·3746	2·4750	68	34	·5591	1·4825	56
10¼	·1779	5·5300	79¾	22¼	·3786	2·4443	67¾	34¼	·5628	1·4686	55¾
10½	·1822	5·3955	79½	22½	·3826	2·4142	67½	34½	·5664	1·4550	55½
10¾	·1865	5·2671	79¼	22¾	·3867	2·3847	67¼	34¾	·5699	1·4414	55¼
11	·1908	5·1445	79	23	·3907	2·3558	67	35	·5735	1·4281	55
11¼	·1950	5·0273	78¾	23¼	·3947	2·3275	66¾	35¼	·5771	1·4149	54¾
11½	·1993	4·9151	78½	23½	·3987	2·2998	66½	35½	·5807	1·4019	54½
11¾	·2036	4·8076	78¼	23¾	·4027	2·2726	66¼	35¾	·5842	1·3890	54¼
Degree.	Cosine.	Tangent.	Degree.	Degree.	Cosine.	Tangent.	Degree.	Degree.	Cosine.	Tangent.	Degree.

TABLE OF NATURAL SINES AND COTANGENTS (*continued*).

Degree.	Sine.	Cotangent.	Degree.	Degree.	Sine.	Cotangent.	Degree.	Degree.	Sine.	Cotangent.	Degree.
36	·5877	1·3763	54	48	·7431	·9004	42	60	·8660	·5773	30
36¼	·5913	1·3638	53¾	48¼	·7460	·8925	41¾	60¼	·8681	·5715	29¾
36½	·5948	1·3514	53½	48½	·7489	·8847	41½	60½	·8703	·5657	29½
36¾	·5983	1·3391	53¼	48¾	·7518	·8769	41¼	60¾	·8724	·5600	29¼
37	·6018	1·3270	53	49	·7547	·8692	41	61	·8746	·5543	29
37¼	·6052	1·3150	52¾	49¼	·7575	·8616	40¾	61¼	·8767	·5486	28¾
37½	·6087	1·3032	52½	49½	·7604	·8540	40½	61½	·8788	·5429	28½
37¾	·6122	1·2915	52¼	49¾	·7632	·8465	40¼	61¾	·8808	·5373	28¼
38	·6156	1·2799	52	50	·7660	·8391	40	62	·8829	·5317	28
38¼	·6190	1·2684	51¾	50¼	·7688	·8316	39¾	62¼	·8849	·5261	27¾
38½	·6225	1·2571	51½	50½	·7716	·8243	39½	62½	·8870	·5205	27½
38¾	·6259	1·2459	51¼	50¾	·7743	·8170	39¼	62¾	·8890	·5150	27¼
39	·6293	1·2348	51	51	·7771	·8097	39	63	·8910	·5095	27
39¼	·6327	1·2239	50¾	51¼	·7798	·8025	38¾	63¼	·8929	·5040	26¾
39½	·6360	1·2130	50½	51½	·7826	·7954	38½	63½	·8949	·4985	26½
39¾	·6394	1·2023	50¼	51¾	·7853	·7883	38¼	63¾	·8968	·4931	26¼
40	·6427	1·1917	50	52	·7880	·7812	38	64	·8987	·4877	26
40¼	·6461	1·1812	49¾	52¼	·7906	·7742	37¾	64¼	·9006	·4823	25¾
40½	·6494	1·1708	49½	52½	·7933	·7673	37½	64½	·9025	·4769	25½
40¾	·6527	1·1605	49¼	52¾	·7960	·7604	37¼	64¾	·9044	·4716	25¼
41	·6560	1·1503	49	53	·7986	·7535	37	65	·9063	·4663	25
41¼	·6593	1·1402	48¾	53¼	·8012	·7467	36¾	65¼	·9081	·4610	24¾
41½	·6626	1·1302	48½	53½	·8038	·7399	36½	65½	·9099	·4557	24½
41¾	·6658	1·1204	48¼	53¾	·8064	·7332	36¼	65¾	·9117	·4504	24¼
42	·6691	1·1106	48	54	·8090	·7265	36	66	·9135	·4452	24
42¼	·6723	1·1091	47¾	54¼	·8115	·7198	35¾	66¼	·9153	·4400	23¾
42½	·6755	1·0913	47½	54½	·8141	·7132	35½	66½	·9170	·4348	23½
42¾	·6788	1·0817	47¼	54¾	·8166	·7067	35¼	66¾	·9187	·4296	23¼
43	·6819	1·0723	47	55	·8191	·7002	35	67	·9205	·4244	23
43¼	·6851	1·0630	46¾	55¼	·8216	·6937	34¾	67¼	·9222	·4193	22¾
43½	·6883	1·0537	46½	55½	·8241	·6872	34½	67½	·9238	·4142	22½
43¾	·6915	1·0446	46¼	55¾	·8265	·6808	34¼	67¾	·9255	·4091	22¼
44	·6946	1·0355	46	56	·8290	·6745	34	68	·9271	·4040	22
44¼	·6977	1·0265	45¾	56¼	·8314	·6681	33¾	68¼	·9288	·3989	21¾
44½	·7009	1·0176	45½	56½	·8338	·6618	33½	68½	·9304	·3939	21½
44¾	·7040	1·0087	45¼	56¾	·8362	·6556	33¼	68¾	·9320	·3888	21¼
45	·7071	1·0000	45	57	·8386	·6494	33	69	·9335	·3838	21
45¼	·7101	·9913	44¾	57¼	·8410	·6432	32¾	69¼	·9351	·3788	20¾
45½	·7132	·9826	44½	57½	·8433	·6370	32½	69½	·9366	·3738	20½
45¾	·7163	·9741	44¼	57¾	·8457	·6309	32¼	69¾	·9381	·3689	20¼
46	·7193	·9656	44	58	·8480	·6248	32	70	·9396	·3639	20
46¼	·7223	·9572	43¾	58¼	·8503	·6188	31¾	70¼	·9411	·3590	19¾
46½	·7253	·9489	43½	58½	·8526	·6128	31½	70½	·9426	·3541	19½
46¾	·7283	·9407	43¼	58¾	·8549	·6068	31¼	70¾	·9440	·3492	19¼
47	·7313	·9325	43	59	·8571	·6008	31	71	·9455	·3443	19
47¼	·7343	·9243	42¾	59¼	·8594	·5949	30¾	71¼	·9469	·3394	18¾
47½	·7372	·9163	42½	59½	·8616	·5890	30½	71½	·9483	·3345	18½
47¾	·7402	·9083	42¼	59¾	·8638	·5831	30¼	71¾	·9496	·3297	18¼
Degree.	Cosine.	Tangent.	Degree.	Degree.	Cosine.	Tangent.	Degree.	Degree.	Cosine.	Tangent.	Degree.

Table of Natural Sines and Cotangents (*continued*).

Degree.	Sine.	Cotangent.	Degree.	Degree.	Sine.	Cotangent.	Degree.	Degree.	Sine.	Cotangent.	Degree.
72	·9510	·3249	18	78¼	·9790	·2080	11¾	84½	·9953	·0962	5½
72¼	·9523	·3201	17¾	78½	·9799	·2034	11½	84¾	·9958	·0918	5¼
72½	·9537	·3152	17½	78¾	·9807	·1989	11¼	85	·9961	·0874	5
72¾	·9550	·3105	17¼	79	·9816	·1943	11	85¼	·9965	·0830	4¾
73	·9563	·3057	17	79¼	·9824	·1898	10¾	85½	·9969	·0787	4½
73¼	·9575	·3009	16¾	79½	·9832	·1853	10½	85¾	·9972	·0743	4¼
73½	·9588	·2962	16½	79¾	·9840	·1808	10¼	86	·9975	·0699	4
73¾	·9600	·2914	16¼	80	·9848	·1763	10	86¼	·9978	·0655	3¾
74	·9612	·2867	16	80¼	·9855	·1718	9¾	86½	·9981	·0611	3½
74¼	·9624	·2820	15¾	80½	·9862	·1673	9½	86¾	·9983	·0567	3¼
74½	·9636	·2773	15½	80¾	·9869	·1628	9¼	87	·9986	·0524	3
74¾	·9647	·2726	15¼	81	·9876	·1583	9	87¼	·9988	·0480	2¾
75	·9659	·2679	15	81¼	·9883	·1539	8¾	87½	·9990	·0436	2½
75¼	·9670	·2632	14¾	81½	·9890	·1494	8½	87¾	·9992	·0392	2¼
75½	·9681	·2586	14½	81¾	·9896	·1449	8¼	88	·9993	·0349	2
75¾	·9692	·2539	14¼	82	·9902	·1405	8	88¼	·9995	·0305	1¾
76	·9702	·2493	14	82¼	·9908	·1360	7¾	88½	·9996	·0261	1½
76¼	·9713	·2446	13¾	82½	·9914	·1316	7½	88¾	·9997	·0218	1¼
76½	·9723	·2400	13½	82¾	·9920	·1272	7¼	89	·9998	·0174	1
76¾	·9733	·2354	13¼	83	·9925	·1227	7	89¼	·9999	·0130	¾
77	·9743	·2308	13	83¼	·9930	·1183	6¾	89½	·9999	·0087	½
77¼	·9753	·2262	12¾	83½	·9935	·1139	6½	89¾	·9999	·0043	¼
77½	·9762	·2216	12½	83¾	·9940	·1095	6¼	90	1·0000	·0000	0
77¾	·9772	·2171	12¼	84	·9945	·1051	6				
78	·9781	·2125	12	84¼	·9949	·1006	5¾				

| Degree. | Cosine. | Tangent. | Degree. | Degree. | Cosine. | Tangent. | Degree. | Degree. | Cosine. | Tangent. | Degree. |

Note.—By using these tables backwards and calling 90°, the upright or 0°, the sines become cosines and the cotangents become tangents.

CHAPTER X. (Section I.)

CALCULATIONS.

Contents.—Useful Tables and Rules—Calculation of Weight of Steel Plate—Solid Stanchion—Hollow Stanchion—Gallons in Fresh-Water Tank—Tons in Coal Bunker—Rectangular Barge's Displacement and "Tons per Inch" Immersion—Simpson's Three Rules and Graphic Explanations—Calculation of Area of Deck or Waterplane—"Tons per Inch" Immersion of Ship's Waterplane—Ship's Displacement—Centre of Gravity of a Waterplane, Longitudinally or Transversely — Centre of Buoyancy, Vertically and Longitudinally—Moment of Inertia — Transverse Metacentre above Centre of Buoyancy—Centre of Gravity—Longitudinal Metacentre above Centre of Buoyancy—Alteration of Trim—Area of Section and Volume and Centre of Gravity of Wedge of Immersion or Emersion—Centre of Effort.

USEFUL TABLES, RULES, AND PRELIMINARY CALCULATIONS.

Useful Tables.

1 cubic foot contains 6¼ gallons.
1 ,, ,, of fresh water weighs 1000 ozs. or 62½ lbs.
1 ,, ,, of salt ,, ,, 1025 ,, 64 ,,

In some localities, where the water is brackish, its weight per cubic foot is between 1000 and 1025 ozs., and in other localities, such as the Red Sea, a specific gravity of over 1·025 is found.

36 cubic feet of fresh water weigh 1 ton.
35 ,, ,, salt ,, ,, ,,
40 to 50 ,, ,, coal ,, ,, ,,
1 ,, foot steel weighs 490 lbs.
1 ,, ,, wrought-iron ,, 480 ,,
1 ,, ,, cast-iron ,, 454 ,,

Steel plates and bars are therefore about 2 per cent. heavier than iron plates.

The thickness of steel plates is usually given in twentieths of an inch, and the thickness of iron plates in sixteenths of an inch.

A square foot of steel plate 1 in. thick weighs 40 lbs. + 2%.
Therefore a square foot of steel plate $\frac{1}{20}$ in. thick weighs 2 lbs. + 2%.
A square foot of iron plate 1 in. thick weighs 40 lbs.
Therefore a square foot of iron plate $\frac{1}{16}$ in. thick weighs 2½ lbs.

Timber—

1 cubic foot of elm		weighs	34	lbs.
1 ,, ,,	red pine	,,	36	,,
1 ,, ,,	pitch pine	,,	41	,,
1 ,, ,,	yellow pine	,,	28	,,
1 ,, ,,	greenheart	,,	$62\frac{1}{2}$	,,
1 ,, ,,	lignum vitæ	,,	83	,,
1 ,, ,,	English oak	,,	52	,,
1 ,, ,,	Riga oak	,,	43	,,
1 ,, ,,	Dantzic oak	,,	47	,,
1 ,, ,,	Indian teak	,,	55	,,
1 ,, ,,	African teak	,,	61	,,

112 lbs. = 1 cwt.
2240 ,, = 1 ton.
20 cwts. = 1 ,,
12 inches = 1 foot ⎫
3 feet = 1 yard   ⎬ Measurement of length.
144 square inches = 1 square foot ⎫
9 ,, feet = 1 ,, yard ⎬ ,, ,, area.
1728 cubic inches = 1 cubic foot ⎫
27 ,, feet = 1 ,, yard ⎬ ,, ,, volume.

Definitions—

Area is measurement in square yards, square feet, or square inches.
Volume ,, ,, cubic ,, cubic ,, cubic ,,
Weight ,, ,, tons, cwts., lbs., etc.

Circular Measure 1 degree = ·01745.

Useful Rules.

To find the area of the following figures:—

I. $L \times B = $ area.

II. *Note.*—Opposite sides parallel.
$A \times B = $ area.

III. *Note.*—Sides A and B parallel.
$\dfrac{A + B}{2} \times C = $ area.

IV. $\dfrac{A \times B}{2} = $ area.

V. $D^2 \times ?$

CALCULATIONS. 273

VI. To find circumference of a circle.
$$D \times 3\cdot1416 \text{ or } D \times 3\tfrac{1}{7} = \text{circumference}.$$

VII. To find the volume of an object, the length of which is given, and the section of which is uniform throughout, the section being like either I., II., III., IV., or V.

Rule—Multiply the area of the section (or one end) by the length—the product gives the cubic contents.

To find the cubic contents of an object of the following elevation, the breadth being constant throughout the length.

Find the area by Rule III. (B and b being parallel), and multiply by the breadth.
$$\frac{B + b}{2} \times L \times \text{breadth} = \text{cubic contents}.$$

VIII. To find the volume of a spherical object, the diameter D being given.
$$D^3 \times \cdot5236 = \text{volume}.$$

USEFUL RULES AND PRELIMINARY CALCULATIONS.

Example I.

Find the weight of a steel plate as per sketch, $\tfrac{9}{20}$ *thick, with a circular hole, 2 feet diameter punched out.*

$\dfrac{5 + 3}{2} \times 6 = 24$, area of plate 6 feet long.

$\dfrac{4 \times 5}{2} = 10$, ,, 4 ,,

$2^2 \times \cdot 7854 = 3\cdot1416$ area of circular hole.
$(24 + 10) - 3\cdot1416 = 30\cdot85$ area of plate.
1 square foot $\tfrac{9}{20} = 18$ lbs. weight.
$\qquad 30\cdot85 \times 18 = 555\cdot3 + 2\% = 555\cdot3 + 11 = 566\cdot3$ lbs. weight.

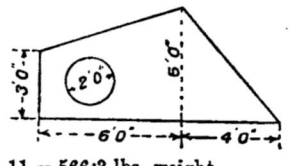

Example II.

To find the weight of a solid iron stanchion 20 feet long, 3 inches diameter.

3 inches = $\cdot25$ of a foot.
$\cdot25^2 \times \cdot7854 \times 20 \times 480 = 471\cdot24$ lbs. weight.

Example III.

To find the weight of a hollow iron stanchion 4 inches outside diameter, 2 inches inside diameter, and 10 feet long.

The mean width of plate which would form this stanchion = 3 inches $\times 3\cdot1416 = 9\cdot42$ inches.

Width. Thickness. Length in inches.
$9\cdot42 \quad \times 1 \times \quad (10 \times 12) = 1130\cdot4$ cubic inches.
$\qquad \dfrac{1130\cdot4}{1728} \times 480 = 314$ lbs. weight.

Example IV.

To find the number of gallons and tons of fresh water a tank will contain. The dimensions are:—Length, 12 feet; breadth, 6 feet; depth 7·3 feet.

$$12 \times 6 \times 7\cdot3 = 525\cdot6 \text{ cubic feet in tank.}$$
$$525\cdot6 \times 6\cdot25 \text{ (gallons in 1 cubic foot)} = 3285 \text{ gallons contained in tank.}$$
$$\frac{525\cdot6}{36} = 14\cdot6 \text{ tons of water contained in tank.}$$

Example V.

How many tons of coal will a coal bunker contain at 45 cubic feet per ton? It is 30 feet long, 10 feet broad, 13 feet deep.

$$\frac{30 \times 10 \times 13}{45} = 86\cdot6 \text{ tons.}$$

Example VI.

What is the displacement of a rectangular barge 60 feet long, 20 feet beam, and 6 feet depth? It draws 4 feet aft, and 3 feet 6 inches forward in fresh water.

$$\frac{4\cdot0 + 3\cdot5}{2} = 3\cdot75 \text{ mean draught.}$$
$$\frac{60 \times 20 \times 3\cdot75}{36} = 125 \text{ tons displacement.}$$

Example VII.

The same barge in a light condition draws 1 foot fore and aft. What weight or cargo is there on board?

$$\frac{60 \times 20 \times 1}{36} = 33\cdot33 \text{ tons, displacement light.}$$

Therefore $125 - 33\cdot33 = 91\cdot67$ tons weight on board.

Example VIII.

How many tons would be required to increase the draught of this barge 2½ *inches?*

$$\frac{60 \times 20}{420} = 2\cdot85 \text{ tons per inch.}$$

$2\cdot85 \times 2\cdot5 = 7\cdot125$ tons to increase draught 2½ inches.

Example IX.

What would be the increase in draught if 12 tons were placed on board?

$$\frac{12}{2\cdot85} = 4\cdot21 \text{ inches increase in draught.}$$

Example X.

To find the area of such a figure as this, which is similar to half of the deck of a ship or a half waterplane. Such areas are found by what are known as Simpson's Rules.

CALCULATIONS.

Simpson's First Rule.—"*Divide the base into any even number of equal lengths,*" say 6, then $\frac{60}{6} = 10$ feet each length, "*through these points draw ordinates to the curve, which ordinates will consequently be odd in number,*" in this case, including the endmost ordinates, there are 7. "*Multiply the length of each of the even ordinates by 4, and each of the odd ordinates by 2, excepting the first and last, which multiply by 1. The sum of these products multiplied by $\frac{1}{3}$ of the common interval between the ordinates will give the area required.*"

The calculation could be arranged in either of the following ways (*the 7th ord. in the diagram is supposed to be 1 foot*):—

	No. of Ord.	Ord.		No. of Ord.	Ord.	S. M.	Products or Functions.
EVEN ORDS.	2	3·8		1	0·0	1	0·0
	4	7·0		2	3·8	4	15·2
	6	4·0		3	6·2	2	12·4
		14·8	Sum of even ordinates.	4	7·0	4	28·0
				5	6·0	2	12·0
		29·6	Twice sum of even ordinates.	6	4·0	4	16·0
ODD ORDS.	1	·0	Half of 1st ord.	7	1·0	1	1·0
	3	6·2					84·6 Sum of products.
	5	6·0		$\frac{1}{3}$ common interval $\frac{10}{3}$ =			3·33
	7	·5	Half of last ord.				2538
							2538
		42·3	Sum of products obtained by treating the ordinates by the half Simpson's Multipliers.				2538
							281·718 Area.

Multiplier for Simpson's whole Multipliers   2

This method is more commonly adopted in actual practice.

84·6 Sum of products.

$\frac{1}{3}$ common interval $\frac{10}{3}$ =   3·33

2538
2538
2538

281·718 Area.

Simpson's Second Rule.—"*Divide the base into equal lengths, so that their number will be a multiple of 3,*" in this case (see figure) 6 lengths. "*Through these points of division draw ordinates to the curve, the total number of which, when divided by 3, gives a remainder of 1.*" There are seven ordinates, $\frac{7}{3} = 2$, and a remainder of 1. "*Call the 4th* (and if there had been more than 7 ordinates), *the 7th, 10th, 13th, etc., ordinates, dividing ordinates, and the others, excepting the first and last ordinates, intermediate ordinates. Add together the first and last ordinates, twice the dividing ordinates, and three times the intermediate ordinates. Multiply the sum by $\frac{3}{8}$ of the common interval* (10 feet), *and the product will be the area of the figure nearly.*"

The calculation is usually arranged as follows:—

No. of Ord.	Ord.	S.M.	Functions or Products.
1	0·0	1	0·0
2	3·8	3	11·4
3	6·2	3	18·6
4	7·0	2	14·0
5	6·0	3	18·0
6	4·0	3	12·0
7	1·0	1	0·0

$\frac{3}{8}$ common interval = $\frac{3}{8}$ of 10 =
74·0
3·75

1500
2625

277·50 Area.

It will be observed that the result is 4 square feet less than by the 1st Rule. The latter is preferable when the number of ordinates is such as to permit of the 1st Rule being applied.

Simpson's Third Rule is for finding the area of a part of a figure such as shown in the adjoining diagram. It is required to find the area of the part C A E F (*the 14 feet ordinate = G H*).

Let A E = 15 feet. Make E G = 15 feet also, and draw G H to the curve.

Rule.—*Add together five times the near end ordinate* (A C) *and eight times the middle ordinate* (E F). *From the sum subtract the far end ordinate* (G H), *and multiply the remainder by $\frac{1}{12}$ of the common interval. The product will give the area required.*

CALCULATIONS.

The calculation is arranged as follows :—

$$8 \times 5 = 40$$
$$12 \times 8 = 96$$

136 sum.
Subtract 14

122 remainder.
$\frac{1}{12}$ common interval = $\frac{15}{12}$ = 1·25

610
244
122

152·50 area required.

By the following graphic method it is hoped that the application of the foregoing Simpson's Rules will be more comprehensive.

First Rule.—Here is a figure, say, a piece of a ship's deck A B is the fore and aft middle line, and D C the curve of the deck. To find its area by Simpson's First Rule, A B is divided into two equal parts at the point E, and E G is drawn to the curve. By this rule, the first and last ordinates, B C and A D, are always multiplied by 1, and the second ordinate (that is, the even ordinate) is multiplied by 4. Then these products are multiplied by $\frac{1}{3}$ of the common interval, that is, $\frac{1}{3}$ of A E—E B.

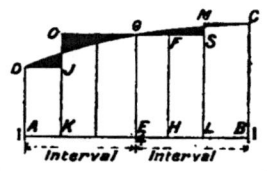

Splitting this calculation up into its separate parts or steps in the mode of procedure, we get :—

A D × $\frac{1}{3}$ of A E = area of A D J K.
E G × $\frac{1}{3}$ of A E = area of G E H F.

But this second ordinate G E, according to the rule, has to be multiplied by 4.

∴ 4 (E G × $\frac{1}{3}$ of A E or E B) = area of the 4 rectangles contained in O K L S.

Then, finally, B C × $\frac{1}{3}$ of E B = area of L B C M.

The sum of these three parts gives the area of the whole figure A B C D.

It will be noticed in this calculation that a piece of the area of the curve is lost over the rectangle A D J K, but a piece is gained within the rectangle at the other extremity M L B C. Similarly, a piece of the area of the curve is lost over the two rectangles G E L S, and a piece gained within the rectangles O K E G.

The areas of the two pieces gained and the two pieces lost approximately balance each other.

While this rule approaches very nearly to the truth, it is not absolutely correct. However, for all practical purposes, and in all ship calculations, when the rule is carefully applied, the error is so slight as to be unnoteworthy.

Where a figure is divided into a considerable even number of equal spaces, the rule applies in exactly the same manner as just described. For the first two spaces the multipliers are 1, 4, 1, and for the second two spaces the multipliers are 1, 4, 1, and so on.

```
1   4   1
    1   4   1
        1   4   1
            1   4   1
_____
1   4   2   4   2   4   2   4   1
```

To give a practical example illustrating what has been explained, let A B = 18 feet, A D = 4 feet, E G = 6 feet, and B C = 7 feet. The common interval between ordinates = $\frac{18}{2}$ = 9 feet, and one-third of the common interval = 3 feet.

$4 \times 3 = 12$, area of A D J K.
$4(6 \times 3) = 72$,, the four rectangles contained in O K L S.
$7 \times 3 = 21$,, L B C M.
─────
105 ,, whole figure.

Arranged in the usual form, and as previously described.

No. of Ord.	Ord.	S. M.	Products.
1	4	1	4
2	6	4	24
3	7	1	7

35
⅓ common interval = $\frac{9}{3}$ = 3
─────
105 Area of whole figure.

Simpson's Second Rule.—With the explanatory notes upon Simpson's First Rule, the reader will be able to follow the graphic explanation here given for Simpson's Second Rule.

Simpson's Multipliers are 1, 3, 3, 1, making altogether 8 oblongs and 3 intervals. 1 oblong is as long as the first ordinate, 3 as long as the second ordinate, 3 as long as the third ordinate, and 1

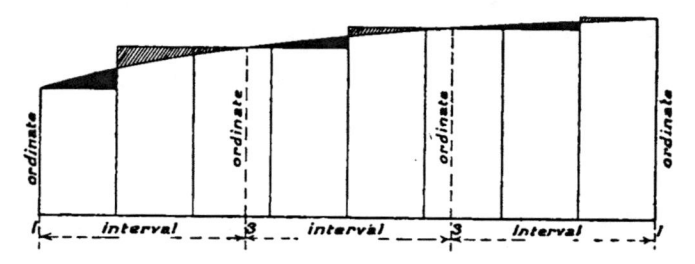

as long as the fourth ordinate. The width of each oblong is $\frac{3}{8}$ of the whole interval.

Simpson's Third Rule.—C D A B is the part of the figure the area of which is found by this rule.

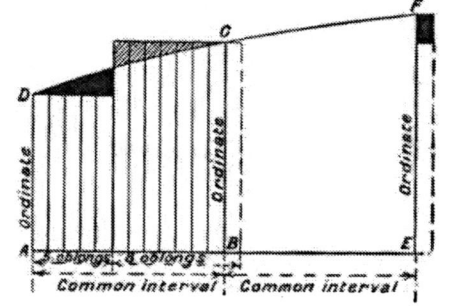

Simpson's Multipliers are 5, 8, and 1, which latter product has to be deducted. There are, therefore, 5 + 8 - 1 = 12 oblongs, 5 of them are of the length of the ordinate A D, 8 of the length of the ordinate B C, which includes 1 oblong more than the required number to cover the area A D C B. The 5 oblongs neglect the black wedge at their upper extremities, while the 8 oblongs gain a wedge outside the curve. By deducting the last ordinate E F, the surplus oblong to the right of B C is corrected, while the excess in length of the ordinate E F over B C produces an area indicated by a black oblong, which together with the lost wedge over the first 5 oblongs approximately neutralises the excessive area obtained over the next 7 oblongs. The width of an oblong is $\frac{1}{12}$ of the whole interval.

A study of the foregoing diagrams, showing the application of Simpson's Rules, indicates clearly that the nearer the ordinates are spaced to each other, or, in other words, the more intervals into which an area is divided, the nearer does the calculation approach to accuracy. Supposing the area of a ship's deck, 200 feet long, be divided into 12 intervals, a study of these rules will further show that, while they apply with great accuracy to the middle

280   KNOW YOUR OWN SHIP.

¾ length or more, the sudden curve of the ends of the deck towards the stem and stern, especially in a bluff cargo vessel, renders the calculation far from even approximately accurate for these end areas.

Let the adjoining figure represent the after 33·33 feet of this deck, covering exactly 2 intervals. The black wedges indicate the area within the deck line which is included by the rule, and the hatched areas indicate the area outside the deck line which is gained by the calculation. The inadaptability of the rule, as previously given, to apply to such an area as this, with any degree of accuracy, is obvious. But by a modification in the application of the rule itself, this inaccuracy can be largely obviated.

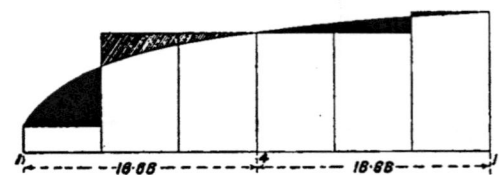

As previously stated, the closer the ordinates are spaced, the greater the degree of accuracy obtained. By this process of application of the first rule, the first interval is subdivided, and an ordinate measured at the point of division. Simpson's Multipliers now become half of what they were originally, viz., ½, 2, ½. In other respects the calculation is carried out in the usual manner. Simpson's Multipliers are now :—

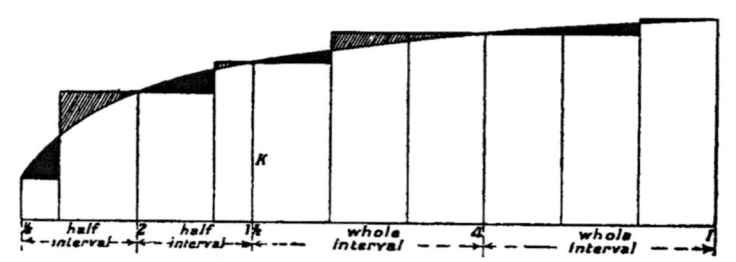

½	2	½			(See accompanying
		1	4	1	diagrams).
½	2	1½	4	1	

By this method of subdivision of the intervals the discrepancy between the excess and loss of area of the deck is reduced, and by means of a still further subdivision the error can be lessened.

Example 11.

To Calculate an Area by Introducing Subdivided Intervals.—The half ordinates for a ship's waterplane at, say, the load line, are 2, 8,* 12, 14, 16, 17, 16·6, 15, 11, 7,* 0, ordinates 8 and 7 being subordinates (half-ordinate means ordinate for half width of waterplane). The common interval is 18 feet. Find area of whole waterplane.

No. of Ord.	½ Ord.	S. M.	Products.
1	2	½	1
1½	8*	2	16
2	12	1½	18
3	14	4	56
4	16	2	32
5	17	4	68
6	16·6	2	33·2
7	15	4	60
8	11	1½	16·5
8½	7*	2	14
9	0	½	0·0

$$\begin{array}{rr} & 314\cdot7 \\ \tfrac{1}{3} \text{ common interval} = \tfrac{18}{3} = & 6 \\ \hline \text{Area of half waterplane} & 1888\cdot2 \\ \text{Multiplier for both sides} & 2 \\ \hline \text{Area of whole waterplane} & 3776\cdot4 \end{array}$$

Example 12.

To Find the "Tons per Inch" Immersion of the Foregoing Vessel at the Load Line.—$\dfrac{3776\cdot4}{420} = 8\cdot99$ "tons per inch" immersion.

By means of Simpson's Rules, the area of any waterplane, deck, or transverse section of the hull of a ship is easily found, whatever may be the number of intervals into which these areas may be divided.

In such an example as that shown by the figure, where there are eleven intervals, neither the First nor the Second Rule will apply for the whole length, but then, by making use of the Third Rule, the area can be found. Thus, by using the First Rule, the area of the whole figure up to the 11th ordinate can be found, and by using the Third Rule the area of the last space between

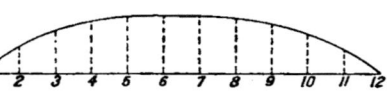

282   KNOW YOUR OWN SHIP.

the 11th and 12th ordinates is found. The sum of the two parts gives the whole area.

EXAMPLE 13.

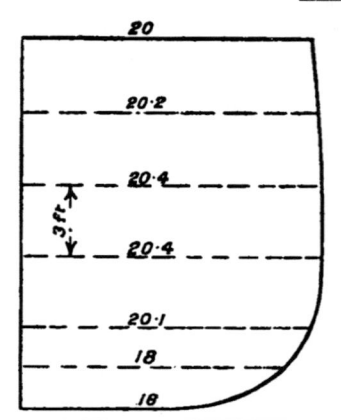

Shows a further Application of Simpson's Rules in ascertaining the Capacity of a Cross Bunker.—The bunker is 20 feet long and of constant section throughout its length. The ordinates of the transverse section are shown upon the sketch. Find the quantity of coal it [will contain at 45 cubic feet per ton. The bottom interval is subdivided.

No. of Ord.	Ord.	S. M.	Products.
1	20	1	20
2	20·2	4	80·8
3	20·4	2	40·8
4	20·4	4	81·6
5	20·1	1½	30·15
5½	18·0	2	36·00
6	16·0	½	8

⅓ of common interval = ⅓ of 3 =   297·35
                                                                                                    1

Multiplier for both sides   297·35
                                                                 2

Area for both sides   594·7
                                                                20 Length of bunker.

                                        45)11894·0 Cubic feet in bunker.
                                          264·31 Tons in bunker.

We have shown how Simpson's Rules may be applied to find the area of a deck, a waterplane, or a transverse section of a ship. In a similar manner, these rules may be used to find the displacement of a ship floating at a given waterline.

Example 14.

To Calculate a Ship's Displacement, 1st Method.—She draws 20 feet of water from the top of the heel. Divide this depth into a number of equal intervals, suitable for the application of Simpson's First or Second Rule, say, five equal intervals, each being 4 feet, with the bottom interval subdivided. The area of each of these waterplanes is now calculated as previously shown.

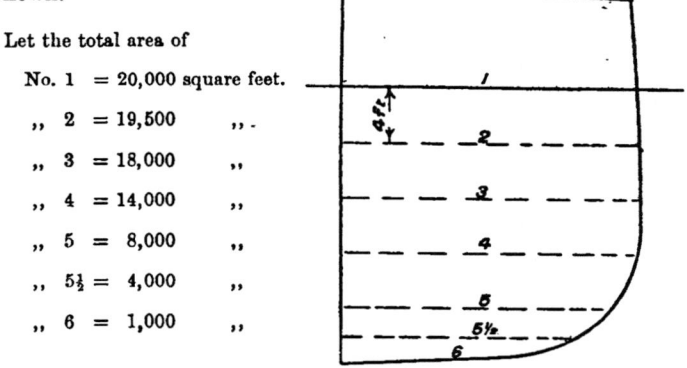

Let the total area of

No. 1 = 20,000 square feet.
,, 2 = 19,500 ,, .
,, 3 = 18,000 ,,
,, 4 = 14,000 ,,
,, 5 = 8,000 ,,
,, 5½ = 4,000 ,,
,, 6 = 1,000 ,,

These areas are treated by Simpson's Multipliers; the sum of the products multiplied by ⅓ of the common interval (when the First Rule is applied) equals cubic feet displacement, and in a similar manner the Second Rule may be applied when the intervals are suitable.

No. of Area.	Area.	S. M.	Products.
1	20,000	1	20,000
2	19,500	4	78,000
3	18,000	2	36,000
4	14,000	4	56,000
5	8,000	1½	12,000
5½	4,000	2	8,000
6	1,000	½	500

$$210,500$$
Common interval 4

Divide by 3 for ⅓ common interval 3)842,000

35)280,666 Cubic feet displacement.

8,019 Tons displacement.

Example 15.

To Calculate a Ship's Displacement, 2nd Method.—Another method of finding the displacement is to divide the length of the vessel into a number of equal intervals suitable for the application of Simpson's Rules. The areas of the transverse sections up to the load waterplane are calculated at each of these stations, and these areas are put through Simpson's Multipliers, and the calculation carried out in a way exactly similar to the foregoing examples.

A vessel, 200 feet long, is divided into twelve intervals, and the areas are calculated at each station. The areas are as follows:—

5, 200, 280, 350, 400, 400, 390, 370, 330, 280, 240, 180, 18.

The common interval between the areas is $\frac{200}{12} = 16\cdot66$ feet.

No. of Section.	Area.	S. M.	Products.
1	5	1	5
2	200	3	600
3	280	3	840
4	350	2	700
5	400	3	1200
6	400	3	1200
7	390	2	780
8	370	3	1110
9	330	3	990
10	280	2	560
11	240	3	720
12	180	3	540
13	18	1	18

$$\frac{3}{8} \text{ of } 16\cdot66 = \quad \begin{array}{r} 9263 \\ 6\cdot24 \end{array}$$

35)57801·12 Cubic feet displacement.

1651·46 Tons.

Example 16.

To Find the Centre of Gravity of a Waterplane or of a Transverse Section of a Vessel's Displacement.—In Chapter II., it has been shown how to find the centre of gravity of a number of weights ranged along a bar, from a given point. The principle holds good in all other calculations for centres of gravity.

Let the adjoining figure represent a ship's deck, the area of which is required. The lengths of the half-ordinates are shown on the diagram, and the intervals between the ordinates. Had 1,

10, 12, 8, 0 been weights upon the deck the centre of gravity of these weights only would have been found by multiplying each by its distance or leverage from one end of the deck, and dividing the sum of the moments obtained by the sum of the weights. However, though the ordinates are not weights, they serve the same purpose as weights, being representative of where the areas are greatest and least. But just as in calculating the area of a deck, Simpson's Rules are applied in order to obtain greater accuracy than would be obtained by simply taking the mean of all the ordinates, in like manner, in obtaining the centre of gravity of the deck we employ Simpson's Rules, and thereby, greater accuracy is obtained by using the ordinates multiplied by their respective multipliers as indices of the fulness or fineness of the deck, than by using the ordinates themselves. The calculation would be done as follows :—

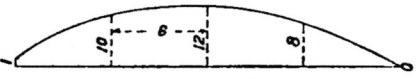

No. of Ords.	½ Ords.	S. M.	Products.	Leverages.	Moments.
1	1	1	1	0	0
2	10	4	40	6	240
3	12	2	24	12	288
4	8	4	32	18	576
5	0	1	0	24	0
			97		1104

$$\frac{1104}{97} = 11\cdot 38$$

The centre of gravity of the deck is 11·38 feet from the left-hand endmost ordinate. In practice, it is usual to use only the number of intervals for leverages, and multiply the sum of the moments so obtained by the interval. This effects a saving in figures. By this method the calculation would be as follows :—

No. of Ords.	½ Ords.	S. M.	Products.	Leverages.	Moments.
1	1	1	1	0	0
2	10	4	40	1	40
3	12	2	24	2	48
4	8	4	32	3	96
5	0	1	0	4	0
			97		184

$$\frac{184 \times 6}{97} = 11\cdot 38 \text{ C. G. from first ordinate.}$$

Example 17.

To Find the Perpendicular Centre of Gravity of a Half Waterplane Transversely from the Fore and Aft Centre Line.—Rule.
—Take the half squares of the ordinates, and treat them by Simpson's Multipliers as though they were the ordinates for a new curve. The area of this hypothetical curve is the moment of the figure relatively to the fore and aft centre line. Divide this moment by the actual area of the half waterplane, and the result is the perpendicular distance of its centre of gravity from the longitudinal middle line of the waterplane.

The following example will serve as an illustration, the common interval between the ordinates being 18 feet :—

No. of Ord.	Ords.	S. M.	Products.	Squares of Ordinates.	S. M.	Products of Squares of Ordinates.
1	2	1	2	4	1	4
2	5	4	20	25	4	100
3	8	2	16	64	2	128
4	4	4	16	16	4	64
5	0	1	0	0	1	0

$\frac{1}{3}$ of common interval $= \frac{18}{3}$ $\frac{54}{6}$

$\overline{324}$

$\frac{1}{2}$ squares $2)\overline{296}$

$\overline{148}$

$\frac{1}{3}$ common interval 6

moment $\overline{888}$

$\frac{888}{324} = 2\cdot74 =$ Perpendicular distance of centre of gravity from longitudinal centre line.

Example 18.

To Find the Centre of Buoyancy or Centre of Gravity of Displacement.—Suppose, first, that it is desired to obtain the height of the centre of buoyancy above the keel.

The draught (measuring from the top of keel) is divided into a number of equal spaces in exactly the same way as is done for displacement. The area of each of these waterplanes is calculated by the application of one of Simpson's Rules. These areas of waterplanes are then put through Simpson's Multipliers. Thus far the calculation has resembled that for displacement. All this information may, therefore, be copied direct from the displacement calculation, or the displacement calculation itself may be used, as is generally done in practice. If the centre of buoyancy is required from the top of the keel, the *products of areas* are multiplied by their respective leverages from the top of the keel. The sum of these moments divided by the sum of the *products of areas* will give the height of the centre of buoyancy above the top of the keel. The steps in this calculation are identical with those followed in the previous example for finding the centre of gravity of a waterplane, the areas of the horizontal waterplanes taking the place of the ordinates of the waterplane.

As an example, suppose it is required to find the centre of buoyancy above the keel of the vessel whose displacement was found in Example 14.

No. of Horizontal Area.	Area.	S. M.	Products.	Leverages.	Moments.
1	20,000	1	20,000	5	100,000
2	19,500	4	78,000	4	312,000
3	18,000	2	36,000	3	108,000
4	14,000	4	56,000	2	112,000
5	8,000	1½	12,000	1	12,000
5½	4,000	2	8,000	½	4,000
6	1,000	½	500	0	000
			210,500		648,000

Common interval 4

210,500)2,592,000

Centre of buoyancy above top of keel = 12·3 feet.

Example 19.

To find the fore and aft centre of buoyancy the same method exactly is adopted, using vertical areas.

As an example, find the centre of buoyancy of the vessel in Example 15.

288 KNOW YOUR OWN SHIP.

No. of Vert. Area.	Area.	S. M.	Products.	Leverages.	Moments.
1	5	1	5	0	0
2	200	3	600	1	600
3	280	3	840	2	1680
4	350	2	700	3	2100
5	400	3	1200	4	4800
6	400	3	1200	5	6000
7	390	2	780	6	4680
8	370	3	1110	7	7770
9	330	3	990	8	7920
10	280	2	560	9	5040
11	240	3	720	10	7200
12	180	3	540	11	5940
13	18	1	18	12	216
			9263		53,946
					16·66

9263)898,740·36

Centre of buoyancy from area 1   97·02 feet.

EXAMPLE 20.

To find the moment of inertia of a waterplane (required in order to arrive at the height of the transverse metacentre above the centre of buoyancy), relatively to the fore and aft axis passing through the centre of gravity of the waterplane. Imagine the waterplane to be divided into an infinitely small number of units of area. Multiply each one of these units of area by the square of its distance from the fore and aft centre line of the waterplane, which is the axis about which the moment of inertia is calculated. The sum of all these products is the moment of inertia required. Let the adjoining figure serve as an illustration. Here we have, say, a midship portion of the area of a waterplane. It is 6 feet long and 9 feet wide, that is, 4½ feet on each side of the fore and aft middle line. Let it be divided into units of area of 1 square foot each.

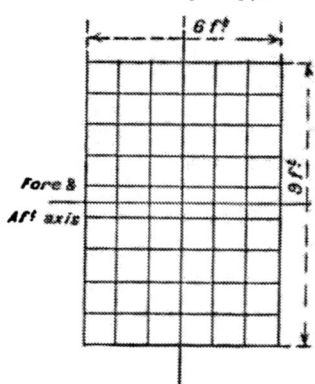

Multiply each of these areas by the squares of their respective distances from the axis, as follows:—

CALCULATIONS. 289

```
6 ( ·5  × ·25²) =     ·1875
6 ( 1  × 1²)  =    6·0000
6 ( 1  × 2²)  =   24·0000
6 ( 1  × 3²)  =   54·0000
6 ( 1  × 4²)  =   96·0000
                 ────────
                 180·1875  Moment of inertia for one half of plane.
                      2   Multiplier for both halves.
                 ────────
                 360·3750  Total moment of inertia.
```

The moment of inertia is first obtained for the half waterplane relatively to the fore and aft axis of the plane, as in the foregoing case, and the result is multiplied by 2 for the whole plane—the other half being exactly similar when the vessel is upright.

Thus, on one side, we have :—

6 half units of area at a distance of ·25 of a foot.
then 6 ,, ,, ,, 1 foot.
,, 6 ,, ,, ,, 2 feet, and so on.

Naturally, as our units of area are not infinitesimally small, the result can only be approximate. The smaller the units of area, and therefore the greater number of them, the more correct is the result. It will be obvious that such a method as this could not be applied to such a huge area as a ship's waterplane.

Rule for Moment of Inertia of a Ship's Waterplane :—

Divide the fore and aft axis into a number of equal intervals suitable for the application of one of Simpson's Rules. Measure the half ordinates at the points of division, and cube each of them. Take one-third of these cubes, and deal with them as though they were ordinates of a curve, the area of which has to be found. Such area would be the moment of inertia of the half waterplane. The moment of inertia of the foregoing figure found by this method would be as follows :—

½ Ords.	Cubes.	S. M.	Products.
4·5	91·125	1	91·125
4·5	91·125	4	364·500
4·5	91·125	1	91·125

```
            ⅓ of cubes  3)546·750
                         ────────
                         182·25
  ⅓ common interval = ⅔ =    1
                         ────────
                         182·25   Moment of inertia for half plane.
                              2   Multiplier for both halves.
                         ────────
                         364·5    Total moment of inertia.
```

T

It is now seen that the previous method was only approximate, being over 4 less than the correct calculation gives.

In an earlier chapter, we have shown that the moment of inertia of a rectangular waterplane is:—

$$\frac{\text{Length of waterplane} \times \text{Breadth}^3}{12}$$

therefore, applying this to the same figure

$$\frac{6 \times 9^3}{12} = 364 \cdot 5,$$

which result is similar to that of the last calculation.

Calculation for Moment of Inertia of an Actual Ship's Waterplane.—To find the moment of inertia of the load waterplane of a vessel whose half ordinates are 0, 2·5, 6, 9, 11, 12, 10, 7, 5, 2, ·5. The common interval between the ordinates is 9 feet.

No. of Ord.	½ Ords.	Cubes of ½ Ords.	S. M.	Moments.
1	·0	·00	1	·00
2	2·5	15·62	4	62·48
3	6·0	216·00	2	432·00
4	9·0	729·00	4	2916·00
5	11·0	1331·00	2	2662·00
6	12·0	1728·00	4	6912·00
7	10·0	1000·00	2	2000·00
8	7·0	343·00	4	1372·00
9	5·0	125·00	2	250·00
10	2·0	8·00	4	32·00
11	·5	·12	1	·12

⅓ of cubes   3)16,638·60

⅓ common interval = ⅓ =   5546·2
                              3

For both halves of waterplane   16,638·6   M of 1 of ½ plane.
                                      2

                                 33,277·2   Moment of inertia.

EXAMPLE 21.

To Find the Height of Metacentre above the Centre of Buoyancy :—

$$\frac{\text{Moment of inertia of waterplane}}{\text{Displacement in cubic feet}} = \text{metacentre above centre of buoyancy.}$$

CALCULATIONS.

As an example, suppose the moment of inertia of the foregoing example is for a vessel of 300 tons displacement at load draught. The height of metacentre above centre of buoyancy would be :—

$$\frac{33,277}{300 \times 35} = \frac{33,277}{10,500} = 3\cdot16 \text{ feet.}$$

EXAMPLE 22.

To Find the Position of the Centre of Gravity of a Vessel in Relation to the Metacentre, by Experiment. — This has been dealt with, and an example worked out in Chapter VI. (Section I.), page 128.

Let W = weight moved across deck.
  d = distance weight is moved.
  D = displacement in tons.
$\frac{W \times d}{D}$ = G G' (shift of centre of gravity transversely).
G G' × cotangent of angle of keel = G M (metacentric height).
Cotangent of angle = $\frac{\text{length of plumb line in inches}}{\text{mean deviation of plumb line in inches.}}$
$\frac{W \times d}{D} \times \frac{\text{length of plumb line in inches}}{\text{mean deviation of plumb line in inches}} = $ G M.

EXAMPLE 23.

To Find the Centre of Gravity of a Ship by Finding the Distance of the Centre of the Weight from, say, the Bottom of the Keel.— Multiply every item of weight in the ship by its distance above the bottom of the keel. The sum of all these products or moments divided by the sum of all the weights (the total of which equals the total weight of the ship), gives the height of the centre of gravity above the bottom of the keel.

EXAMPLE 24.

To Find the Height of the Longitudinal Metacentre above the Centre of Buoyancy.—The principle of this calculation is identical with that already given for height of transverse metacentre above centre of buoyancy.

The moment of inertia of the waterplane is found relatively to a transverse axis passing through the centre (centre of gravity) of the particular waterplane. When this moment of inertia is found, and divided by the displacement in cubic feet, the result is the height of the longitudinal metacentre above the centre of buoyancy.

Here, again, the moment of inertia is the sum of the products of each of the units of area in the waterplane, multiplied by the square of their respective distances from the transverse axis. Were the waterplane rectangular, the moment of inertia would be found by identically the same method as in the first two calculations in Example 20, excepting that what was in that case the length now becomes breadth, and the breadth becomes length.

$$\frac{\text{Breadth} \times \text{Length}^3}{12} = \frac{9 \times 6^3}{12} = 162 \text{ moment of inertia.}$$

Owing, however, to the shape of an actual ship's waterplane, a modification in the application of the rule is necessary.

The method adopted is as follows :—

1st. Find the Moment of Inertia of the Waterplane, Relatively to One End of the Waterplane.—Rule. Divide the fore and aft middle line into a number of equal intervals suitable for the application of one of Simpson's Rules. Through these points draw ordinates to the curve. Then multiply each ½ ordinate by its proper multiplier. Each of these products is next multiplied by the square of the number of whole intervals it is distant from the end of the waterplane. The sum of these moments multiplied by ⅓ or ⅜, the cube of a whole interval (according to the Simpson's Rule applied) will give the moment of inertia of the ½ waterplane relatively to the first ordinate. By multiplying by 2 the M of I for whole plane is obtained.

As we require, however, the moment of inertia of the waterplane, relatively to the centre of the waterplane, a correction is necessary.

2nd. Find the Moment of Inertia of the Waterplane, relatively to a Transverse Axis passing through the Centre of the Waterplane.—Rule. Multiply the area of the waterplane by the square of its distance from the first ordinate. This product subtracted from the moment of inertia relatively to the first ordinate gives the moment of inertia relatively to the axis passing through the centre of gravity of the waterplane.

By applying this rule to the figure at the beginning of Example 20, we shall see that it agrees with $\dfrac{\text{Length}^3 \times \text{Breadth}}{12}$.

CALCULATIONS.

No. of Ord.	½ Ords.	S. M.	*Products.	Squares of Intervals.	Moments.
1	4·5	1	4·5	0	0
2	4·5	4	18	1	18
3	4·5	1	4·5	4	18

$$\frac{\text{Common interval cubed}}{3} = \frac{3^3}{3} = 9 \qquad \begin{array}{r} 36 \\ 9 \end{array}$$

$$\begin{array}{r} 324 \\ \text{Both halves} \quad 2 \end{array}$$

Moment relatively to 1st ordinate   648
Area of waterplane × (distance of centre of gravity from 1st ordinate)² = (9 × 6) × 3² =   486

162 =
Moment of inertia relatively to transverse axis passing through centre of W P.

As already stated, the moment of inertia divided by the displacement in cubic feet gives the height of the longitudinal metacentre above the centre of buoyancy.

Calculation for Longitudinal Metacentre above the Centre of Buoyancy.—As an example, the half ordinates of a ship's waterplane are, 0, 4, 7, 9, 10, 8, 5, 2, 0. The common interval is 12 feet, the displacement, 150 tons.

No. of Ord.	½ Ords.	S. M.	Prods.	Leverages.	Prods. of Moments.	Levers.	Prods. for M. of I.
1	0	1	0	0	0	0	0
2	4	4	16	1	16	1	16
3	7	2	14	2	28	2	56
4	9	4	36	3	108	3	324
5	10	2	20	4	80	4	320
6	8	4	32	5	160	5	800
7	5	2	10	6	60	6	360
8	2	4	8	7	56	7	392
9	0	1	0	8	0	8	0

⅓ Longt. interval ⅓ = $\frac{136}{4}$ $\frac{508}{12}$ $\frac{(\text{Com. int.})^3}{3}$ = $\frac{2268}{576}$

$\frac{544}{\text{Both halves} \quad 2}$ 136)6096 $\frac{1306368}{\text{2 Both halves.}}$

Area of waterplane 1088   C.G of W.P from ord. 1.= 44·82   2612736 Moment of inertia relatively to ord. 1.

The result at the foot of the fourth column is area of waterplane. Instead of multiplying the products obtained after using

Simpson's Multipliers, by the square of the number of intervals in one operation, it is done in two operations, as shown in the fifth and seventh columns. This enables us to use the sixth column in order to ascertain the position of the centre of gravity of the waterplane, as shown in the calculation.

Area of waterplane × square of the distance of C.G of waterplane from first ordinate = 1088 × (44·82)² = 2185609.

2612736 Moment of inertia relatively to ordinate 1.
2185609

150 tons × 35 = 5250)427127 M. of I. about axis passing through C.G of W.P.

81·35 feet = Height of longt. M.C above C.B.

Example 25.

Alteration of Trim.—The longitudinal metacentric height is chiefly used in order to ascertain the alteration to trim caused by loading, discharging, or shifting weights or cargo on board a vessel. The principle upon which Trim calculations are worked is fully explained in Chapter VII., to which the reader should refer at this stage. A variety of examples are dealt with at the end of that chapter.

The moment altering the trim 1 inch, when a vessel is floating at any particular draught, has been shown to be

$$\frac{D \times GM}{WL \times 12}$$

D = displacement.
G M = metacentric height.
W L = length of waterplane.

Suppose, in the previous example where the longitudinal metacentre was found to be 81·35 feet above the centre of buoyancy, that a weight of 8 tons in the fore hold, and at a distance of 30 feet forward of the centre of gravity of the waterplane, is discharged. What would be the alteration in trim?

The moment altering the trim is 8 × 30 = 240 foot tons.

The "tons per inch" is $\frac{1088}{420}$ = 2·59 tons.

The decrease in draught is therefore $\frac{8}{2·59}$ = 3·08 inches.

The moment to change trim 1″ =
$\frac{150 \times 82·35 \text{ G.M (supposing the C.G is found to be 1 foot below C.B)}}{96 \times 12}$ =

$\frac{4117·5}{38·4}$ = 10·72

Change in trim = $\frac{240}{10·72}$ = 22·38 inches.

CALCULATIONS.

Suppose the vessel to have been floating upon even keel at a draught of 6 feet before the weight was discharged. By discharging the 8-ton weight from forward, the vessel has changed trim 11·19-inches at the stem, and 11·19-inches at the stern.

	Draught forward.	Draught aft.
	6' 0"	6' 0"
	less 11·19	add 11·19
	5' – 0"·81	6' – 11"·19
The mean draught has decreased by 3·08"	3·08	3·08
Draught after weight discharged	4' – 9"·73	6' – 8"·11 aft.

EXAMPLE 26.

Alteration of Trim caused by Damage to Fore Peak, owing to Collision. — As a further example of change of trim, suppose this vessel has a fore-peak watertight bulkhead, at a distance of 15 feet from the stem, extending a considerable height above the waterline (the length from stem to stern being 96 feet). Let the draught be as before, 6 feet on even keel. Owing to collision, this fore peak is damaged, and water is admitted from the sea. A very important question in such a case is "What will be the change in trim after the ship has again come to rest?"

First of all, it is clear, that, owing to the loss of buoyancy in the fore peak, the mean draught must have increased. Immediately the fore peak is perforated, and free communication established with the water outside, the waterplane of the ship terminates at the after side of the fore-peak bulkhead, the part before the bulkhead providing neither buoyancy nor moment of inertia of waterplane.

The loss of buoyancy in the fore peak must be compensated for by the vessel increasing in draught, and obtaining the amount lost from the reserve buoyancy abaft the peak bulkhead.

The amount of lost buoyancy must be found. This is, of course, the volume of the fore peak below the 6-feet draught level. Though there may be considerable space in the fore peak above the 6-feet level, this never afforded buoyancy, being entirely reserve buoyancy. Now that the peak is damaged, it is no longer reserve buoyancy, so that the buoyancy to be dealt with is simply that below the 6-feet level. This volume could easily be calculated by Simpson's Rules, as shown previously for displacement, in fact, it is simply a calculation of the displacement of the fore peak. Let the capacity be 210 cubic feet or 6 tons. The centre of gravity of the waterplane between the stern post and the fore-peak bulkhead has next to be found, and also the centre of buoyancy of the fore-peak displacement. Then add together the distance of the centre of gravity of the waterplane from the fore-peak bulkhead, say, 38 feet, and the distance of the centre of buoyancy of the fore peak from the bulkhead, say, 5 feet. The sum 43 is the leverage used in ascertaining the moment altering trim.

Moment to alter trim = 6 × 43 = 258 foot tons.

The length of the waterplane is now 96 − 15 = 81 feet.

The area of the waterplane will be less, let it be, say, 1030 square feet.

The shortening of the length of the waterplane and the reduction of its area will have reduced the longitudinal metacentric height. Let it be now 74 feet.

The moment to alter trim 1 inch will be:—

$$\frac{150 \times 74}{81 \times 12} = \frac{3700}{324} = 11\cdot42$$

Total change in trim $\frac{258}{11\cdot42}$ = 22·6 inches = 11·3 inches, at each end of intact waterplane.

The "tons per inch" = $\frac{1030}{420}$ = 2·45

The increase in mean draught = $\frac{6}{2\cdot45}$ = 2·45 inches.

The new draught will be:—

```
         Aft.              Forward.
       6'  0"              6'  0"
           2·45                2·45
       ─────               ─────
       6   2·45           6   2·45
    −     11·3         +     11·3
       ─────               ─────
       5   3·15           7   1·75
```

But the draught 7 feet 1¾ inches forward is at the collision bulkhead, and we require the draught at the stem 15 feet forward.

The length from C G of waterplane to stem is 38 + 15 = 53 feet

$\frac{11\cdot3'' \times 53}{38}$ = 15·7 inches.

```
                          6'  2·45"
                          1   3·7
                          ─────
The draught at stem will be  7' − 6·15"
```

CALCULATIONS. 297

EXAMPLE 27.

To find the area of a wedge-shaped figure, such as shown by A B C in the adjoining figure, which may be taken to represent one of the wedges of immersion or emersion of a rolling vessel, B C being a plain curve. Let the whole angle at A = 40 degrees. Divide the angle at A into a number of equal angular intervals, so that the whole number of radii may be suitable for the application of one of Simpson's Rules.

Rule.—Measure the length of each of the radii, and find their half squares. Treat these half squares as if they were ordinates of a curve, by the application of Simpson's Rules. If the first of Simpson's Rules be applied, then the sum of the half squares will require to be multiplied by $\frac{1}{3}$ of the common angular interval, which must be taken in circular measure.

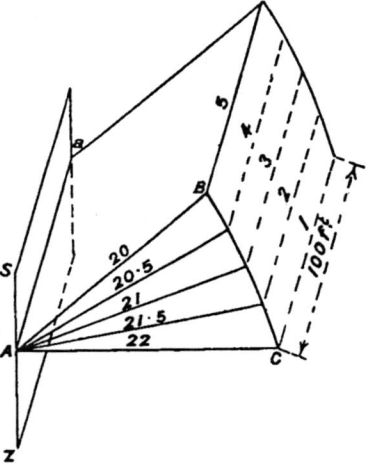

Note.—Circular measure for 1 degree is ·01745.

The calculation would be arranged as follows:—

No. of Radius.	Radii.	Squares of Radii.	S. M.	Products.
1	20·0	400·0	1	400·0
2	20·5	420·2	4	1680·8
3	21·0	441·0	2	882·0
4	21·5	462·2	4	1848·8
5	22·0	484·0	1	484·0

Divide by 2 for half sqs. 2)5295·6
               2647·8
$\frac{1}{3}$ of circular measure for 10° = $\frac{·1745}{3}$ = ·0581
             Area 153·8

Example 28.

To Find the Volume of a Wedge of Immersion or Emersion.—Suppose that the figure in the previous example had been one of the sections in a wedge of immersion 100 feet long, then, by dividing the length of the wedge into a suitable number of equal intervals, and finding the area of the sections at each of these intervals, and treating these areas as though they were ordinates of a new curve by the application of one of Simpson's Rules, the volume of the whole wedge would be ascertained.

Example 29.

To Find the Longitudinal Centre of Gravity of a Wedge, such as that just dealt with.—*Rule.* Multiply the area of each sectional area by its distance from one *extremity* (call it x) of the wedge. Treat the products as though they were ordinates of a new curve of the same length as the wedge (100 feet), by the application of one of Simpson's Rules; the result so found will be the moment of the wedge relatively to the *extremity* x. This moment, divided by the volume of the wedge, will give the longitudinal distance of the centre of gravity from the *extremity* x.

Example 30.

To Find the Perpendicular Distance of the Centre of Gravity of the Fore-Mentioned Wedge (see fig. in Example 27), *relatively to the Longitudinal Plane Z A S, which is perpendicular to the Radius A C.* —*Rule.* Divide the wedge into a number of longitudinal planes, radiating from the edge A *a* at equi-angular intervals. Find the moment of inertia of each of these longitudinal planes (5 in number, see the fig.) relatively to A *a*, as explained in Example 20, page 290. Multiply each of these moments of inertia by the cosines of the angles (see page 268) made by their respective planes with the plane A C, and apply to these results the multipliers for Simpson's First Rule. The sum of these products must be multiplied by $\frac{1}{3}$ of the common interval (using circular measure). The result is the moment of the wedge relatively to the plane Z A S. This moment, divided by the volume of the wedge, gives the distance of the centre of gravity of the wedge from the longitudinal plane Z A S.

When finding the centre of gravity of the wedge, as described in this example, the volume of the wedge is usually found as follows (example 31):—

EXAMPLE 31.

To Find Volume of Wedge of Immersion or Emersion.—The wedge, as has been explained, is divided into 5 longitudinal planes radiating from A *a*, which planes have already been divided into a number of equal intervals longitudinally, suitable for the application of Simpson's Multipliers (this has been done in order to find the moment of inertia in the previous example). Measure the ordinates, and compute their half squares. Treat these half squares as though they were ordinates of a new curve by the application of Simpson's Multipliers, and find the hypothetical area in the usual way. The results thus obtained are moments for each of the longitudinal planes.

Use now these moments of planes as though they were ordinates of another new curve by the application of Simpson's Multipliers. The sum of products thus obtained, multiplied by $\frac{1}{3}$ of the angular interval, gives the volume of the wedge which ought to agree with the result obtained by the method previously described in Example 28, where the transverse sections were dealt with.

EXAMPLE 32.

Calculation for Finding the Position of the Centre of Effort Relatively to the Centre of Lateral Resistance.—In order to work out an actual example, the Three-masted Schooner-rigged Vessel, page 213, is used. The sails are numbered as shown upon the diagram. The steps in the process of the calculation are fully explained in Chapter VI. (Section VII.).

No. of Sail.	Areas. Sq. Feet.	Distances of Centres of Sails from Centre of Lateral Resistance. Before. Feet.	Abaft. Feet.	Moments. Before.	Abaft.	Height of Centre of Sails above Centre of Lateral Resistance. Feet.	Vertical Moments.
1	480	45		21,600		25	12,000
2	560	35		19,600		24	13,440
3	580	30		17,400		22	12,760
4	1000	12		12,000		28	28,000
5	300	15		4,500		50	15,000
6	1100		8		8,800	28	30,800
7	350		4		1,400	51	17,850
8	1500		30		45,000	26	39,000
9	340		25		8,500	50	17,000
Sum	6210			Sum 75,100	63,700		Sum 185,850

As the moments before the centre of lateral resistance preponderate, it is evident that the centre of effort lies forward of the centre of lateral resistance, and the distance is:—

$$\frac{75100 - 63700}{6210} = 1\cdot 83 \text{ feet.}$$

The height of the centre of effort above the centre of lateral resistance is:—

$$\frac{185850}{6210} = 29\cdot 92 \text{ feet.}$$

CHAPTER X. (Section II.)

A SET OF SHIP CALCULATIONS AS WORKED FROM ACTUAL DRAWINGS.

CONTENTS.—Displacement—Longitudinal Centre of Buoyancy—Vertical Centre of Buoyancy—Transverse Metacentre above Centre of Buoyancy, showing two Methods of Arrangement—Tons per Inch Immersion—Wetted Surface and Shell Displacement—Longitudinal Metacentre above Centre of Buoyancy—Results of Calculations upon Curves—Stability Calculation.

IN Section I. of this chapter, a series of tables, rules, and worked examples have been given, all of which are met with in working out actual ship calculations from the actual drawings.

In this chapter it is proposed to work through an actual set of ship calculations, taking all dimensions, etc., from the actual drawing which figs. 137 and 141 provide.

Though in appearance and amount of labour involved, these ship calculations will be larger than anything we have yet undertaken, yet in practice they present nothing more difficult than what has already been encountered in the preliminary calculations in Section I. of this chapter.

As we proceed, notes and explanations, as far as such are necessary, will accompany each calculation, by which means it is hoped that no reader, though possessed of even the most elementary mathematical attainments, will be prevented from intelligently following the steps in the processes involved. And, as encouragement, it is repeated that these larger calculations are only composed of combinations of the smaller ones already dealt with at an earlier stage. The calculations are arranged in the order in which they are usually worked in actual practice.

DISPLACEMENT AND LONGITUDINAL CENTRE OF BUOYANCY CALCULATIONS.
(See Drawing, fig. 137.)
1st Method of Arrangement.

DISPLACEMENT.

S.M	Top of Keel	1' 0" W.L.	2' 0" W.L.	3' 0" W.L.	4' 6" W.L.	6' 0" W.L.	8' 0" W.L.	10' 0" W.L.	12' 6" W.L.

Water Lines.
Simpson's Multipliers.

LONGITUDINAL CENTRE OF BUOYANCY.

	Functions of Vertical ½ Areas.	S.M.	Multiples of Functions.	Leverages.	Moments.
	6·00	1	6·00	0	0·00
	65·14	4	260·56	1	260·56
	123·45	2	246·90	2	493·80
	163·30	4	653·20	3	1959·60
	185·00	2	370·00	4	1480·00
	195·72	4	782·88	5	3914·40
	198·02	2	396·04	6	2376·24
	194·26	4	777·04	7	5439·28
	181·89	2	363·64	8	2909·12
	155·93	4	623·72	9	5613·48
	119·43	2	238·86	10	2388·60
	62·54	4	250·16	11	2751·76
	3·28	1	3·28	12	39·36
			4972·28		29626·7

The horizontal rows of black figures are the ⅓ ordinates of the transverse sections, the bottom two intervals of which are subdivided. These are treated by the Simpson's Multipliers on the top row, and the products arranged vertically below the black figures. The sum of these products gives the *Functions of vertical half areas*. These Functions, treated by Simpson's Multipliers, give the sum of which, multiplied by the number 13·2 obtained before, gives the displacement in tons, which practically agrees with that already found (the difference being accounted for by the effect of the 1½ Simpson's Multiplier), and therefore constitutes a valuable check upon the calculation. The rest of the operation in arriving at the longitudinal centre of buoyancy will be clear after noting Example 19 in Section I. of this chapter.

Sum of functions Horl.⎫	3·72	168·22	218·24	248·58	268·54	297·10	316·64	336·42	348·50		2962620 = Sum of Moments.
zontal areas Simpson's Multipliers⎫	1	2	4	2	4	4	2	4	1		12 = Common Interval.
Multiples Func-⎫ tions	1·86	336·44	109·12	497·16	134·27 497·16 109·12 447·42	1188·40	316·64 1188·40 268·54 1187·97	1345·68	348·50 1345·68 316·64 2961·55	Sum of Multiples of Functions = Multiplier for Displacement =	4972·28)355514·40 ·152 71·49 = Centre of Buoy-ancy from No. I., Section. 994456 2486140 497228
Sum of Multiples of Functions Multiplier for Displacement =			447·42 ·152			1187·97 ·152	2961·55 ·152		4972·37 ·152		Tons . 755·78656 Displacement
			89·484 223710 44742			237594 593985 118797	592310 1490775 296155		994474 2496185 497237		
See page 304, Shell Displacement =	Tons Tons	68·00764 8·7		Tons Tons	180·67144 3·7	Tons	450·16560 5·5	Tons	755·80024 7·3		
	Total at 4' 0" waterline.	184·27 Ts.		Total at 8' 0" waterline.		Total at 8' 0" waterline.	455·65 Ts.	Total at 12' 0" waterline.	763·10 Ts.		

Notes on Displacement Calculations.—The vertical columns of black figures are the half ordinates of the successive horizontal waterplanes, which are measured from the displacement sections shown on fig. 137. These are treated by Simpson's Multipliers, the products being arranged in the adjoining vertical column. The sum of these products gives the *Sum of Functions of Horizontal Half Areas*. In order to construct a curve of displacement, as is usually done, the displacement is calculated in stages, according as Simpson's Rules will apply. It will be noted in fig. 137, that the interval between the horizontal areas is 2 feet, the bottom two intervals being subdivided in order to obtain greater accuracy, and thus, in finding the displacement, the "sums of functions" of these areas have to be treated by Simpson's Half Multipliers. In this example the displacement is calculated in four separate stages—to the 2 ft. W.L., the 4 ft. W.L., the 8 ft. W.L., and the 12 ft. W.L. After the "sums of functions of horizontal half areas" are treated by Simpson's half multipliers, the "sum of multiples of functions" is found, which in turn is multiplied by the number 152. This number is obtained as follows:—In the first place, the sum of functions of horizontal half areas" should be multiplied by ⅓ of the longitudinal interval for area = ⅓ of 12, and the sum of the multiples of functions should be multiplied by ½ of the vertical interval for volume = ½ of 2. Then volume has to be divided by 35 to obtain tons displacement, and this displacement is only for one-half of the vessel, since only the half ordinates of waterplanes were taken. Therefore the multiplier for total displacement is:—

$$\frac{12 \times 2 \times 2}{3 \times 3 \times 35} = \frac{16}{105} = \cdot 152.$$

As this displacement is only taken over the frames of the vessel, an addition has to be made for the displacement of the shell. See page 304 for Shell Displacement calculation.

Though in practice it is not always done, as the result is practically unnoteworthy, yet, for greater accuracy, the volume of the keel, stem, and stern frame, and the exceedingly small immersed appendages at the ends of the vessel, may also be computed and added to the displacement.

WETTED SURFACE AND SHELL DISPLACEMENT.

\multicolumn{3}{c	}{Up to 4' 0" Waterline.}			\multicolumn{3}{c	}{Up to 8' 0" Waterline.}			\multicolumn{3}{c}{Up to 12' 0" Waterline.}		
No. of Sec.	Half Girths.	S.M.	Products.	No. of Sec.	Half Girths.	S.M.	Products.	No. of Sec.	Half Girths.	S.M.	Products.
1	4·1	1	4·1	1	8·0	1	8·0	1	12·5	1	12·5
2	4·5	4	18·0	2	9·4	4	37·6	2	15·1	4	60·4
3	7·2	2	14·4	3	12·3	2	24·6	3	16·7	2	33·4
4	10·1	4	40·4	4	14·7	4	58·8	4	18·8	4	75·2
5	11·8	2	23·6	5	16·1	2	32·2	5	20·1	2	40·2
6	12·9	4	51·6	6	17·0	4	68·0	6	21·0	4	84·0
7	13·2	2	26·4	7	17·3	2	34·6	7	21·3	2	42·6
8	12·9	4	51·6	8	17·0	4	68·0	8	21·0	4	84·0
9	11·9	2	23·8	9	16·1	2	32·2	9	20·1	2	40·2
10	10·0	4	40·0	10	14·3	4	57·2	10	18·4	4	73·6
11	7·9	2	15·8	11	12·2	2	24·4	11	16·2	2	32·4
12	5·2	4	20·8	12	9·4	4	37·6	12	13·5	4	54·0
13	0·0	1	0·0	13	4·0	1	4·0	13	8·0	1	8·0

330·5	487·2	640·5
⅓ longt. interval 4	4	4
1322·0	1948·8	2562·0
Both sides 2	2	2
Total Area of Immersed Surface, which is the Wetted Surface, 2644 sq. ft.	Area 3897·6	Area 5124

Let the average thickness of shell plating up to the load line be 8/20.

As the strakes of plating are alternately in and out the **average thickness** from the frames to the outside of the plating will be :—

$$\frac{8+4}{20} = \frac{12}{20} \text{ of an inch.} \quad \frac{12}{20} \times \frac{1}{12} = \frac{1}{20} \text{ of a foot.}$$

$$\frac{2644}{1} \times \frac{1}{20} = \frac{2644}{20} \text{ cubic feet displacement.}$$

$$\frac{2644}{20} \div 35 = 3·7 \text{ tons displacement at the 4 ft. waterline.}$$

$$\frac{3897}{20} \div 35 = 5·5 \quad ,, \quad ,, \quad ,, \quad 8 \text{ ft.} \quad ,,$$

$$\frac{5124}{20} \div 35 = 7·3 \quad ,, \quad ,, \quad ,, \quad 12 \text{ ft.} \quad ,,$$

Notes on Wetted Surface Calculation.—The half girths of the frames, measured from the top of the keel to the height of the particular waterlines, up to which the areas are required, are taken at each section. These are treated as the ordinates of a new curve, and the area found, by the application of one of Simpson's Rules, in the usual way. The area of the immersed keel, stem, and stern frame, may be added to this.

Shell Displacement Calculation.—The area of the immersed surface multiplied by the average thickness, from the frames to the outside of the plating, gives the volume of displacement.

CALCULATION FOR HEIGHTS OF VERTICAL CENTRES OF BUOYANCY ABOVE TOP OF KEEL.

1st Method of Arrangement.

Sum of Functions of Horizontal Half Areas (taken from Displacement Calculation).

Top of Keel.	1' 0" W.L.	2' 0" W.L.	3' 0" W.L.	4' 0" W.L.	6' 0" W.L.	8' 0" W.L.	10' 0" W.L.	12' 0" W.L.	
3·72 0	168·22 ½	218·24 1	248·58 1½	268·54 2	297·10 3	316·64 4	336·42 5	348·50 6	= Leverages.
0·00 ½	84·11 2	218·24 ½	372·87 2	537·08 1	891·3 4	1266·56 1	1682·1 4	2091·00 1	Simpson's = Multipliers.
0·00	168·22	109·12 168·22 0·00	745·74	268·54 745·74 109·12 277·34	3565·2	1266·56 3565·20 537·08 1400·74	6728·4	2091·00 6728·40 1266·56 6769·58	= Moments.

Sum of Multiples of Functions } 447·42)277·34   1187·97)1400·74   2961·55)6769·58   4972·37)16855·64

Vertical Interval   ·61
2   1·17
2   2·28
2   3·38
2

   1·22   2·34   4·56   6·76

Height of Centre of Buoyancy above } Keel for 2' 0" Waterline.   2' 0" W.L.   4' 0" W.L.   8' 0" W.L.   12' 0" W.L.

Notes on Vertical Centre of Buoyancy Calculation.—As described in Example 18, the areas of waterplanes (or "functions of horizontal half areas") are treated by Simpson's Multipliers, and by the leverages (number of intervals). The sum of moments so obtained multiplied by the vertical interval, and the product divided by the "sum of the multiples of functions," gives the height of centre of buoyancy above keel. In the above calculation the operation of multiplying by the vertical interval was effected at the end of the calculation.

CALCULATION FOR HEIGHTS OF TRANSVERSE METACENTRES ABOVE CENTRES OF BUOYANCY.

1st Method of Arrangement.

2' 0" Waterline.

No. of Ord.	⅓ Ords.	Cubes of Ords.	S.M.	Functions of Cubes.
1	·12	0·00	1	0·00
2	1·08	1·25	4	5·00
3	3·65	48·62	2	97·24
4	6·60	287·49	4	1149·96
5	8·65	647·21	2	1294·42
6	9·90	970·29	4	3881·16
7	10·22	1067·46	2	2134·92
8	9·90	970·29	4	3881·16
9	8·70	658·50	2	1317·00
10	6·85	321·41	4	1285·64
11	4·72	105·15	2	210·30
12	2·23	11·08	4	44·32
13	0·10	0·00	1	0·00

⅓ of Cubes 3)15301·12
                 5100·37
⅓ of Longt. Interval 12/3 = 4
Moment of Inertia for ⅓ W.P. } 20401·48
Multiplier for both sides = 2
Displacement* 2880)40802·96
Metacentre above the } = 17·14
Centre of Buoyancy } ft.

4' 0" Waterline.

⅓ Ords.	Cubes of Ords.	S.M.	Functions of Cubes.
·12	0·00	1	0·00
1·89	6·75	4	27·00
5·70	185·19	2	370·38
8·80	681·47	4	2725·88
10·51	1160·93	2	2321·86
11·44	1497·19	4	5988·76
11·51	1524·84	2	3049·68
11·25	1423·82	4	5695·28
10·38	1118·38	2	2236·76
8·48	609·80	4	2439·20
6·05	221·44	2	442·88
3·17	31·85	4	127·40
0·00	0·00	1	0·00

              3)25425·08
                 8475·02
                    4
                33900·08
                    2
         6819·95)67800·16
                 10·72
                  ft.

8' 0" Waterline.

⅓ Ords.	Cubes of Ords.	S.M.	Functions of Cubes.
·14	·002	1	0·00
4·60	97·330	4	389·32
8·80	681·470	2	1362·94
10·85	1277·280	4	5109·12
11·65	1581·160	2	3162·32
11·98	1719·370	4	6877·48
12·00	1728·000	2	3456·00
11·85	1664·000	4	6656·00
11·31	1446·730	2	2893·46
9·85	955·670	4	3822·68
7·60	438·970	2	877·94
4·24	76·220	4	304·88
·30	·020	1	0·02

              3)34912·16
                11637·38
                    4
                46549·52
                    2
      15755·25)93099·04
                 5·90
                  ft.

12' 0" Waterline.

⅓ Ords.	Cubes of Ords.	S.M.	Functions of Cubes.
2·00	8·00	1	8·00
8·71	660·77	4	2643·08
10·65	1207·94	2	2415·88
11·40	1481·54	4	5926·16
11·75	1622·23	2	3244·46
11·90	1685·15	4	6740·60
11·98	1719·37	2	3438·74
11·85	1664·00	4	6656·00
11·51	1524·84	2	3049·68
10·50	1157·62	4	4630·48
8·46	605·49	2	1210·98
4·98	123·50	4	494·00
·44	·08	1	·08

              3)40458·14
                13486·04
                    4
                53944·16
                    2
     26458·00)107888·32
                 4·07
                  ft.

This calculation has been fully described in Examples 20 and 21, Section 1, of this chapter.
* Displacement = cubic feet.

CALCULATIONS.

DISPLACEMENT CALCULATION.
2nd Method of Arrangement.

No. of Ord.	Top of Keel.	1' 0" W.L.	2' 0" W.L.	3' 0" W.L.	4' 0" W.L.	6' 0" W.L.	8' 0" W.L.	10' 0" W.L.	12' 0" W.L.	
2	·12	·65	1·08	1·42	1·89	3·00	4·60	6·78	8·71	
4	·12	4·45	6·60	7·92	8·80	10·14	10·85	11·26	11·40	
6	·12	7·80	9·90	10·87	11·44	11·90	11·98	11·95	11·90	
8	·12	7·92	9·90	10·80	11·25	11·71	11·85	11·90	11·85	
10	·12	6·05	6·85	7·86	8·48	9·34	9·85	10·23	10·50	
12	·00	1·35	2·23	2·80	3·17	3·80	4·24	4·65	4·98	
a	·60	28·22	36·56	41·67	45·03	49·89	53·37	56·77	59·34	
2a	1·20	56·44	73·12	83·34	90·06	99·78	106·74	113·54	118·68	
(½)1	·06	·06	·06	·06	·06	·06	·07	·30	1·00	
3	·12	2·15	3·65	4·76	5·70	7·40	8·80	9·88	10·65	
5	·12	6·70	8·65	9·80	10·51	11·36	11·65	11·76	11·75	
7	·12	8·40	10·22	11·05	11·51	11·90	12·00	12·00	11·98	
9	·12	7·00	8·70	9·74	10·38	11·02	11·31	11·45	11·51	
11	·12	3·36	4·72	5·54	6·05	6·93	7·60	9·10	8·46	Sums of Functions
(½)13	·00	0·00	0·00	0·00	0·00	·10	·15	·18	·22	= of Horizontal ½ Areas.
	1·86	84·11	109·12	124·29	134·27	148·55	158·32	168·21	174·25	
Simpson's ½ Multipliers	⅓	1	⅓	1	⅓		2	½	2	½
	·46	84·11	27·28	124·29	33·56	297·10	79·16	336·42	87·12	
			84·11		124·29		297·10		336·42	
			·46		27·28		67·13		79·16	
					111·85		296·98		740·37	
Sum of Multiples of Functions			111·85		296·98		740·37		1243·07	
Multiplier for Displacement	·608				·608		·608		·608	
			89480		237584		592296		994456	
			671100		1781880		4442220		7458420	
		Tons	68·00480		180·56384		450·14496		755·78656	

Shell displacement to add. See page 304.

CALCULATION FOR "TONS PER INCH IMMERSION."

	2' 0" W.L.	4' 0" W.L.	8' 0" W.L.	12' 0" W.L.	
$\frac{2 \times 2 \times 12}{3}$ =	109·12 / 16	134·27 / 16	158·32 / 16	174·25 / 16	= { Sum of Functions of Horizontal ½ Areas.
	420)1745·92	420)2148·32	420)2533·12	420)2788	Areas of Waterlines
Tons per Inch	4·15	5·11	6·03	6·63	

Notes on Displacement Calculation, 2nd Method.—This calculation involves somewhat less labour than the first method. It is simply a method of using Simpson's ½ Multipliers throughout. All the even ordinates are put in the top column, and the sum of them, a, is multiplied by 2, and placed in the 2a line. In the column below, the odd ordinates are placed together with half of the endmost ordinates, and these are added together, including the 2a line. By this process we have simply used, ⅓, 2, 1, 2 and so on, as Simpson's Multipliers, instead of 1, 4, 2, 4, etc. The sum of these horizontal functions are treated by Simpson's ½ Multipliers, and the displacements found as before. It will be noted that the multiplier for displacement is ·608, this being ·152 × 2 (for horizontal ½ Simpson's Multipliers) × 2 (for vertical ½ Simpson's Multipliers) = ·608.

Tons per Inch.—The area of each particular waterplane is obtained by multiplying the sum of the functions of the horizontal half areas by 2 for ½ Simpson's Multipliers, by 2 for the other side, and by ⅓ of the longitudinal interval =

$$\frac{2 \times 2 \times 12}{3} = 16.$$

CALCULATION FOR HEIGHTS OF VERTICAL CENTRES OF BUOYANCY ABOVE TOP OF KEEL.

2nd Method of Arrangement.

Sum of Functions of Horizontal Half Areas (taken from Displacement Sheet).

Top of Keel.	1' 0" W.L.	2' 0" W.L.	3' 0" W.L.	4' 0" W.L.	6' 0" W.L.	8' 0" W.L.	10' 0" W.L.	12' 0" W.L.	
1·86 0	84·11 ½	109·12 1	124·29 1½	134·27 2	148·55 3	158·82 4	168·21 5	174·25 6	= Leverage.
0·00 ½	42·05 1	109·12 4	186·43 1	268·54 2 2/3	445·65 3	633·28 4 2/3	841·05 2	1045·50 4 2/3	= Simpson's Multipliers.
0·00	42·05	27·28 42·05 0·00	186·43	67·13 186·43 27·28 69·33	891·30	316·64 891·30 134·27 350·17	1682·10	522·75 1682·10 316·64 1692·38	= Moments.
Sum of Multiples of Functions	{ 111·85)69·33		296·98)350·17		740·37)692·38		1243·07)4213·87	
Vertical Interval	·61 2		1·17 2		2·28 2		3·38 2	
Height of C.B. above top of Keel for 2' 0" Waterline } = 1·22 ft.		2·84 ft.		4·56 ft.		6·76 ft.	
	4' 0" W.L.		8' 0" W.L.		12' 0" W.L.	

See notes, page 305.

308 KNOW YOUR OWN SHIP.

CALCULATION FOR LONGITUDINAL CENTRES OF BUOYANCY.

2nd Method of Arrangement.

N. of Ord.	Levers	Top of Keel	1' 0" W.L.	2' 0" W.L.	3' 0" W.L.	4' 0" W.L.	6' 0" W.L.	8' 0" W.L.	10' 0" W.L.	12' 0" W.L.	
2	1	·12	·65	1·08	1·42	1·89	3·00	4·60	6·78	8·71	
4	3	·36	13·35	19·80	23·76	26·40	30·42	32·55	33·78	34·20	
6	5	·60	39·00	49·50	54·35	57·20	59·50	59·90	59·75	59·50	
8	7	·84	55·44	69·30	75·60	78·75	81·27	82·95	83·30	82·95	
10	9	1·08	54·45	61·65	70·74	76·32	84·06	88·65	92·07	94·30	
12	11	0·00	14·85	24·53	30·80	34·87	41·80	46·64	51·15	54·78	
a		3·00	177·74	225·86	256·67	275·43	300·75	315·29	326·83	334·64	
2a		6·00	355·48	451·72	513·34	550·86	601·50	630·58	653·66	669·28	
1	0	0·00	0·00	0·00	0·00	0·00	0·00	0·00	0·00	0·00	
3	2	·24	4·30	7·30	9·52	11·40	14·80	17·60	19·76	21·30	
5	4	·48	26·80	34·60	39·20	42·04	45·44	46·60	47·04	47·00	
7	6	·72	50·40	61·32	66·30	69·06	71·40	72·00	72·00	71·38	
9	8	·96	56·00	69·60	77·92	83·04	88·16	90·48	91·60	92·08	
11	10	1·20	33·60	47·20	55·40	60·50	69·30	76·00	91·80	84·60	
13	12	0·00	0·00	0·00	0·00	0·00	1·20	1·80	2·16	2·64	
		9·60	526·58	671·74	761·68	816·90	891·80	935·06	977·22	988·78	Simpson's Multipliers.
			½	1	½	1	½	2	½	2	½
		2·40	526·58	167·93	761·68	204·22	1783·60	467·53	1954·44	494·39	
				526·58		761·68		1783·60		1954·44	
				2·40		167·93		408·45		467·53	
						696·21		1830·74		4490·32	
Sum of Multiples of Functions			111·85)696·91		296·98)1830·74		740·37)4490·32		1243·07)7406·68	Moments.
				6·23		6·16		6·06		5·95	
Longitudinal Interval, =				12		12		12		12	
Centre of Buoyancy from No. 1. Section		=		74·76 ft.		73·92 ft.		72·7 ft.		71·40 ft.	

Longitudinal Centre of Buoyancy.—In this calculation, the ordinates (see Displacement Sheet) for the various waterplanes are multiplied by their respective leverages (intervals) from the first ordinate, and the moments so obtained are treated by Simpson's Multipliers. In the above calculation, the centre of buoyancy is worked for four separate draughts, giving four separate results.

The sums of moments are next treated by Simpson's Multipliers, and the new sum is divided by the "sum of multiples of functions" (from the Displacement Sheet). This result, multiplied by the common interval, gives the position of the longitudinal centre of buoyancy relatively to the first section.

CALCULATION FOR HEIGHTS OF TRANSVERSE META-CENTRES ABOVE CENTRES OF BUOYANCY.

2nd Method of Arrangement.

No. of Ord.	2' 0" W.L. Cubes of ½ Ordinates.	4' 0" W.L. Cubes of ½ Ordinates.	8' 0" W.L. Cubes of ½ Ordinates.	12' 0" W.L. Cubes of ½ Ordinates.	
2	1·25	6·75	97·83	660·77	
4	287·49	681·47	1277·28	1481·54	
6	970·29	1497·19	1719·37	1685·15	
8	970·29	1423·82	1664·00	1664·00	
10	321·41	609·80	955·67	1157·62	
12	11·08	31·85	76·22	123·50	
a	2561·81	4250·88	5789·87	6772·58	
2a	5123·62	8501·76	11579·74	13545·16	
(½) 1	0·00	0·00	0·00	4·00	
3	48·62	185·19	681·47	1207·94	
5	647·21	1160·93	1581·16	1622·23	
7	1067·46	1524·84	1728·00	1719·37	
9	658·50	1118·38	1446·73	1524·84	
11	105·15	221·44	438·97	605·49	
(½) 13	0·00	0·00	0·01	·04	
Functions of Cubes,	7650·56	12712·54	17456·08	20229·07	Multiplier for Moment of Inertia.
	5·33	5·33	5·33	5·33	
	2295168	3813762	5236824	6068721	
	2295168	3813762	5236824	6068721	
	3825280	6356270	8728040	10114535	
	2380*)40777·4848	6319·60*)67757·8382	15754·90*)93040·9064	26465·80*)107820·9431	Moment of Inertia.
M.C. above C.B. = 17·13ft.		10·72ft.	5·90ft.	4·07ft.	

* = displacement in cubic feet.

With the previous explanations the only point in this calculation needing comment is the multiplier 5·33.

$$\frac{12 \times 2 \times 2}{3 \times 3} = 5 \cdot 33.$$

Where 12 is the longitudinal interval

    2 is for half Simpson's Multipliers.
    2 is for both halves of waterplane.

      3 is for ⅓ of cubes.
      3 is for ⅓ of longitudinal interval.

CALCULATION FOR HEIGHT OF LONGITUDINAL METACENTRE ABOVE THE CENTRE OF BUOYANCY.

4' 0" WATERPLANE.

No. of Ord.	Half Ordinates.	S.M.	Functions or Products.	Levers.	Products for Moments.	Mults. for M. of I.	Products for Moment of Inertia.
1	·12	1	·12	0	0·00	0	0·00
2	1·89	4	7·56	1	7·56	1	7·56
3	5·70	2	11·40	2	22·80	2	45·60
4	8·80	4	35·20	3	105·60	3	316·80
5	10·51	2	21·02	4	84·08	4	336·32
6	11·44	4	45·76	5	228·80	5	1144·00
7	11·51	2	23·02	6	138·12	6	828·72
8	11·25	4	45·00	7	315·00	7	2205·00
9	10·38	2	20·76	8	166·08	8	1328·64
10	8·48	4	33·92	9	305·28	9	2747·52
11	6·05	2	12·10	10	121·00	10	1210·00
12	3·17	4	12·68	11	139·48	11	1534·28
13	0·00	1	0·00	12	0·00	12	0·00
	89·30		268·54		1638·80		11704·44

⅓ Long. Int. = $\frac{12}{3}$ = 4

Both sides

1074·16
2

Area of Waterplane 2148·32
sq. ft.

268·54) 19605·60 (12
78·00 ft.
C. G. of Waterplane from Ord. 1.

676

7022664
8193108
5852220

2) 67417574·44

73 × 73 = 5329   13483514·88
5329 × 2148·32 = 11448397·28

Displacement in cubic feet = 180·57 × 85 = 6319·85) 2035117·60 (322·01

feet.

= ⅓ of Longitudinal Interval, cubed = $\frac{12^3}{3}$

= Both sides.

= Moment of Inertia of Waterplane about 1st Ordinate.

= { Moment of Inertia of Waterplane about the Axis passing through the Centre of Gravity of Waterplane.

Longitudinal Metacentre above Centre of Buoyancy.

CALCULATION FOR HEIGHT OF LONGITUDINAL METACENTRE ABOVE THE CENTRE OF BUOYANCY.

8' 0" WATERPLANE.

No. of Ord.	½ Ord.	S.M.	Products.	Levers.	Products for Moments.	Mults. for M. of I.	Products for Moment of Inertia.
1	·14	1	·14	0	0·00	0	·00
2	4·60	4	18·40	1	18·40	1	18·40
3	8·80	2	17·60	2	35·20	2	70·40
4	10·85	4	43·40	3	130·20	3	390·60
5	11·65	2	23·30	4	93·20	4	372·80
6	11·98	4	47·92	5	239·60	5	1198·00
7	12·00	2	24·00	6	144·00	6	864·00
8	11·85	4	47·40	7	331·80	7	2322·60
9	11·31	2	22·62	8	180·96	8	1447·68
10	9·85	4	39·40	9	354·60	9	3191·40
11	7·60	2	15·20	10	152·00	10	1520·00
12	4·24	4	16·96	11	186·56	11	2052·16
13	·30	1	·30	12	3·60	12	43·20

316·64

⅓ Long. Int. = 12/3 = 4

Both sides 1266·56 × 2

Area of Waterplane 2533·12 sq. ft.

1870·12 × 12

316·64) 22441·44
70·87 ft.
C. G. of Waterplane from Ord. 1.

Displacement in cubic feet = 450·15 × 35 = 15755·20) 2819169·15

13491·24
576
—————
8094744
9443868
6745620
—————
7770864·24 × 2
—————
15541908·48
2533·12 × (70·87)² = 12722789·38
—————
2819169·15

178·93 ft.

= ⅓ of Longitudinal Interval, cubed = 12³/3

Moment of Inertia of Waterplane about 1st Ordinate.

{ Moment of Inertia of Waterplane about the Axis passing through the Centre of Gravity of Waterplane.

Longitudinal Metacentre above Centre of Buoyancy.

CALCULATION FOR HEIGHT OF LONGITUDINAL METACENTRE ABOVE THE CENTRE OF BUOYANCY.

12′ 0″ WATERPLANE.

No. of Ord.	Half Ordinates.	S. M.	Products.	Levers.	Products for Moments.	Mults. for M. of I.	Products for Moment of Inertia.
1	2·00	1	2·00	0	0·00	0	0·00
2	8·71	4	34·84	1	34·84	1	34·84
3	10·65	2	21·30	2	42·60	2	85·20
4	11·40	4	45·60	3	136·80	3	410·40
5	11·75	2	23·50	4	94·00	4	376·00
6	11·90	4	47·60	5	238·00	5	1190·00
7	11·98	2	23·96	6	143·76	6	862·56
8	11·85	4	47·40	7	331·80	7	2322·60
9	11·51	2	23·02	8	184·16	8	1473·28
10	10·50	4	42·00	9	378·00	9	3402·00
11	8·46	2	16·92	10	169·20	10	1692·00
12	4·98	4	19·92	11	219·12	11	2410·32
13	·44	1	·44	12	5·28	12	63·36

Sum of Products: 348·50   1977·56   14322·56

⅓ Long. Int. = 12/3 = 4   12

Both sides   2   348·50)23730·72   8698386
                                                                        10026792
1394·00   68·09 ft.   7161280

Area of Waterplane   2   C. G. of Waterplane from Ord. 1.   8249794·56
sq. ft. 2788   2

$2788 \times (68 \cdot 09)^2 = 16499589·12$ $= 19925859·70$

Displacement in cubic feet = $755·8 \times 35 = 26453$) $3573729·42$

135·09 ft.

$= \tfrac{2}{3}$ of Longitudinal Interval, cubed $= \tfrac{12^3}{3}$

= Moment of Inertia of Waterplane about 1st Ordinate.

= { Moment of Inertia of Waterplane about the Axis passing through the Centre of Gravity of Waterplane.

= Longitudinal Metacentre above Centre of Buoyancy.

The steps in these calculations have been fully described in Example 24 in Section 1 of this chapter.

314 KNOW YOUR OWN SHIP.

By means of the results given in the foregoing calculations, curves may be constructed, see figs. 138 and 139 as described and illustrated in the earlier chapters of this book, by which means any of the quantities or distances may be ascertained for any intermediate draught or waterline.

Care should be taken in noting that, while the draughts are measured from the bottom of the keel, the calculations have been for waterlines measured from the top of the keel, and need to be set off accordingly in constructing the curves. When a vessel trims by the head or the stern, the mean draught is that worked to, in taking any particulars from the curves

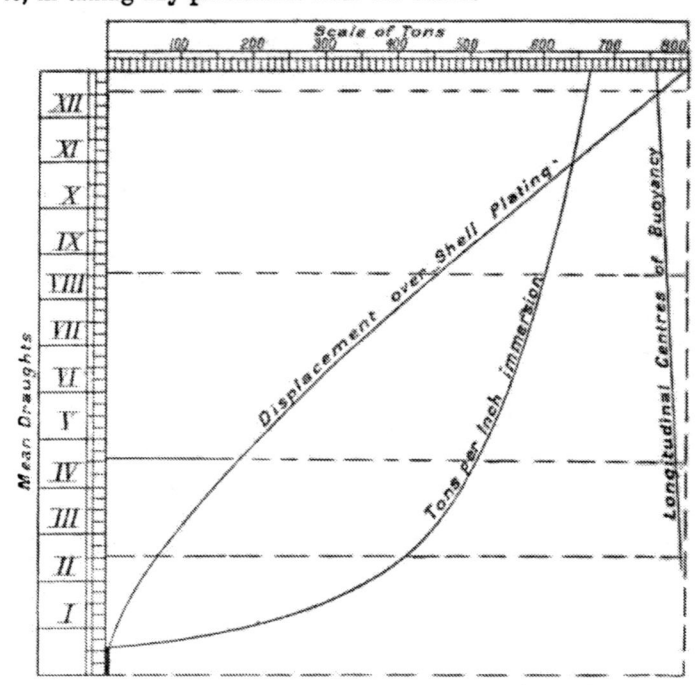

Fig. 138.—Curves of Displacement—Tons per Inch Immersion—Longitudinal Centres of Buoyancy set off from the Aft Side of the After Stern Post.

Note.—100 on the horizontal scale of tons equals 1 ton for the tons per inch immersion, and 10 feet for the longitudinal centres of buoyancy.

Stability Calculations.—Fig. 140 shows our vessel keeled

CALCULATIONS. 315

to an angle of 14°, at which angle the stability is computed in the stability calculation, page 320.

In fig. 140, V = the volume of the wedge of immersion, or, the wedge of emersion.

BR is what the righting arm of stability would be if the centre of gravity coincided with the centre of buoyancy in the upright condition. It is perpendicular to the vertical line through B'.

BR may therefore be called the lever of stability produced by design or form. It is usually called the lever of statical surface stability.

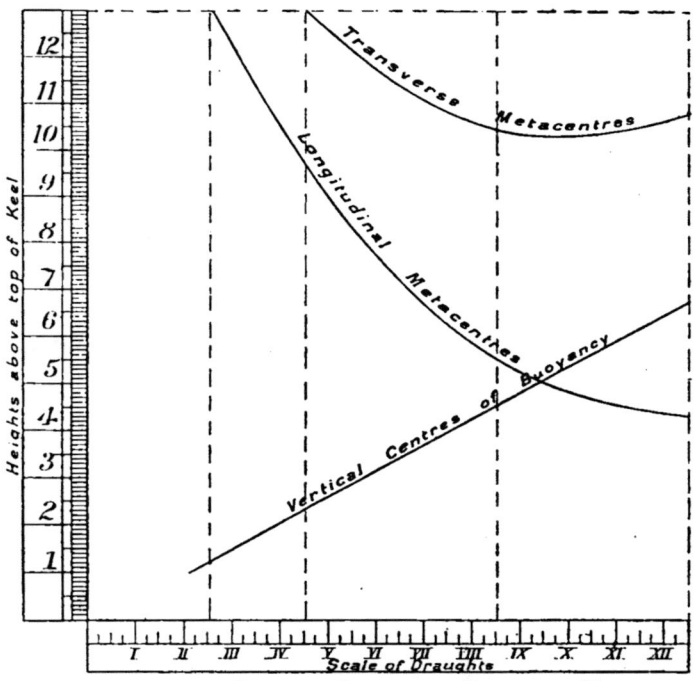

Fig. 139.—Curves of Vertical Centres of Buoyancy—Transverse Metacentres—Longitudinal Metacentres (note, 3 feet on the vertical scale represents 100 feet).

G is the centre of gravity of the ship, its position being governed by the loading and position of weights carried.

The lever of statical stability is therefore :—

$$BR - BC = GZ \qquad BC = BG \times \text{sine of angle of } 14°.$$

The moment of statical surface stability is BR × displacement in tons.
The moment of statical stability is GZ × D (displacement in tons).
gh and $g'h'$ are perpendicular to W'L'.
B' is the position of the centre of buoyancy at 14° of inclination.
g, g' are the centres of gravity of the wedges of immersion and emersion.

$$BB' = \frac{V \times gg'}{D} \qquad\qquad BR = \frac{V \times hh'}{D}$$

$$GZ = \frac{V \times hh'}{D} - BG \times \text{sine of } 14°$$

Righting moment of stability $= D \times \left\{ \dfrac{V \times hh'}{D} - (BG.\ \text{sine of } 14°) \right\}$

Fig. 140.

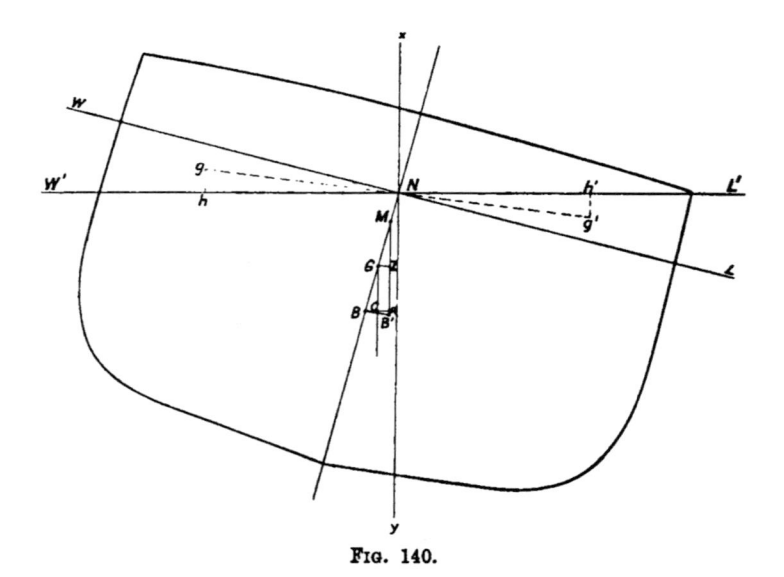

Fig. 141 shows the sections of our vessel prepared for the stability calculation. The calculation is divided into two parts, the first being of a preliminary nature in order to obtain results which are transferred to the combination table in the second part.

From the stability sections, the ordinates are measured for the upright, and the two inclined, waterplanes, for both the immersed and emerged wedges, and inserted in their proper columns in the preliminary tables.

CALCULATIONS. 317

The three results obtained in the preliminary tables almost explain themselves, after the examples worked in Section I. of this chapter.

(1) By treating the ordinates by Simpson's Multipliers, the total area of the new horizontal waterplane is obtained, W′L′ (fig. 140),—the immersed and emerged sides being added together. This is only necessary for the third waterplane in the calculation. See Example 10. Section 1 of this chapter.

(2) By treating the half squares of ordinates by Simpson's Multipliers, it is found on which side of the inclined waterplane there is a preponderance of moment. In this case, the immersed is the greater. The centre of gravity of the waterplane, therefore, lies towards this side. By subtracting the emerged side from the immersed, and dividing by 2 for ½ squares, and multiplying by ⅓ of the longitudinal interval, the preponderating moment is ascertained. This, divided by the area of the same inclined waterplane, gives the distance the centre of gravity is out from the longitudinal middle line of the waterplane on the immersed side. See Example 17.

It will now be seen that the area was necessary in order to find the transverse centre of gravity of the whole inclined waterplane.

(3) By cubing the ½ ordinates of the three planes, and ultimately taking one-third of them, we obtain the moment of inertia which is necessary in order to find the moments of the wedges. See example 30.

In fig. 141 the wedges of immersion and emersion have been drawn as though the waterplanes for all angles of inclination intersected at the original fore and aft centre line of the upright waterplane. In reality, these waterplanes do not, at least for considerable inclinations, for it is well known to the reader that the volume of the wedge of immersion so obtained, would in all probability, differ from the volume of the wedge of emersion. So a correction becomes necessary in order to arrive at an accurate result. If the volume of the wedge of immersion be larger than that of the wedge of emersion, it is clear that the vessel is drawing more water in the diagram than she does in reality. By dividing the difference in volume of the two wedges as calculated, by the area of the waterplane, the thickness of the correcting layer to be deducted from the draught is ascertained. Had the immersed wedge been less than the emerged, the difference in volume of the two wedges divided by the area of the waterplane, would give the thickness of the layer to be added to the draught. In the combination table, the volume of the immersed and emerged wedges are calculated as explained in Example 31, and the thickness of the correcting layer is obtained.

In our vessel we see that the volume of the wedge of immersion exceeds that of the wedge of emersion by 89·75 cubic feet, and the thickness of the correcting layer is ·03 feet. And as the centre of gravity of the inclined waterplane which represents the centre of gravity of the correcting layer is ·25 feet (see preliminary table) towards the immersed side, the moment of the wedges is in excess, and the moment of the layer must be deducted from the moment of the wedges. Had the centre of gravity of this layer lain towards the opposite side, its moment would have had to be added.

The last operation in the combination table is to find the moment of the wedges. This process has been fully explained in Example 30. The sums of the functions of the cubes of the ordinates for the waterplanes of both wedges, are transferred from the preliminary tables. These functions are treated by Simpson's Multipliers, and also by the *cosines* of the angles of inclination, obtaining "functions of cubes for moments of wedges." The sum of these functions for moments is divided by 3, as $\frac{1}{3}$ of cubes of ordinates is required for moment of inertia of waterplanes. The result is multiplied by $\frac{1}{3}$ of the angular interval, and this by $\frac{1}{3}$ of the longitudinal interval giving moment of the wedges relatively to a plane passing through $x\,N\,y$ perpendicular to W', N, L' (see fig. 140). From the moment of wedges is subtracted the correcting moment for the layer, and the remainder divided by the displacement gives the distance BR (fig. 140) produced by the transference of the wedge WNW' to $L'NL$, a distance hh'. To obtain the righting lever GZ, subtract BC from BR ($BC = BG$ multiplied by the sine of the angle BGC, which is 14°).

$GZ \times$ displacement = Righting moment in foot tons.

As a check upon the GZ, when calculated for small angles of inclination (before the deck edge is immersed), not exceeding 10° or 15°.

$GM \times$ Sine of angle $= GZ$.
2 feet × ·2419 = ·4838 which is approximately correct.

Note.—The metacentre was found to be 4·07 feet above the centre of buoyancy and the metacentric height was *assumed* to be 2 feet.

In order to construct a curve of stability a succession of calculations, identical to that we have just described, would have to be made for the vessel inclined to a succession of angles of inclination in order to find the GZ's at each inclination.

CALCULATIONS.

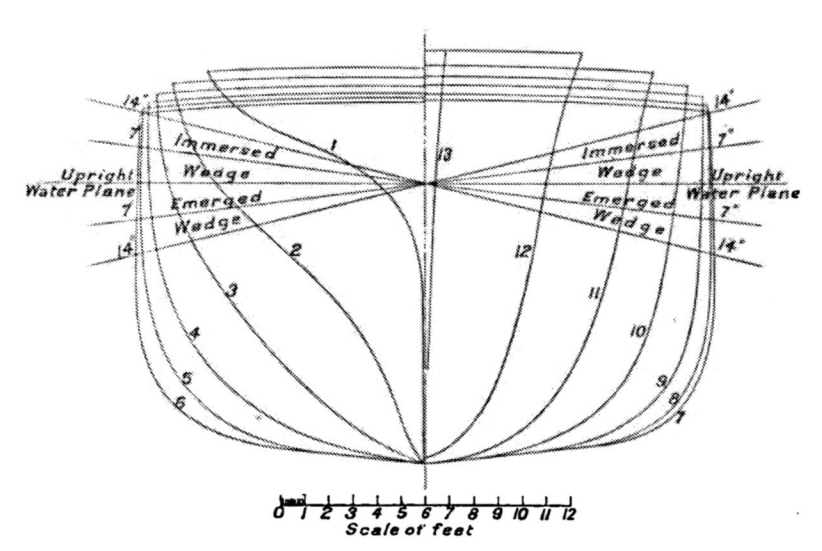

FIG. 141.—SECTIONS FOR CALCULATION OF STATICAL STABILITY.

CALCULATION FOR THE STABILITY OF A VESSEL
See fig. 141

PRELIMINARY

Upright

IMMERSED WEDGE.

No. of Section.	Ordinates.	Multipliers.	Functions of Ordinates.	Squares of Ordinates.	Multipliers.	Functions of Squares.	Cubes of Ordinates.	Multipliers.	Functions of Cubes.
1	2·00			4·00	1	4·00	8·00	1	8·00
2	8·71			75·86	4	303·44	660·77	4	2643·08
3	10·65			113·42	2	226·84	1207·94	2	2415·88
4	11·40			129·96	4	519·84	1481·54	4	5926·16
5	11·75			138·06	2	276·12	1622·23	2	3244·46
6	11·90			141·61	4	566·44	1685·15	4	6740·60
7	11·98			143·52	2	287·04	1719·37	2	3438·74
8	11·85			140·42	4	561·68	1664·00	4	6656·00
9	11·51			132·48	2	264·96	1524·84	2	3049·68
10	10·50			110·25	4	441·00	1157·62	4	4630·48
11	8·46			71·57	2	143·14	605·49	2	1210·98
12	4·98			24·80	4	99·20	123·50	4	494·00
13	·44			·19	1	·19	·08	1	·08
						3693·89			40458·14
									2
									80916·28 For both wedges.

PRELIMINARY

Waterplane inclined

IMMERSED WEDGE.

1	2·4			5·7	1	5·7	13·8	1	13·8
2	9·5			90·2	4	360·8	857·3	4	3429·2
3	11·0			121·0	2	242·0	1331·0	2	2662·0
4	11·6			134·5	4	538·0	1560·8	4	6243·2
5	11·8			139·2	2	278·4	1643·0	2	3286·0
6	12·0			144·0	4	576·0	1728·0	4	6912·0
7	12·0			144·0	2	288·0	1728·0	2	3456·0
8	11·9			141·6	4	566·4	1685·1	4	6740·4
9	11·6			134·5	2	269·0	1560·8	2	3121·6
10	10·7			114·4	4	457·6	1225·0	4	4900·0
11	8·7			75·6	2	151·2	658·5	2	1317·0
12	5·2			27·0	4	108·0	140·6	4	562·4
13	·4			·1	1	·1	·0	1	·0
						3841·2			42643·6 Immersed wedge.
									40197·0 Emerged wedge.
									82840·6 Sum for both wedges.

CALCULATIONS.

INCLINED TO AN ANGLE OF 14 DEGREES.
for Drawings.

TABLE I.
Waterplane.

			EMERGED WEDGE.						
No. of Section.	Ordinates.	Multipliers.	Functions of Ordinates.	Squares of Ordinates.	Multipliers.	Functions of Squares.	Cubes of Ordinates.	Multipliers.	Functions of Cubes.

SAME AS IMMERSED WEDGE.

TABLE II.
to an angle of 7°.

			EMERGED WEDGE.						
1	1·8			3·2	1	3·2	5·8	1	5·8
2	7·9			62·4	4	249·6	493·0	4	1972·0
3	10·3			106·0	2	212·0	1092·7	2	2185·4
4	11·4			129·9	4	519·6	1481·5	4	5926·0
5	11·9			141·6	2	283·2	1685·1	2	3370·2
6	12·1			146·4	4	585·6	1771·5	4	7086·0
7	12·1			146·4	2	292·8	1771·5	2	3543·0
8	12·0			144·0	4	576·0	1728·0	4	6912·0
9	11·5			132·2	2	264·4	1520·8	2	3041·6
10	10·4			108·1	4	432·4	1124·8	4	4499·2
11	8·4			70·5	2	141·0	592·7	2	1185·4
12	4·9			24·0	4	96·0	117·6	4	470·4
13	·4			·1	1	·1	·0	1	·0
						3655·9			40197·0

CALCULATION FOR THE STABILITY OF A VES$

PRELIMIN

(Waterplane incl;

IMMERSED WEDGE.

No. of Section.	Ordinates.	Multipliers.	Functions of Ordinates.	Squares of Ordinates.	Multipliers.	Functions of Squares.	Cubes of Ordinates.	Multipliers.	Functions of Cubes.
1	3·1	1	3·1	9·6	1	9·6	29·7	1	29·7
2	10·3	4	41·2	106·0	4	424·0	1092·7	4	4370·8
3	11·4	2	22·8	129·9	2	259·8	1481·5	2	2963·0
4	11·8	4	47·2	139·2	4	556·8	1643·0	4	6572·0
5	12·0	2	24·0	144·0	2	288·0	1728·0	2	3456·0
6	12·0	4	48·0	144·0	4	576·0	1728·0	4	6912·0
7	12·2	2	24·4	148·8	2	297·6	1815·8	2	3631·6
8	12·2	4	48·8	148·8	4	595·2	1815·8	4	7263·2
9	11·9	2	23·8	141·6	2	283·2	1685·1	2	3370·2
10	11·1	4	44·4	123·2	4	492·8	1367·6	4	5470·4
11	9·2	2	18·4	84·6	2	169·2	778·6	2	1557·2
12	5·4	4	21·6	29·1	4	116·4	157·4	4	629·6
13	·4	1	·4	·1	1	·1	·0	1	·0

Sum of functions { Immersed, 368·1
of Ordinates, { Emerged, 346·4
          Total, 714·5
½ of longitudinal interval, . 4
Total area of Waterplane, 2858·0

Immersed side, . 4068·7
Emerged side, . 3707·2
For ½ squares, . 2) 361·5 difference.
          180·75 preponderance of ½
          squares on immersed side.
⅓ longitudinal interval, 4
Area of Waterplane, } 2858)723·00 moment.
Centre of gravity = ·25 of a foot towards immersed side.

46225·7 Immer:
      wedg
41419·2 Emerg:
      wedg
87644·9 Sum fo
      both
      wedg:

CALCULATIONS.

INCLINED TO AN ANGLE OF 14 DEGREES—*continued.*

TABLE III.

to an angle of 14°.)

No. of Section.	Ordinates.	Multipliers.	Functions of Ordinates.	Squares of Ordinates.	Multipliers.	Functions of Squares.	Cubes of Ordinates.	Multipliers.	Functions of Cubes.
				EMERGED WEDGE.					
1	1·6	1	1·6	2·5	1	2·5	4·0	1	4·0
2	7·3	4	29·2	53·2	4	212·8	389·0	4	1556·0
3	10·0	2	20·0	100·0	2	200·0	1000·0	2	2000·0
4	11·5	4	46·0	132·2	4	528·8	1520·8	4	6083·2
5	12·1	2	24·2	146·4	2	292·8	1771·5	2	3543·0
6	12·4	4	49·6	153·7	4	614·8	1906·6	4	7626·4
7	12·4	2	24·8	153·7	2	307·4	1906·6	2	3813·2
8	12·2	4	48·8	148·8	4	595·2	1815·8	4	7263·2
9	11·8	2	23·6	139·2	2	278·4	1643·0	2	3286·0
10	10·5	4	42·0	110·2	4	440·8	1157·6	4	4630·4
11	8·3	2	16·6	68·8	2	137·6	571·7	2	1143·4
12	4·9	4	19·6	24·0	4	96·0	117·6	4	470·4
13	·4	1	·4	·1	1	·1	·0	1	·0
			346·4			3707·2			41419·2

COMBINATION TABLE FOR STABILITY AT 14° INCLINATION.

	IMMERSED WEDGE.			EMERGED WEDGE.		
Angles of Inclination.	Functions of Squares of Ordinates.	Multipliers.	Functions of Squares of Ordinates for Volume of Wedge.	Functions of Squares of Ordinates.	Multipliers.	Functions of Squares of Ordinates for Volume of Wedge.
0°	3693·8	1	3693·8	3693·8	1	3693·8
7°	3841·2	4	15364·8	3655·9	4	14623·6
14°	4068·7	1	4068·7	3707·2	1	3707·2

Immersed wedge,	.	.	.	23127·3	22024·6
Emerged ,,	.	.	.	22024·6	
For half squares,	.	.	.	2) 1102·7 difference.	
				551·35 preponderance of half squares in immersed wedge.	
⅓ of angular interval,	.	.		·0407	
				385945	
				2205400	
				————	*Correcting Layer.*
				22·439945	89·759 × ·25 (centre of gravity of waterplane towards immersed side. See Preliminary Table III.).
⅓ of longitudinal interval,	.			4	
Excess in volume of immersed wedge,	.	.	.	89·759780	=22·439 Moment for Layer.
Area of waterplane,	.	.	2858)89·759		
			·03 of a foot Thickness of Layer.		

BOTH WEDGES.

Angles of Inclination.	Sums of Functions of Cubes of Ordinates.	Multipliers.	Products of Functions of Cubes.	Cosines of Angles of Inclination.	Functions of Cubes for Moments of Wedges.
0°	80916·28	1	80916·28	·9702	78504·9
7°	82840·60	4	331362·40	·9925	328877·1
14°	87644·90	1	87644·90	1·0000	87644·9

⅓ of cubes, ...		3)495026·9
		165008·9
⅓ of angular inclination, . . .		·0407
		6715·8
⅓ of longitudinal interval, . . .		4
Moment of wedges, ...		26863·2
(Subtract) Correction for layer, . .		−22·4
Disp. in cubic feet, . . .		26452·3)26840·8
	BR=	1·01
BG × Sine of Angle = 2·07 × ·2419 =		·50
	Righting Arm GZ =	·505

Righting Moment of Stability = GZ × Displacement in Tons = ·505 × 755·78 = 381·66 foot tons.

Righting Moment when inclined to an angle of 14° = 381·66 foot tons.

APPENDIX A.

The Author desires, in conclusion, to call the attention of such of his readers as have some mathematical attainments to an able Paper, read before the Institute of Marine Engineers, on Jan. 12, 1892, by John A. Rowe, Esq., Surveyor to the Board of Trade.* From this Paper, which was kindly placed at his disposal, he has made the following extract, which will be found of much interest :—

DYNAMIC STABILITY AND OSCILLATIONS AMONG WAVES.

"Most of you are aware that in computing the rolling period of a ship—that is, the time in seconds she will take to roll from the vertical and back again—she is somehow or other regarded as a pendulum. This is a correct view to take of the matter if the subject is approached in a proper direction. But many able men have been perplexed by what has appeared to be the contradiction between theory and practice. For instance, most of you are aware that the period T of a bob-pendulum in seconds is :

$$T = 3\cdot1416 \times \sqrt{\frac{\text{Length in feet}}{\text{Gravity}}} = 3\cdot1416 \times \sqrt{\frac{L}{32}} = \cdot554 \times \sqrt{L}$$

L is the length of the pendulum in feet.

In this formula it is clear that the period of a pendulum varies as the square root of its length.

For the smooth water period of a ship the formula is somewhat different, and is as follows :—

$$T = 3\cdot1416 \sqrt{\frac{\text{Radius of Gyration squared}}{\text{Gravity} \times \text{Metacentric Height}}} = \cdot554 \times \sqrt{\frac{R^2}{GM}}$$

The radius of gyration and the metacentric height to be in feet.

An examination of this formula reveals the fact that a vessel's rolling period varies directly as the radius of gyration, and inversely as the square root of the metacentric height. If G M be increased in length, the vessel's period will be shortened, and she will become a quicker roller than before. But practical men looking at the usual diagrams, have reasoned thus :— 'The length G M is the distance between the vessel's point of oscillation and her centre of gravity. If there is any pendulum-like motion in the ship it is of necessity about G or M, at a length G M.'

But the formula $T = \cdot554 \times \sqrt{\frac{R^2}{GM}}$ shows clearly enough that whatever be the equivalent pendulum length it is not G M but is something entirely different. The following fig. has been constructed by the writer in the hope of simplifying some points not generally understood.

The vessel is shown upright on the wave-slope. In this position the force of buoyancy, *which acts at right angles to the wave surface,* as it acts at right angles to the surface of smooth water, creates a righting arm G Z. This length in feet, multiplied by the ship's weight in tons, is what we have called the vessel's righting moment in foot-tons. In the position we

* Now Chief Examiner of Engineers to the Board of Trade.

326 APPENDIX.

have shown her, it is obvious that this power, usually regarded as the power of recovery, starts the vessel rolling; and if, after she had acquired a position at right angles to the wave-slope, we could instantly give a smooth

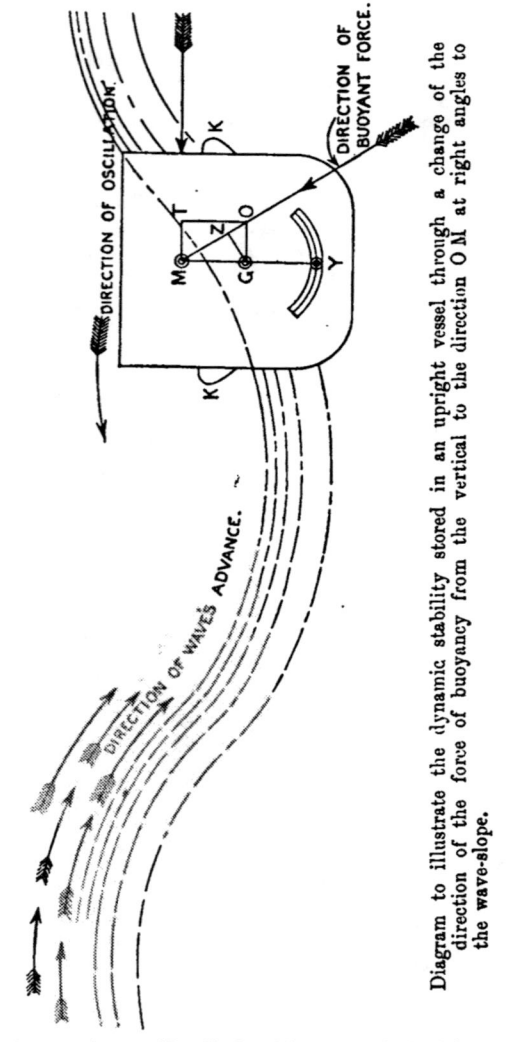

Diagram to illustrate the dynamic stability stored in an upright vessel through a change of the direction of the force of buoyancy from the vertical to the direction O M at right angles to the wave-slope.

surface to the sea she would roll through the angle G M O on **each side of** the vertical, and gradually extinguish the range of oscillation by the fluid resistance offered to the immersed portion of the hull.

Again, if G O and M T be drawn at right angles to G M, and T O drawn parallel to G M, we obtain a parallelogram of forces, whose resultant M O may be regarded as the buoyant force which equals the weight of the ship, and whose components are G M and M T. As G M is acting upward through the vessel's centre line we may disregard it, and direct our attention exclusively to the component M T, whose direction is shown by arrow, and whose amount is M O × sine of the angle of inclination.

Let the effective angle of the wave-slope be 9°, the vessel's weight 10,000 tons, her metacentric height 6 feet. Find the turning moment about G, the vessel's centre of gravity.

The component M T = M O × sine of 9° = 10,000 tons × ·156 = 1560 tons.

This force of 1560 acts at the end of the lever G M = 6 feet; therefore, the righting, or in this case, the turning moment

= 1560 tons × 6 feet = 9360 foot-tons.

But the righting moment is the weight of the ship multiplied by the righting arm. What is the product of these quantities?

$$\text{G Z, the righting arm} = \text{G M} \times \text{sine of } 9° = 6 \text{ Ft.} \times ·156 = ·936 \text{ Foot.}$$

$$\text{Righting moment } 10,000 \text{ Tons.} \times ·936 \text{ Foot.} = 9360 \text{ foot-tons.}$$

Both calculations declare the righting moment to be 9360 foot-tons.

In other words, the weight of the ship into the righting arm G Z = the component M T into the metacentric height. If now we plot Y as the centre of gyration, we shall be able to realise the nature of the force tending to produce motion, and the character of the resistance offered to it.

Let us for a moment suppose that G is a fixed point—the ship's fulcrum. Let us also regard the ship as a portion of a huge wheel (a portion of a flywheel) its radius of gyration being G Y.

By an examination of the fig. it will be seen that the greater G M is (with a given horizontal force M T) the greater is the turning moment. And the smaller the radius of gyration G Y, the smaller will be the resistance and the quicker will be the motion of oscillation.

To obtain great stability and quick motion, we must increase the leverage G M, and reduce the pendulum length G Y; to obtain moderate stability, but a slow angular motion, and, therefore, a comfortable vessel at sea, and one offering a steady gun-platform, we must diminish the leverage G M and increase the length of G Y.

With regard to G Y, which is obtained by dividing the vessel's moment of inertia about G by the sum of the weights, and extracting the square root, it is evident that it can be of great length only in a large vessel. In a small vessel, G Y can be increased by placing movable weights towards the bulwarks, but no such change as this will be sufficient to make the radius of gyration great enough to give rise to a slow rate of oscillation. An easy motion in small crafts may be obtained by shortening G M, but this may give rise to want of stability. Hence the difficulty of builders to make a perfect ship. They have to steer between Scylla and Charybdis. Worse still, they strive to please shipowners, who know but little of the difficulties of naval architecture; and to please themselves, with the result that they sometimes please neither. . . .

Wave action upon ships, stores wave energy *in* ships to an extent depending on their weight and length of righting arm; and the manner of ascertaining the amount of work put into the before-mentioned ship is as follows:—The force $MT = 1560$ tons, becomes nil when the vessel's deck is parallel to the wave-slope or to a smooth sea. Therefore the *mean* force acting to turn the ship at M about the centre G is $\frac{MT}{2} = \frac{1560}{2} = 780$ tons. This *mean* force acts through the space $MT = MG \times$ tangent of the angle of inclination $= 6$ feet $\times \cdot 158 = \cdot 948$ foot. And 780 tons $\times \cdot 948$ foot $= 739\cdot 44$ foot tons dynamic stability. . . .

Storm-waves produce violent rolling in the largest of floating structures, and these structures are occasionally brought to rest by a sudden and complete expenditure of their stored energy. And the greater the energy *in* the vessel—*i.e.*, the heavier the *ship*, and the quicker the motion, the more tremendous is the blow she can inflict upon an approaching wave. But, unhappily, when the momentum of an ocean wave is not only resisted by a vessel's hull, but is increased by the dynamic energy of the ship, a climax occurs, the severity of the blow is manifested by the vessel ceasing to roll (her energy being expended), and by the wave bursting high above the decks and sweeping them from end to end. This condition of things, as about to happen, the writer wishes to convey in the rough sketch, by arrows, showing the direction of the ship's oscillation and the wave's advance. The fig. is not by any means to scale." The projections K K on each side of the diagram are short lengths of troughs which, in Mr. Rowe's opinion, would prevent rolling. They are open ended, and the dimensions would vary with the weight of the ship and the metacentric height. They would probably render torpedo boats habitable in choppy seas and stormy weather.*—(From paper on "Stability and Motions of a Vessel among Waves," by John A. Rowe. Part ii., p. 10, *et seq.*)

APPENDIX B.

TEST QUESTIONS.†

CHAPTER I.

1. What is displacement? What is a displacement curve? Explain its construction and use.
2. What is deadweight? What is a deadweight scale, and how is it constructed?
3. What is meant by "tons per inch" immersion? Give and explain the rule for "tons per inch" immersion.
4. Explain how a curve of "tons per inch" immersion is constructed, and show clearly its use.
5. What is a coefficient of displacement? State approximately the coefficient for an average cargo steamer, and a fine passenger steamer.
6. What is the weight of 1 cubic foot of salt water and 1 cubic foot of fresh water? How many cubic feet of salt water, and also of fresh water, are there in 1 ton?
7. Find the displacement of a box ship floating light in sea water at a

---

* By instantly exhausting the energy derived from each wave. These troughs would prevent the *accumulation* of energy and therefore limit the effect of wave action to that due to the passing of one wave only under the ship's bottom.

† Many of these questions are taken from Science and Art Examination Papers.

draught of 5 feet forward and 6 feet aft. The length is 100 feet; the breadth 20 feet (use mean draught). *Ans.* 314·2 tons.

8. When loaded, the box ship in the previous question draws 12 feet of water fore and aft. What is the deadweight? *Ans.* 371·5 tons.

9. A ship of 1000 tons displacement, loaded, is floating in sea water. What will be the change in draught in passing into river water? The "tons per inch" at the load line is 9. *Ans.* Draught increases 1·7 inches.

10. A steamer on a voyage burns 200 tons of coal. The "tons per inch" is 24. What is the approximate change in draught? *Ans.* Draught decreases 8·3 inches.

11. A vessel is 200 feet long, 30 feet broad, and of 16 feet depth to top of weather deck at amidships. The freeboard is, say, 2 feet, and the coefficient of fineness ·65. What is her displacement in salt water? *Ans.* 1560 tons.

CHAPTER II.

12. What is meant by saying that a vessel has a righting or a capsizing moment?

13. Define the term "centre of gravity."

14. A ship has a displacement of 2000 tons when floating at a certain draught; 100 tons are then placed on deck at a height of 9 feet above the centre of gravity of the ship as it was before the weight was placed on board. Find the alteration in the position of the centre of gravity. *Ans.* Centre of gravity is raised ·42 foot.

15. A ship, with a displacement of 2000 tons, has a weight of 100 tons, already on board on the centre of the upper deck, moved 8 feet to the port side. Find the distance the centre of gravity has shifted. *Ans.* ·4 foot.

16. The centre of gravity of a ship is 12 feet from the bottom of the keel. In this condition her displacement is 2500 tons. She is then loaded in the following manner:—100 tons are placed 9 feet above the bottom of the keel, 300 tons 14 feet, and 500 tons 12 feet. Find the new position of the centre of gravity from the bottom of the keel. *Ans.* 12·08 feet.

CHAPTER III.

17. What is meant by buoyancy, reserve buoyancy, centre of buoyancy?
18. How do water pressures act? Which of them afford support?
19. What is meant by sheer? and explain its use.
20. Of what value are deck erections as regards buoyancy?
21. What is the vertical centre of buoyancy? also longitudinal centre of buoyancy?
22. Show how curves of longitudinal and vertical centres of buoyancy are constructed.
23. Supposing a ship's longitudinal centres of buoyancy to be in the middle of the length at every draught when floating on even keel, and she is loaded in the following manner:—

10 tons are placed	30 feet forward of the centre of buoyancy
120	,,	50	,,	,,	,,
500	,,	60	,,	,,	,,
400	,,	70 feet aft	,,	,,
15	,,	60	,,	,,	,,
150	,,	25	,,	,,	,,

Where would a weight of 200 tons need to be placed on board to bring her again on even keel ? *Ans.* 18·25 feet aft of the centre of buoyancy.

24. Why does a ship increase in draught on a comparatively small compartment being damaged below the water level, into which the sea enters ?

25. What is meant by camber, and why is it given to a vessel ? State the rule for the minimum.

26. A box ship is 100 feet long, 20 feet broad, and floats at 6 feet draught. Calculate the amount of upward water pressure in lbs. *Ans.* 768,000 lbs.

CHAPTER IV.

27. Enumerate in order of importance the principal strains to which a ship may be subject.

28. State clearly what strains a ship may experience when floating light and in calm water.

29. Show how strains may be decreased or enormously increased in the operation of loading.

30. Describe the strains experienced by a ship among waves—fore and aft and athwartships.

31. Explain the term "unequal distribution of weight and buoyancy."

32. What is a compressive strain and a tensile strain ?

33. Show in any graphic way how to combine and arrange the material used in the construction of ships so as to give greatest resistance to bending.

34. What kind of ships offer greatest resistance to longitudinal bending, and which offer least ?

35. Where are the fore and after strains greatest ? Show why.

36. What is the tendency of strains due to rolling motion ?

37. What strains are supposed to be provided for in vessels built to the requirements of the recognised classifying societies ? What strains may a vessel experience which such rules do not profess to cover ?

CHAPTER V.

38. Enumerate the parts of a ship's structure known as transverse framing. What is the function of transverse framing ?

39. Describe carefully and in detail how the parts which make up a complete transverse frame are connected with one another, and also the various forms of material which may be used.

40. What is meant by compensation in ship construction ? Give illustrations.

41. Give rules for beam knees.

42. Which are the best beams to fit under iron or steel decks, and also under wooden decks ? Give illustrations.

43. What depth must a ship be to require two tiers of beams, and also three tiers of beams ?

44. How may the lowest of these tiers be dispensed with, and state clearly the compensation made for the loss ?

45. What are web frames ? When and where are they fitted ?

46. State which parts of the transverse framing specially resist the tendency to "working," produced by rolling motion.

47. Mention the parts comprising longitudinal framing. What is the function of longitudinal framing ?

48. What means are adopted to secure a good connection between the longitudinal and the transverse framing ?

APPENDIX. 331

49. What is a bar keel? How are the parts comprising this keel connected?
50. What is a keelson and a stringer?
51. Sketch roughly the different forms of keels and centre keelsons, and state which (if any) combination is preferable.
52. What is the garboard strake? The sheer strake?
53. How is it that most of the material used in a ship's construction is reduced in thickness towards the ends?
54. Describe any method adopted to compensate for cutting down a centre keelson or reducing a stringer plate in width.
55. Describe what provision may be made to resist panting strains, and also strains from masts due to wind pressure.
56. What are the special characteristics of a three-deck, a spar-decked, an awning-decked, and a quarter-decked vessel?
57. Why are bulkheads fitted? Show how they are connected to the shell plating. To what height are bulkheads carried? How is a recessed bulkhead made watertight?
58. Show how a bulkhead is made watertight where a keelson or stringer passes through it.
59. What is a rimer and a drift punch, and their use? Which is the best form of rivet, and why?

CHAPTER VI.

60. Define the term "stability."
61. What are the two factors producing moment of stability?
62. What is the metacentre? What is metacentric height?
63. What is meant by a righting moment of stability?
64. State the conditions under which a vessel will float in stable equilibrium.
65. What is the condition of a ship which is said to be "stiff" or "tender"?
66. Having given the metacentric height, how can the righting lever be found, and when is it unsafe to adopt this method?
67. Give the formula for finding the height of the metacentre above the centre of buoyancy.
68. What features in the design are most important in influencing the height of metacentre?
69. How can stiffness be obtained?
70. What is a curve of stability, and how is it constructed?
71. It is usual for the metacentre to fall when the draught is increasing from light towards the load draught, but on approaching the load it is often found to rise again. What explanation can be given for this?
72. Describe clearly the steps of the operation for finding the metacentric height by experiment.
73. Describe clearly the effect of beam, freeboard, and height of centre of gravity upon the maximum levers and range of stability.
74. What is tumble home?
75. State the relation between metacentric height and transverse rolling motions in still water.
76. Enumerate the resistances to rolling motions.
77. What are bilge keels, and why are they fitted? What is the danger among waves of great metacentric height?
78. What methods may be adopted to obtain steadiness among waves, and state clearly the condition of a vessel affected by each of these methods?
79. What is meant by synchronism? How is it produced and how averted? What conditions of loading are most liable to produce it?

80. State how it is that similar metacentric heights for load and light or ballast conditions do not produce similar stability and behaviour at sea.

81. What are the most important considerations in ballasting as regards the amount, position, and securing the ballast?

82. What is the condition and danger developed by a shifted cargo?

83. What means may be adopted to prevent a cargo shifting?

84. State under what conditions a vessel's behaviour and stability may vary upon a single voyage.

85. State under what circumstances a ship will sink or capsize, owing to the entry of water into the interior, either through an opening in the deck or a hole in the side or bottom below the water level.

86. What are the necessary features or conditions of a vessel in order to be able to carry large sail area?

87. Enumerate the chief resistances to propulsion.

CHAPTER VII.

88. What is meant by the terms "trim," "moment to alter trim," and "moment to alter trim one inch"?

89. How would you distribute the cargo in the holds of a vessel so as to produce no alteration of trim in immersing her from the light to the load draught?

90. Explain how you would arrange the cargo in a vessel so as to obtain a definite condition of trim.

91. How is the change of trim estimated, owing to the filling of a fore peak tank?

92. Explain how to estimate the change of trim caused by an empty compartment in the double bottom becoming damaged, and the sea filling the compartment.

93. Give the formula for "moment to alter trim one inch."

94. Why is there so great a difference between the height of the longitudinal metacentre and the vertical metacentre above the centre of buoyancy at any particular draught?

CHAPTER VIII.

95. What is meant by gross and under-deck tonnage, and what spaces are included in each?

96. Enumerate the deductions from the gross tonnage for register tonnage.

97. How is the propelling space deduction in steamers obtained?

98. When are deep water-ballast tanks allowed as deductions, and when are they not?

99. How are deck cargoes reckoned as regards tonnage?

100. What are the important differences between the ordinary tonnage and Suez Canal tonnage?

CHAPTER IX.

101. What is meant by the term "freeboard"?

102. What are the leading considerations in determining the freeboard for any particular vessel?

103. Why have spar-decked vessels more freeboard than three-deckers, and awning-decked vessels more freeboard than spar-deckers?

104. What effect have sheer, camber, length, and deck erections upon freeboard?

APPENDIX A.

105. Describe Rowe's anti-rolling troughs.

INDEX.

A

AMIDSHIPS strength, 60.
Ardency, 214.
Atmospheric pressure, 25.
Awning-deck vessels, 74, 95, 97; Scantlings of, 102, 103.

B

BALLAST, Amount and arrangement of, 166-185; Means to prevent shifting of, 169; Minimum ballast draught, 180; Testing ballast tanks, 38; Water, 169-185.
Beam knees, 54.
Beams, 53, 74, 89; Compensation for dispensing with, 54; Compensation for loss of, in engine and boiler space, 76.
Behaviour, Effect of bilge keels upon, 150, 163; at sea, how affected, 160; Alteration of, on a voyage, 161; how affected by arrangement of weights fore and aft, 163; how affected by arrangement of weights transversely, 161; how affected by metacentric height, 162.
Bilge, Strengthening of, 53, 56.
,, keels, Effect of, upon rolling motions, 150.
Boiler stools, 78.
Bosom piece, 51.
Breadth, Extreme, 62.
,, Moulded, 62.
Breast hook, 89, 90.
Bridge over half midship length, 76.
Bridges, Value of, 61.
Bulb angle, 52, 53.
Bulkheads, 97-109; Height of, 104, 210; liners, 107; longitudinal, 107; Number of, 97; recessed, and means of making watertight, 105; stiffening, 107; watertight doors, 108.

Buoyancy, 19, 113; afforded by cargo, 203; Centre of, 23, 27; Curves of centres of, 28-33; effect of camber, 38; Effect of entry of water upon, 36; Effect of longitudinal bulkheads upon, 206; Reserve, 22, 26; Wedges of, 33-36; forces among waves, 153-160.
Butts, 51, 58.
Buttstrap, 109.

C

CALCULATIONS (see Contents, pages 271, 301).
Camber, 38.
Cargoes, Homogeneous, 188; Shifting, 193.
Cement washing, 76.
Centre of effort, 214.
,, gravity, 14; Height of, 128, 132.
Centrifugal force, 154.
Channel bar, 53.
Coefficients, 10, 11.
Collars, 107.
Compensation for dispensing with hold beams, 54.
Connection of longitudinal and transverse framing, 54-56.
Cosines, Table of, 268-270.
Cotangents, Table of, 268-270.
Countersinking, 110.
Cylinders, Stability of, 121.

D

DEADWEIGHT, 6.
,, scale, 5; Relation of, to type, 91-97.
Deck cargoes, Means to support, 91.
,, erections, Value of, 27.
Decks, Steel, 75.

Deck weights, Means to support, 91.
Depth, Lloyd's, 62–74; moulded, 62; spar- and awning-deck vessels, 62.
Diamond plate, 55.
Displacement, Coefficient of, 10; curve, Construction of, 3–5; Definition of, 1; scale, Vertical, 5.
Double bottom for water ballast, 80, 170.
Draught, after lying aground, 25.
„ Salt and fresh water, 9.
Drift punch, 109.

E

ENGINE, Foundation plate under, 79; seat, 76–78; space, Strengthening of, 76; trough, 78.
Equilibrium, Condition of, 114; neutral, 115; stable, 115; unstable, 114.

F

FAYING surface, 109.
Flam, 164.
Flare, 164, 200.
Floors, 53; Depth of, 53, 82, 89; Thickness of, under engines and boilers, 76.
Foot-ton, 12, 112.
Frame bar, 52.
„ heel, 61.
„ spacing, 51.
Framing, Longitudinal, 40, 57.
„ Transverse, 40, 51.
Freeboard, 7; Corrections upon, for erections on deck, 258; for length, 256; for round of beam, 258; for sheer, 257; Definition of, 251; examples showing how ascertained, 262–267.

G

GARBOARD strake, 61.
Girder, ship, Strengthening flanges of, 61.
Gravity, 14, 113; force among waves, 154.
Gunwale, Strengthening of, 61.

H

HEEL piece, 38.
Hold beams, 74.

I

IMPORTANT terms, Definition of, 62.
Intercostal plates, 78, 81.

K

KEEL, 57–60.
Keel scarph, 57.
Keelsons, centre, 58–60; Position of, 60; Function of, 60; Number of, 75; compensation for reduced depth, 81.
Kinetic energy, 149.

L

LATERAL resistance, Centre of, 213.
Leeway, 214.
Length between perpendiculars, 62; Extreme, 64–73; Lloyd's, 62; Standard of, 75, 76.
Leverage, 12.
Loading, 186; Effect of, upon behaviour, 160.
Lug piece, 59, 60.

M

MAST partner, 91.
Masts, Strengthening and fixing of, to resist strain, 91.
Metacentre, Transverse, 114; above centre of buoyancy, Rule for, 115; Curves of, 126, 127; Relation of design to height of, 118; Variations in height of, 200.
Metacentre, Longitudinal, 232; Curve of, 235.
Metacentric height, Effect of loading upon, 120; how found, 128; Relation of, to wind pressure, 215.
Metacentric heights, similar, Effect of, at different draughts, 106; stability, 115.
Metal chocks, 106.
Midship sections, 51, 64–72, 80, 98–102.
Moments, 12; Calculation of, 13–15; righting and capsizing, 12.

N

NEUTRAL axis, 45.
Numerals, Lloyd's, 63.

O

ORLOP beams, 75.

P

PANTING beams, 89; Means to prevent, 88, 89; stringers, 88.
Pillars, 56, 91.
Pitching, 163; how affected by arrangement of weights, 164.

Q

QUESTIONS, Test, 329.

R

RAISED quarter deck, 94, 95.
Repose, Angle of, 195.
Resistance, Eddymaking, 217; Frictional, 216; Wavemaking, 217.
Reverse bar, 52.
,, frames, Height of, 74.
Rimer, 109.
Riveting, 109.
Rivets, Forms of, 110.
Rolling, 148; among waves, 151; in still water, 149; motion of cylinder, 153; of raft, 152, 156; Resistance to, 149; Unresisted, 163.

S

SAIL area, 213; Nature of design to carry large area, 216.
Scantlings, 64-73, 98-103; Reduction in, 74-76.
Self-trimming vessels, 191, 192.
Sheer, 27.
Sheer strake, 61.
Shell bar, 60.
Sines, Table of, 268-270.
Slackness, 214.
Snap cup, 111.
Spar-deck vessels, 74, 95, 96; Scantlings of, 100, 101.
Stability Curves of, for actual ships, 143, 146, 174, 179, 182, 183, 219; Curve of, for cylinder, 124; Definition of, 112; Effect of beam upon, 134; Effect of freeboard upon, 135; Effect of position of centre of gravity upon, 139; Effect of tumble home upon, 144; Effect of sheer upon, 146; Effect of wedges of immersion and emersion upon, 125; Moment of, 112, 114; Necessary information upon, 219-221; of cylinder, 121; of different types of vessels, 147, 192; Range of, 123; Righting lever of, 112; Variations in, on a voyage, 190, 195.
Stern frame, Fixing of, 82, 83.
Stiff, 115.
Stiffness, Danger of, 151.
Strain, 40.
Strains from deck cargoes, 48; in dry dock, 50; in loading cargoes aground, 49; from permanent heavy weights, 48; due to loading, 42; due to rolling, 47; due to propulsion by steam and sail, 48; when floating light, 41; from panting, 48; from shipping seas, 49; from water pressure, 42; on waves, 46; on vessel partly waterborne, 49; Compressive and Tensile, 45; Types of vessels subject to greatest, 46.
Strength, distribution of material to resist bending, 43-46.
Strength, Relation of, to dimensions, 74.
Strengthening at aft end of shafting, 81; Local, 76; Longitudinal, 75; Special, 75, 76.
Stringer plate, Compensation for reduction in width of, 81.
Stringers, Function of, 60.
Struts for twin-screws, 82-88; Method of dispensing with, 87, 89.
Synchronism, 158.

T

TANGENTS, Table of, 268-270.
Tender, 115.
Three-deck vessels, 63, 70-73, 76, 94, 98-99.
Tonnage, deep water ballast tanks, 245; Examples of, 245-247; Gross, 241; Importance of, 239; of deck cargoes, 245; Register, 242; regis-

ter, Deductions for, 242–244; Royal Thames Yacht Club, 250; Suez Canal, 247; under deck, 239; Yacht, 250.

Tons per inch immersion, Definition of, 7; curve, 8; Rule for, 7; use and application, 8, 9.

Transome plate, 82.

Trim, 226; Definition of, 226; Examples showing how to estimate change of, 236–238; Moment to alter, 227; Moment to alter one inch, 234.

Trunk steamer, 191, 192.

Tumble home, 120.

Turret steamer, 191, 192.

Twin screws, Struts for supporting, 82–88.

Types of vessels, 91, 191, 192.

W

WATER, Admission of, above a watertight flat, 205.

Water, Admission of, below a watertight flat, 204; below water-level, 198; into an end compartment, 209; through deck opening, 207.

Waterlogged vessels, 211.

Water on deck, 206.

Water ports, Value of, 206.

Water pressures, 19–26.

Waves, Theory of, 154–160.

Web frames, 55.

Web stringer, 56.

Weight and buoyancy, Relation between, 41; Unequal distribution of, 41.

Weight of material in structure to strength, Relation of, 40.

Weights, Effect of moving, upon ship's centre of gravity, 16, 17.

Wind pressure, 167, 214.

Z

Z-BAR, 53.

PRINTED BY NEILL AND CO., LTD., EDINBURGH.

Printed in the United States
92883LV00006B/239/A